Lecture Notes in Computer Science 12801

More information about this subseries at http://www.springer.com/series/7409

Elisabeth Métais · Farid Meziane ·
Helmut Horacek · Epaminondas Kapetanios (Eds.)

Natural Language Processing and Information Systems

26th International Conference on Applications
of Natural Language to Information Systems, NLDB 2021
Saarbrücken, Germany, June 23–25, 2021
Proceedings

Springer

Editors
Elisabeth Métais
Conservatoire National des Arts et Métiers
Paris, France

Farid Meziane ⓘ
University of Derby
Derby, UK

Helmut Horacek
German Research Center for Artificial
Intelligence
Saarbrücken, Germany

Epaminondas Kapetanios
University of Hertfordshire
Hatfield, UK

ISSN 0302-9743   ISSN 1611-3349 (electronic)
Lecture Notes in Computer Science
ISBN 978-3-030-80598-2   ISBN 978-3-030-80599-9 (eBook)
https://doi.org/10.1007/978-3-030-80599-9

LNCS Sublibrary: SL3 – Information Systems and Applications, incl. Internet/Web, and HCI

This Springer imprint is published by the registered company Springer Nature Switzerland AG
The registered company address is: Gewerbestrasse 11, 6330 Cham, Switzerland

Preface

This volume contains the papers presented at NLDB 2021, the 26th International Conference on Applications of Natural Language to Information Systems held during June 23–25, 2021, as a video conference at the German Research Center for Artificial Intelligence in Saarbrücken, Germany. We received 82 submissions for the conference. Each paper was assigned to three reviewers, taking into account preferences expressed by the Program Committee members as much as possible. After the review deadline, Program Committee members were asked to complete missing reviews. On the basis of these reviews, the Conference Organization Committee members and the Program Committee Chair decided to accept papers with an average score closer to *acceptance* than *weak acceptance* as full papers and papers with an average score around *weak acceptance* as short papers. In borderline cases, credit was given to experimentally-oriented papers with novel and ambitious concepts.

The final acceptance rate counting the number of full papers according to NLDB tradition was 23 percent (19 out of 82), similarly competitive in comparison to previous years. In addition, 14 submissions were accepted as short papers, and no posters since NLDB had to be held as a video conference like last year. Full papers were allowed a maximum of 12 pages and short papers a maximum of 8 pages. Originally, one more long paper and three more short papers were accepted, but the authors preferred to retract their submissions for personal reasons.

Similar as last year, the popular topics of classification and sentiment analysis have been addressed by many papers, and successful tools are reused and adapted, such as the transformer BERT. Following the trends of previous years, there is more diversification in the topics and specific issues addressed even in comparison to NLDB 2020. Several papers focus on methodological issues per se, prominently on perspectives on learning.

In addition to the reviewed papers, there were two invited talks at NLDB 2021:

- *Manfred Stede*, University of Potsdam, Germany
- *Elke Teich*, Saarland University, Germany

Moreover, a panel was organized by the recently built consortium NL4XAI (Interactive Natural Language Technology for Explainable Artificial Intelligence) to introduce their activities.

The accepted contributions (long and short papers) covered a wide range of topics, which we classified in nine topic areas, each covering a section in this volume:

- The Role of Learning
- Methodological Approaches
- Semantic Relations
- Classification
- Sentiment Analysis
- Social Media

- Linking Documents
- Multimodality
- Applications

The Role of Learning

One long and two short papers were categorized in this section. The long paper addresses limitations of learning, which may arise due to insufficient conceptual coverage of the data. One of the short papers deals with the combination of learning models, and the other short paper shows a machine in a learning environment inspired by human learning.

Methodological Approaches

Four long and one short paper were categorized in this section. The first two long papers deal with making frequently used techniques more fine-grained, namely word embedding and auto encoding. The short paper emphasizes modularity in a speech-to-speech translation model. Among the other two long papers, one addresses a specific form of entities, when numerals are included, and the other one deals with transfer across domains.

Semantic Relations

Three long papers were categorized in this section. The first one concentrates on causality relations, elaborated for a medical subdomain. The second one shows how precision about the entities involved can improve recognition in business domains. The third paper also addresses the role of preciseness of information, to detect combinations of relations which constitute events.

Classification

This traditional topic constitutes the longest section in the proceedings, with three long and four short papers. Some of the papers feature technical enhancements, including the use of word embeddings, features of multiword expressions, and applying a BERT capsule model. The majority of the contributions is oriented on the needs of the area of application. Some cover certain kinds of speech, such as hate speech, figurative language - irony and sarcasm - and technical language. Others are dedicated to specific domains: law and COVID-19 messages.

Sentiment Analysis

This is another popular topic, providing the second largest section, with three long and three short papers. As in the preceding section, there are papers featuring methodological issues and others oriented on the intended application. Methodological enhancements include multi-step transfer learning, cross-lingual learning, and importance weighting. Application topics vary greatly, they range from low resource language, opinions about vaccines, and aspects of structured product reviews, to headline stance and even literary artifacts.

Social Media

Three long papers were categorized in this section. They cover quite diverse tasks in analyzing texts in social media. These tasks are the detection of claims and evidence in arguments, attribution of authorship, and prediction of mental disorders.

Linking Documents

One long and one short paper were categorized in this section. They both address linking of some sort of a master document to a set of enhancing documents. One paper uses the supplementary material for background information in news, whilst the other one does this for citation context of scientific documents.

Multimodality

One long and one short paper were categorized in this section. The long one addresses the recognition of essential elements from an image and its textual description, and the short one enhances the categorization of writers by incorporating visual data about them in the categorization process.

Applications

Two short papers were categorized in this section. They both address human computer interaction aspects of some sort. One of them attempts to enhance intent classification by multiple models, whilst the other one aims at retrieving graphical user interface prototypes on the basis of natural language specifications.

The conference organizers are indebted to the reviewers for their engagement in a vigorous submission evaluation process. We would also like to thank, for the organization help, some members of the DFKI GmbH.

June 2021

Epaminondas Kapetanios
Helmut Horacek
Elisabeth Métais
Farid Meziane

Organization

Conference Organization

Elisabeth Métais   Conservatoire des Arts et Métiers, France
Farid Meziane   University of Derby, UK
Helmut Horacek   German Research Center for AI, Germany

Program Chair

Epaminondas Kapetanios   University of Hertfordshire, UK

Program Committee

Jacky Akoka   CNAM & TEM, France
Hidir Aras   FIZ Karlsruhe, Germany
Nicolas Béchet   IRISA, France
Gloria Bordogna   IREA, CNR, Italy
Sandra Bringay   LIRMM, France
Christian Chiarcos   University Frankfurt am Main, Germany
Raja Chiky   ISEP, France
Philipp Cimiano   University of Bielefeld, Germany
Mohsin Farid   University of Derby, UK
Vladimir Fomichov   Moscow Aviation Institute, Russia
Flavius Frasincar   Erasmus University Rotterdam, The Netherlands
André Freitas   The University of Manchester, UK
Yaakov Hacohen-Kerner   Jerusalem College of Technology, Israel
Michael Herweg   IBM, Germany
Helmut Horacek   German Research Center for AI, Germany
Dino Ienco   IRSTEA, France
Ashwin Ittoo   HEC, University of Liège, Belgium
Epaminondas Kapetanios   University of Hertfordshire, UK
Zoubida Kedad   UVSQ, France
Eric Kergosien   GERiiCO, University of Lille, France
Christian Kop   University of Klagenfurt, Austria
Valia Kordoni   Humboldt University Berlin, Germany
Jochen Leidner   Refinitiv Labs and University of Sheffield, UK
Deryle W. Lonsdale   Brigham Young University, USA
Cédric Lopez   Emvista, France
Natalia Loukachevitch   Moscow State University, Russia
John P. Mccrae   National University of Ireland, Ireland
Elisabeth Métais   Conservatoire des Arts et Métiers, France
Farid Meziane   University of Derby, UK

Luisa Mich	University of Trento, Italy
Balakrishna Mithun	Limba Corp., USA
Nada Mimouni	Conservatoire des Arts et Métiers, France
Rafael Muñoz	Universidad de Alicante, Spain
Karen O'Shea	University of Derby, UK
Davide Picca	Columbia University, USA
Mathieu Roche	Cirad, TETIS, France
Paolo Rosso	Technical University of Valencia, Spain
Bahar Sateli	Concordia University, Canada
Khaled Shaalan	The British University in Dubai, UAE
Lucia Siciliani	University of Bari Aldo Moro, Italy
Max Silberztein	Université de Franche-Comté, France
Vijay Sugumaran	Oakland University Rochester, USA
Maguelonne Teisseire	Irstea, TETIS, France
Maria Trușcă	Bucharest University of Economic Studies, Romania
Dan Tufis	Academia Romana, Romania
Nicolas Turenne	INRA UPEM, France
Luis Alfonso Ureña	Universidad de Jaén, Spain
Sunil Vadera	University of Salford, UK
Csaba Veres	UIB, Norway
Isabelle Wattiau	ESSEC & CNAM, France
Wlodek Zadrozny	UNCC, USA

Webmaster

Christian Willms	German Research Center for AI, Germany

Additional Reviewers

Rémy Decoupes	Martin Lentschat
Mohammed Erritali	Nils Müller
Javi Fernández Koustava Goswami	Atul Kr. Ojha
Seethalakshmi Gopalakrishnan	Priya Rani
Philipp Heinisch	Raniah Rawass
Jude Hemanth	Lucile Sautot
Roberto Interdonato	Rima Türker
Rodrique Kafando	Lei Zhang

Contents

Classification

Sentiment Analysis

Social Media

Linking Documents

Multimodality

Applications

The Role of Learning

You Can't Learn What's Not There: Self Supervised Learning and the Poverty of the Stimulus

Csaba Veres[1](✉) and Jennifer Sampson[2]

[1] University of Bergen, Bergen, Norway
csaba.veres@uib.no
[2] Equinor, London, UK
jensam@equinor.com

Abstract. *Diathesis alternation* describes the property of language that individual verbs can be used in different subcategorization frames. However, seemingly similar verbs such as *drizzle* and *spray* can behave differently in terms of the alternations they can participate in (drizzle/spray water on the plant; *drizzle/spray the plant with water). By hypothesis, primary linguistic data is not sufficient to learn which verbs alternate and which do not. We tested two state-of-the-art machine learning models trained by self supervision, and found little evidence that they could learn the correct pattern of acceptability judgement in the locative alternation. This is consistent with a poverty of stimulus argument that primary linguistic data does not provide sufficient information to learn aspects of linguistic knowledge. The finding has important consequences for machine learning models trained by self supervision, since they depend on the evidence present in the raw training input.

1 Introduction

Language models which employ self supervised learning represent a major breakthrough for machine learning. These algorithms construct a large statistical model of language by performing tasks such as masked language modeling on a vast quantity of unlabelled text [7,22]. The idea that linguistic knowledge can be acquired entirely from primary linguistic data (PLD) has been questioned for many years, with the poverty of stimulus (POS) argument. The term itself was introduced in [5], but has been part of Chomsky's arguments since at least 1965 [4]. The main problem raised by the POS argument is that the PLD does not contain the kinds of sentences that would help learners falsify (at least some of) the incorrect hypotheses about the grammar of their language [3,6]. The consequences for machine learning are the same: if POS is correct, self supervised

Notes and Comments. This research was supported by the Project News Angler, which is funded by the Norwegian Research Council's IKTPLUSS programme as project 275872.

models will not have sufficient data for a complete understanding of linguistic structure.

Warstadt [23] developed the Corpus of Linguistic Acceptability (CoLA) to test the POS argument with machine learning models. They argued that if grammatical acceptability judgements can be learned to human level with no in-built language specific principles, then this argues against POS. Their results showed that state-of-the-art recurrent neural network models could not achieve human level performance, suggesting that grammatical knowledge can not be learned in its entirety from linguistic input alone. A similar conclusion was reached a year later with a BiLSTM model using the GLUE benchmark, which included CoLA [20]. However, with the advent of the transformer architecture [18] and ensuing implementations, performance improved dramatically and the subsequent iteration of the benchmark, SuperGLUE, did not include the CoLA suite citing better than human performance by the XLNet-Large architecture [21,24].

Does this result mark the end of the POS hypothesis? Veres and Sandblast [19] argue that it does not, because the CoLA does not pose a sufficiently strong test of the hypothesis. The corpus includes a wide variety of grammatical violations, but no attempt is made to show that any of them are potentially unlearnable in the suggested way. In fact [3] argue that "responsible nativists" try to account for acquired linguistic knowledge with the minimum language specific component of learning, so CoLA may not have focused on the critical test cases. Veres and Sandblast [19] propose a new benchmark which is composed of grammatical violations related to Baker's paradox [1], which are not learnable from linguistic data alone [15]. Instead they rely on knowledge of linking rules between lexico-semantic features and their syntactic expressions. Their preliminary results supported the POS argument, that learning requires knowledge about language which is not directly discernible from the primary linguistic data. This paper reports additional experiments to provide stronger evidence that self supervised models are not able to learn certain aspects of linguistic knowledge.

2 Learnability and Semantics

The learning problem involves verb subcategorisation frames and the possibilities for alternative frames involving the same verb, also called diathesis alternations. For example the verb *load* can appear in the following construction (examples taken from [16]).

(1)   Hal is loading hay into the wagon

In sentence (1) the grammatical subject (Hal) of the verb is the loader, the object is the contents being moved (the hay), and the object of the preposition *into* expresses the container into which the hay is being moved (the wagon). This is called the content-locative construction, or V-locatum-location, because the focus of the sentence is the locatum (hay). The same meaning can be expressed by sentence (2) where the object of the verb is now the container, changing the focus of the sentence. This is called the container-locative construction, or V-location-locatum.

(2)   Hal loaded the wagon with hay

A possible generalisation for the learner is that verbs appearing in content-locative constructions can also appear in container-locative constructions.

However the generalisation does not hold, as there are many other verbs which result in unacceptable sentences. Examples (3) and (4) show that *pour* does not accept the container-locative, and *fill* does not allow content-locative. There does not seem to be a clear way to distinguish the verbs that do, and the ones that don't allow the generalisation. In these examples *pour*, *fill*, and *load* are all verbs which describe someone moving something somewhere.

(3)   a.   Amy poured water into the glass.
       b.   *Amy poured the glass with water.

(4)   a.   *Bobby filled water into the glass.
       b.   Bobby filled the glass with water.

The fact that adult speakers of English can make these distinctions is a learnability paradox. Four conditions lead to the paradox: (a) language speakers generalise from observations, (b) they avoid some possible generalisations, (c) they are not corrected for erroneous generalisations, (d) there is no systematic difference between verbs that allow generalisation and those which do not. Clearly at least one of these statements cannot be correct.

Pinker argues that the fourth condition is where the solution to the paradox lies, and in fact systematic differences do exist. However the differences are described in terms of nonobvious descriptions of semantic structure in the form of broad- and narrow- range semantic rules. Broad range rules provide necessary conditions, and the narrow range rules provide sufficient conditions [15,16].

A necessary condition for a verb to participate in the locative alternation is that it specifies both a type of motion and an end state. For example when someone *smears grease onto a bearing*, or *smears a bearing with grease*, then we know the kind of activity the person is engaged in and how the bearing will end up looking. On the other hand, the non alternating verb *fill* specifies only an end state. If I *fill the bottle with water* (not *fill water into the bottle*) then it is not clear how I filled it; what is clear is that the bottle is full. Conversely, if I *pour water into the bottle* (not *pour the bottle with water*) then the action I perform is more clear, but the end state of the bottle, less so. Note also that this contrast explains the subtle shift in meaning observed with alternating verbs. For example *The farmer loaded the cart with apples* suggests the cart is full, whereas *The farmer loaded apples into the cart* does not [11]. The V-location-locatum diathesis carries the semantic interpretation involving an end state.

The necessary conditions in themselves do not capture the full range of grammatical facts. For example, why is *I dripped the floor with water* not acceptable? In what way does it not entail an end state where the floor is covered with drops of water? If it did, then the construction should be grammatical under the present hypothesis. To explain these facts [15] further proposes a set of narrow range rules which provide fine-grained criteria which are sufficient to license the alternation for a given verb. The rules involve a range of language specific semantic properties which constrain the interpretation of concepts with respect to their expression, and particular argument structures are licensed by these semantic properties.

The broad and narrow range rules together are in fact *rules of construal* which are needed because cognition is too flexible to determine which syntactic device is most suited in expressing the communicative intent of the message. For example if someone in the real world *pours water into a glass*, are they affecting the water by causing it to move from one location to another (V-locatum-location) or are they affecting the glass by causing it to be less empty (V-location-locatum)? The broad range rule makes this determination for us. As far as language is concerned, *pour* is a verb that describes an action performed on the locatum rather than the state of the location. This principle is meant quite generally, such that the role of language is to funnel an infinitely flexible cognition into a more rigid and fixed system suitable for expression.

The narrow range rules provide specific constraints to determine the interpretation of narrow conflation classes. Returning to the example of *drip*, why does *I sprayed the plant with water* entail and end state but **I dripped the plant with water* does not? By hypothesis, the fine grained semantic description of *drip* verbs (which also includes *dribble, drizzle, dump, pour, ...*) is something like "a mass is enabled to move via the force of gravity." On the other hand *spray* verbs (which also includes *splash, splatter, sprinkle, squirt, ...*) are verbs where "force is imparted to a mass, causing ballistic motion in a specified spatial distribution along a trajectory" ([15], p.126). It is therefore a distinction between **enabling** and **causing** the motion of a mass, where the causation implies some element of control over the end state. "Dripping" does not entail an end state because we have no direct control over the end state.

This proposal is called the Grammatically Relevant Subsystem (GRS) approach, because the classification of verbs with respect to their subcategorization options is a matter for the specialised semantics embodied in the narrow range rules, rather than some more general classification problem. The semantic features are a part of the conceptual - linguistic linking system, and can not directly be inferred from the general properties of the observed linguistic input. Diathesis alternations are controlled by lexico-semantic facts that are not directly observable from the strings of the language, cannot be inferred from the statistical distribution of those strings, and should not be learned by systems that depend entirely on such distributions.

3   Related Work

There is a growing body of research whose goal is to investigate the nature of knowledge acquired by machine learning models, beyond the commonly used

NLP benchmark results. Many of these studies draw similar conclusions about the limitations of machine learning.

Bender et al. [2] take a somewhat general view of the limits of machine learning, arguing that text corpora can only provide linguistic *form*, which is not sufficient to capture *meaning*, or more precisely, *communicative intent*. While this is not strictly speaking a POS argument for acquisition of knowledge *about* language, it is a reminder that exposure to written sentences is not sufficient to model the use of language for communication.

Kassner and Schütze [10] test for more specific aspects of linguistic knowledge. They investigate pretrained language models (PLMs) for evidence of specific factual knowledge. They conclude that PLMs have difficulty with learning about negation. Given the statement "The theory of relativity was *not* developed by [MASK]" they are just as likely to predict "Einstein" as if the statement was "The theory of relativity was developed by [MASK]." In addition, PLMs can be misled in a novel technique called *mispriming*, inspired by psycholinguistic studies, where a question framed as "Talk? Birds can [MASK]", can prompt the erroneous response "Birds can talk."

In another set of experiments designed to test linguistic capacities rather than performance on popular NLP tasks, [8] show that BERT [7] lacks knowledge of negation and it struggles with some difficult inference and role-based event prediction tasks.

Turning now to the question of learnability from PLD, [14] propose a hierarchical Bayesian framework which is able to model many aspects of learning verb constructions, including those involved in Baker's paradox. They showed that diathesis alternations could be predicted by distributional evidence alone. They used a hierarchical Bayesian model which regarded deviations from expected frequencies as a form of negative evidence for resolving Baker's paradox. The model uses a hierarchy of inductive constraints, or *overhypotheses*, based on the distributional evidence. The model learns the distribution of verb constructions across all verbs in a language, as well as the degree to which any individual verb tends to be alternating or non-alternating. This way it can learn prior probabilities that can be used to predict the alternation patterns of verbs in the corpus. One limitation of the study is that the model is built to detect the non occurrence of just the right sentences, that is, the lack of the ungrammatical alternation. But this is unnatural because it assumes that, of the potentially infinite non occurring sentences containing a particular verb, language learners are tuned to focus on just the right ones.

The critical point which emerges from prior work is that statistical models can potentially learn the relevant grammatical generalisations from PLD, but only if they include built in assumptions about expected distributions in text. Transformer models make no prior assumptions and therefore it is important to determine if they can learn the relevant generalisations from the input data alone.

4   Dataset

The preliminary studies of [19] showed a mixed set of results for the 24 different types of diathesis alternations selected from [11]. Amongst the poorest performers were the as-, locative-, reciprocal-, and fulfilling- alternations. The locative alternation we have been describing is one of the best understood, and it was used in the sentences in this study.

We constructed a set of 274 sentences in total, 137 alternating and 137 non alternating. The 137 alternating sentences were all grammatical, but half of the non alternating sentences were ungrammatical. Table 1 shows the conditions with a sample sentence in each.

5   Results

5.1   Acceptability Experiments

We use the Hugging Face implementation of BERT (Bidirectional Encoder Representations from Transformers) [7]. The pre-trained model has been trained on vast amounts of general language data and can be fine-tuned by further training on downstream NLP tasks such as named entity recognition, classification, question answering, and acceptability judgement.

BERT is distinguished from other transformer-based networks by the input encoding it uses while training and the problems it was trained to solve during training: masked language modelling (MLM) and next sentence prediction (NSP).

Since acceptability judgement is a form of classification, we used BERT-ForSequenceClassification classifier using bert-base pretrained model, fine tuned on the CoLa dataset. The validation accuracy was 0.70 with validation loss = 0.61.

The common metric for acceptability judgement is the Matthews correlation coefficient which measures the agreement between classification scores and human judgement. The measure is thought to be particularly meaningful because

Table 1. Example sentences from the six different types in the experiment. The asterisk (*) denotes ungrammatical strings. The treatment conditions are named for the verb frame in which the example sentences are judged acceptable.

	V locatum location	V location locatum
Alternating	The farmer had to load all the apples into the cart	The farmer had to load the cart with all the apples
With only	*The final step is to coat chocolate on the cake	The final step is to coat the cake with chocolate
Into/Onto/On only	Carla poured lemonade into the pitcher	*Carla poured the pitcher with lemonade

it takes into account true and false positives and negatives, unlike the F measure typically used in many other tasks [13].

Table 2 shows the Matthews correlation coefficient for the two sentence types, compared to the experimenter's judgement of grammatical acceptability.

Table 2. Matthews correlation coefficient for acceptability judgement obtained with BERT.

	Matthews correlation
With only	0.27
Into/Onto only	0.05

The results show almost no sensitivity to grammatical acceptability for Into/onto only sentences that are ungrammatical in the V location locatum construction. Table 3 shows the reason for this is low accuracy for ungrammatical constructions (e.g. *Carla poured the pitcher with lemonade*). On the other hand there is a weak positive correlation for with-only sentences, but accuracy for unacceptable sentences (e.g. *The final step is to coat chocolate onto the cake*) is still low.

Table 3. Accuracy of acceptability judgement obtained with BERT.

	Grammatical	Ungrammatical
With only	1.0	0.16
Into/Onto	0.9	0.14

Since there was a weak correlation in one condition We repeated the analysis using RoBERTa, a newer model based on BERT with a robustly optimized pretraining approach [12] which uses a much larger training set, and modifies the training regime by dropping the next sentence prediction task.

We used the RobertaForSequenceClassification classifier from Hugging Face based on the roberta-base pretrained model. The classifier was fine tuned on the CoLA task as before, obtaining a higher validation accuracy = 0.86 and loss = 0.43. We submitted our results to Kaggle for test validation and achieved a result of 0.62[1]. Compare this to 0.678 for the Facebook implementation on gluebenchmark.com, where the current leader for this task is StructBERT from Alibaba with a score of 0.753 [22].

Table 4 shows the Matthews correlations. Surprisingly the with-only condition shows a slightly worse performance, but now the Into/Onto condition shows a stronger, moderate correlation.

---

[1] https://www.kaggle.com/c/cola-in-domain-open-evaluation/leaderboard.

Table 4. Matthews correlation coefficient for acceptability judgement obtained with RoBERTa.

	Matthews correlation
With only	0.17
Into/Onto only	0.4

The increased correlation in the Into/Onto condition is due to increased accuracy in the ungrammatical condition, as seen in Table 5. The with-only accuracy is still low, as expected from the correlation score.

Table 5. Accuracy of acceptability judgement obtained with RoBERTa.

	Grammatical	Ungrammatical
With only	1.0	0.09
Into/Onto only	0.97	0.45

5.2 Embeddings

It is generally believed that embeddings capture aspects of word semantics, though the nature of the semantic properties is not well understood [17]. If verb alternations depend on subtle semantic distinctions, then the word embeddings should contain elements of such semantics.

Figure 1 shows a 2-dimensional principal components projection of the vector embeddings for the verbs in the experimental conditions. The with-only verbs are shown with a plus sign "+" in the figure, and the Into/Onto-only verbs with the filled circles. Each verb appears more than once because embeddings are contextualised, and a given verb has a slightly different vector representation in different sentences. There is a very pronounced separation between the two sets of verbs, suggesting that something of the semantic difference was captured in the embedding space.

Closer inspection of the verbs, however, reveal a possible confound. It appears that the Into/Onto-only verbs in the cluster on the right of Fig. 1 appear with various liquids, while the with-only ones on the left can not. So, for example, pour/dribble/slop/slosh are actions one can perform with water but bandage/bind/decorate/dirty/bombard are not. This is just distributional semantics learned from the context in which words appear, where the distributional hypothesis [9] implies that words which cluster together are words which can be used interchangeably in relevant contexts. Sahlgren [17] calls this a *paradigmatic relation*. To control for the confound, we are currently collecting Into/onto-only sentences which do not include liquids. If we are correct then this should reduce classification accuracy to 0.

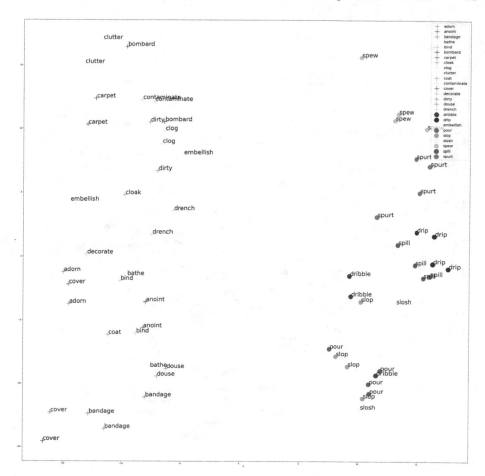

Fig. 1. 2-dimensional PCA projection of "Into/Onto verbs" (filled circles) and "With" verbs ("+") signs

There is an alternative test we can perform with the current sentences, which is to see if verbs that allow alternation cluster differently to ones which do not. Figure 2 shows these verbs as upside down triangles. We can see that the alternating verbs are spread throughout the non alternating ones. This is important for Pinkers's hypothesis since he writes: "The exact differentiation of the non-alternating classes from one another is not crucial as long as the criteria distinguishing them from the alternating classes are clear" ([15], p.237). Clearly they are not distinguished in the verb embeddings. Further, alternating verbs which co-occur with liquids overlap with the non alternating verbs that co-occur with liquids. For example squirt/sprinkle/spray appear next to dribble/pour/spew. This strengthens the hypothesis that the semantics captured in RoBERTa is limited to distributional co-occurrence.

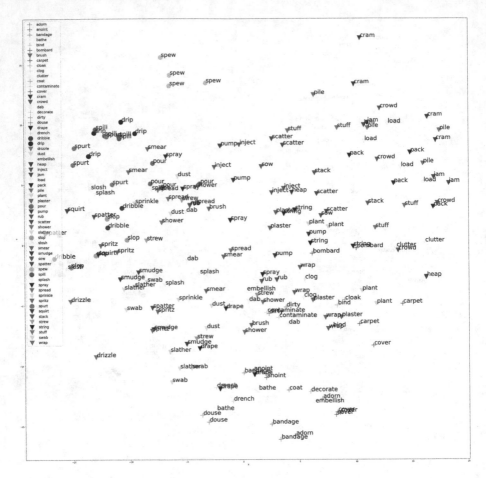

Fig. 2. 2-dimensional PCA projection of "Into/Onto verbs" (filled circles), "With" verbs ("+") signs, and alternating verbs (upside down triangle)

6 Discussion

We began by considering Baker's paradox which concerns problems with the learning of syntactic diathesis alternations from primary linguistic data. The proposed solution involved a number of lexical semantic features that constrain the syntactic behaviour of individual verbs. We then asked if these features could be learned by modern machine learning architectures trained on massive text corpora. The results show that neither BERT nor RoBERTa were able to reliably differentiate the verbs on semantic grounds.

However, RoBERTa achieved moderate performance for recognizing the acceptability of Into/Onto-only verbs, and embeddings from both systems showed an appreciable separation between the two non alternating verb classes. We argued that this result was an artefact because the Into/onto-only verbs in our test sentences tended to have liquids as objects while the with-only verbs did not. We should then be able to abolish the model's classification accuracy with a new test that included Into/onto-only verbs with non liquid objects, for example *"He coiled the chain around the pole"/ "*He coiled the pole with the chain."*

7   Conclusion

The results reported in this paper show that current state-of-the art machine learning systems cannot learn the necessary knowledge to be able to correctly judge the acceptability of the locative alternation, from text input alone. It is suggested that the poverty of the stimulus is a fundamental limitation for statistical learning from text corpora, and practitioners should be aware that their models could have unpredictable "blind spots".

References

1. Baker, C.: Syntactic theory and the projection problem. Linguist. Inquiry **10**, 533–581 (1979)
2. Bender, E.M., Koller, A.: Climbing towards NLU: on meaning, form, and understanding in the age of data. In: Proceedings of the 58th Annual Meeting of the Association for Computational Linguistics, pp. 5185–5198. Association for Computational Linguistics, July 2020. https://www.aclweb.org/anthology/2020.acl-main.463
3. Berwick, R.C., Pietroski, P., Yankama, B., Chomsky, N.: Poverty of the stimulus revisited. Cogn. Sci. **35**(7), 1207–1242 (2011). https://doi.org/10.1111/j.1551-6709.2011.01189.x. ISSN 1551-6709
4. Chmomsky, N.: Aspects of the Theory of Syntax. MIT Press, Cambridge (1965)
5. Chomsky, N.: Rules and Representations. Columbia Classics in Philosophy. Columbia University Press (1980). https://books.google.no/books?id=KdYOYJwjFo0C. ISBN 9780231048279
6. Cowie, F.: Innateness and language. In: Zalta, E.N. (ed.) The Stanford Encyclopedia of Philosophy. Metaphysics Research Lab, Stanford University, Fall 2017 edn (2017)
7. Devlin, J., Chang, M.-W., Lee, K., Toutanova, K.: BERT: pre-training of deep bidirectional transformers for language understanding. In: Burstein, J., Doran, C., Solorio, T. (eds.) NAACL-HLT (1), pp. 4171–4186. Association for Computational Linguistics (2019). http://dblp.uni-trier.de/db/conf/naacl/naacl2019-1.html#DevlinCLT19. ISBN 978-1-950737-13-0
8. Ettinger, A.: What BERT is not: lessons from a new suite of psycholinguistic diagnostics for language models. Trans. Assoc. Comput. Linguist. **8**, 34–48 (2020). https://doi.org/10.1162/tacl_a_00298
9. Harris, Z.S.: Distributional structure. WORD **10**(2–3), 146–162 (1954). https://doi.org/10.1080/00437956.1954.11659520. ISSN 0043-7956
10. Kassner, N., Schütze, H.: Negated and misprimed probes for pretrained language models: birds can talk, but cannot fly. In: Proceedings of the 58th Annual Meeting of the Association for Computational Linguistics, pp. 7811–7818. Association for Computational Linguistics, July 2020. https://www.aclweb.org/anthology/2020.acl-main.698

11. Levin, B.: English Verb Classes and Alternations: A Preliminary Investigation. The University of Chicago Press, The University of Chicago (1993). ISBN 0-226-47532-8
12. Liu, Y., et al.: Roberta: a robustly optimized BERT pretraining approach. CoRR, abs/1907.11692 (2019). http://arxiv.org/abs/1907.11692
13. Matthews, B.W.: Comparison of the predicted and observed secondary structure of T4 phage lysozyme. Biochimica et Biophysica Acta (BBA) - Protein Struct. **405**(2), 442–451 (1975). https://doi.org/10.1016/0005-2795(75)90109-9. http://www.sciencedirect.com/science/article/pii/0005279575901099. ISSN 0005-2795
14. Perfors, A., Tenenbaum, J.B., Wonnacott, E.: Variability, negative evidence, and the acquisition of verb argument constructions. J. Child Lang. **37**(3), 607–642 (2010). https://doi.org/10.1017/S0305000910000012
15. Pinker, S.: Learnability and Cognition: The Acquisition of Argument Structure (1989/2013), New edn. MIT Press, Cambridge (2013)
16. Pinker, S.: The Stuff of Thought : Language as a Window Into Human Nature. Viking, New York (2007)
17. Sahlgren, M.: The distributional hypothesis. Italian J. Linguist. **20**, 33–53 (2008)
18. Vaswani, A., et al.: Attention is all you need. In: Proceedings of the 31st International Conference on Neural Information Processing Systems, NIPS 2017, Red Hook, NY, USA, pp. 6000–6010. Curran Associates Inc. (2017). ISBN 9781510860964
19. Veres, C., Sandblåst, B.H.: A machine learning benchmark with meaning: learnability and verb semantics. In: Liu, J., Bailey, J. (eds.) AI 2019. LNCS (LNAI), vol. 11919, pp. 369–380. Springer, Cham (2019). https://doi.org/10.1007/978-3-030-35288-2_30
20. Wang, A., Singh, A., Michael, J., Hill, F., Levy, O., Bowman, S.R.: GLUE: a multi-task benchmark and analysis platform for natural language understanding. In: The Proceedings of ICLR (2019)
21. Wang, A., et al.: SuperGLUE: A Stickier Benchmark for General-Purpose Language Understanding Systems (2019)
22. Wang, W., et al.: Structbert: Incorporating language structures into pre-training for deep language understanding (2019)
23. Warstadt, A., Singh, A., Bowman, S.R.: Neural Network Acceptability Judgments (2018)
24. Yang, Z., Dai, Z., Yang, Y., Carbonell, J.G., Salakhutdinov, R., Le, Q.V.: XLNet: generalized autoregressive pretraining for language understanding. CoRR, abs/1906.08237 (2019). http://arxiv.org/abs/1906.08237

Scaling Federated Learning
for Fine-Tuning of Large Language
Models

Agrin Hilmkil[1](✉), Sebastian Callh[1], Matteo Barbieri[1], Leon René Sütfeld[2],
Edvin Listo Zec[2], and Olof Mogren[2]

[1] Peltarion, Stockholm, Sweden
https://peltarion.com
[2] RISE Research Institutes of Sweden, Gothenburg, Sweden

Abstract. Federated learning (FL) is a promising approach to distributed compute, as well as distributed data, and provides a level of privacy and compliance to legal frameworks. This makes FL attractive for both consumer and healthcare applications. However, few studies have examined FL in the context of larger language models and there is a lack of comprehensive reviews of robustness across tasks, architectures, numbers of clients, and other relevant factors. In this paper, we explore the fine-tuning of large language models in a federated learning setting. We evaluate three popular models of different sizes (BERT, ALBERT, and DistilBERT) on a number of text classification tasks such as sentiment analysis and author identification. We perform an extensive sweep over the number of clients, ranging up to 32, to evaluate the impact of distributed compute on task performance in the federated averaging setting. While our findings suggest that the large sizes of the evaluated models are not generally prohibitive to federated training, we found that not all models handle federated averaging well. Most notably, DistilBERT converges significantly slower with larger numbers of clients, and under some circumstances, even collapses to chance level performance. Investigating this issue presents an interesting direction for future research.

Keywords: Distributed · Federated learning · Privacy · Transformers

1 Introduction

Transformer-based architectures such as BERT have recently lead to breakthroughs in a variety of language-related tasks, such as document classification, sentiment analysis, question answering, and various forms of text-mining [1,2,10,21,23,25]. These models create semantic representations of text, which can subsequently be used in many downstream tasks [2]. The training process for Transformers typically includes two phases: pre-training and task-specific

© Springer Nature Switzerland AG 2021
E. Métais et al. (Eds.): NLDB 2021, LNCS 12801, pp. 15–23, 2021.
https://doi.org/10.1007/978-3-030-80599-9_2

fine-tuning. During pre-training, the model learns to extract semantic representations from large, task-independent corpora. The pre-training is followed by task-specific fine-tuning on a separate dataset to optimize model performance further. In this paper, we study the effects of fine-tuning large language models in a federated learning (FL) setting. In FL, models are trained in a decentralized fashion on a number of local compute instances, called clients, and intermittently aggregated and synchronized via a central server. As such, FL is a solution which provides a level of privacy with regards to the sharing of personal or otherwise sensitive data. Model aggregation is commonly performed via averaging of the weights of the individual client models, called Federated Averaging (FEDAVG) [16]. Depending on the application, the number of clients in an FL setting can differ wildly. In instances where smartphones are used as clients, their number can reach into the millions [5], whereas settings with higher compute requirements and more data per client will often range between a handful and a few dozens of clients. Here, we focus on the latter. A potential application of this is the medical field, in which automated analyses of electronic health records yield enormous potential for diagnostics and treatment-related insights [26].

Our contribution is a comprehensive overview of the applicability of the federated learning setting to large language models. To this end, we work with a fixed computation budget for each task, and use a fixed total amount of data while varying the number of clients between which the data is split up. This way, we isolate the effects of distributing data over several clients for distributed compute. We leave comparisons with a fixed amount of data per client, and varying non-i.i.d. data distributions between clients for future work. The main contributions of this paper are the following: (1) We provide a comparison of three popular Transformer-based language models in the federated learning setting, using the IMDB, Yelp F, and AG News datasets. (2) We analyze how the number of clients impacts task performance across tasks and model architectures. Finally, we share our code publicly[1].

2   Related Work

Federated optimization was first introduced in [8]. The key challenges in this paradigm are communication efficiency when learning from many clients, privacy concerns with respect to leakage of client data, and variability in data distributions between clients (non-i.i.d. setting). FEDAVG [16] solves the federated optimization problem by building a global model based on local stochastic gradient descent updates and has been shown to work on non-i.i.d. data in some circumstances. Since then, many adaptations have arisen [7,11,18]. [4] proposes a one-shot FL algorithm, learning a global model efficiently in just one communication round. [6,28] and [13] study effects of FEDAVG and non-i.i.d. client

---

[1] https://github.com/Peltarion/scaling_fl.

data. [17] and [5] train large recurrent language models with user-level differ-ential privacy guarantees and for mobile keyboard prediction. [3] use federated learning for named entity recognition in heterogeneous medical data.

Most architectures used in FL to date are relatively small (e.g., CIFG for mobile keyboard prediction: 1.4M parameters [5]), compared to BERT-based language models with hundreds of millions of parameters. How these very large models behave under FEDAVG remains underexplored. To the best of our knowl-edge, [12] and [14] are the first ones to train large Transformer models in a fed-erated setting. [14] trained BERT on a medical corpus and showed that both pre-training and fine-tuning could be done in a federated manner with only minor declines in task performance. Nonetheless, the study is mainly a proof-of-concept and does not explore many of the factors that can be expected in real-world scenarios. For instance, the authors only used five clients, and eval-uated them only on i.i.d. data. [12] introduces FedDF, an ensemble distillation algorithm for model fusion. The authors train a central model through unlabeled data on the client models outputs, and perform fine-tuning on a pre-trained Dis-tilBERT [20] in a federated setting as a baseline. To the best of our knowledge, no systematic variation of the number of clients and other relevant factors has previously been explored in this context.

3   Method

3.1   Federated Learning

Federated learning aims to solve the optimization problem

$$\min_{\theta \in \mathbb{R}^d} \frac{1}{K} \sum_{k=1}^{K} F_k(\theta), \tag{1}$$

where $F_k(\theta) = \mathbb{E}_{x \sim \mathcal{D}_k} [\ell_k(\theta; x)]$ is the expected loss on client k and $\mathcal{D}_k$ is the data distribution of client k. In FEDAVG, a global model f_θ is initialized on a central server and distributed to all K clients, each of which then trains its individual copy of the network using SGD for E local epochs with local batch size B. The clients' updated parameters are then averaged on the central server, weighted by the local data size at each client. The averaged model is distributed to the clients again, and the process is repeated for a defined number of communication rounds. We implement FEDAVG using distributed PyTorch [19]. For each experiment we start from a pre-trained model, and fine-tune it with federated averaging on the current task.

3.2   Models

We include BERT with 110M parameters, 12 layers [2], ALBERT with 11M parameters, 12 layers [9] and DistilBERT with 65M parameters, 6 layers [20]. This allows us to study the effect that both the parameter count and the number of layers have on FEDAVG. All models are the corresponding base models pre-trained on (cased) English. In particular, it should be noted that while the models have similar architectures, they have some key differences. ALBERT introduces factorized embedding parameterization and cross-layer parameter sharing, while the DistilBERT model is a student network trained with knowledge distillation from BERT. We use the weights and implementations of the models available in the Huggingface Transformers library [24].

3.3   Datasets

We performed experiments on three datasets to assess the performance of the proposed approach on different tasks. All of them pose classification problems with a different number of target categories and dataset sizes. **IMDB** [15] contains of a collection of 50,000 movie reviews and their associated binary sentiment polarity labels (either "positive" or "negative"), which is used to train a sentiment classifier. **Yelp F** [27] contains reviews of local businesses and their associated rating (1–5). The task is posed as a text classification task, from the review text to its associated rating. **AG News**[2] consists of over one million categorized news articles gathered from more than 2,000 news sources. We used the common subset [27] of the whole dataset, consisting of 120,000 samples equally divided in four categories.

3.4   Experiments and Hyperparameters

We construct several experiments to evaluate how well Federated Learning scales with an exponentially increasing number of clients. In all experiments, the respective dataset is divided into a number of subsets equal to the number of clients. Data points are uniformly sampled on each client (i.i.d.). Results with a single client are considered centralized training baselines for each model and dataset. We run the baselines for a fixed number of rounds based on our compute budget. The test set performance for the baselines are then compared against varying number of participating clients at the same number of rounds. Finally, since runs with a larger number of clients converge more slowly, we allow those runs to continue to a second threshold and report the number of rounds required to reach 90% of the baseline performance, similar to [16]. Runs not reaching 90% of the baseline performance within the second threshold are reported as failures. We run the baseline for 100 rounds for both IMDB and AG News while setting the second threshold to 200 rounds. However, we only run Yelp F baselines for

---

[2] http://groups.di.unipi.it/~gulli/AG_corpus_of_news_articles.html.

50 rounds due to its large size and set the second threshold at 100 rounds. Like [12], we avoid momentum, weight decay, and dynamic learning rates for simplicity. Instead, all experiments are performed with SGD. Based on [22] we choose a constant learning rate of $2 \cdot 10^{-5}$, a maximum sequence length of 128 and a batch size (B) of 32. Furthermore, the number of local epochs (E) is set to 2 per round.

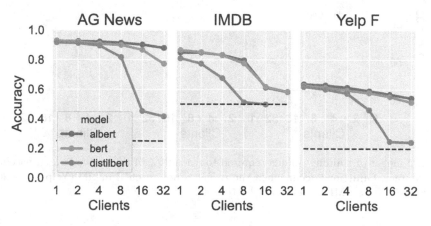

Fig. 1. Test accuracy at a fixed compute budget of 100 rounds for AG, IMDB, and 50 rounds for Yelp F. The expected accuracy of a random classifier for each task has been highlighted in the dashed line. Higher is better.

4 Results

4.1 Fixed Compute Budget

In Fig. 1, we study the effect of increasing the number of clients. It shows the final accuracy after 100 rounds for IMDB and AG News, and 50 rounds for the much larger Yelp F., with an exponentially increasing number of clients. Both ALBERT and BERT are well behaved and exhibit a gradual decrease with an increasing number of clients. However, DistilBERT shows a much steeper decline when moving past 4 clients for all datasets, down to the random classifier baseline (IMDB, Yelp F).

4.2   Rounds Until Target Performance

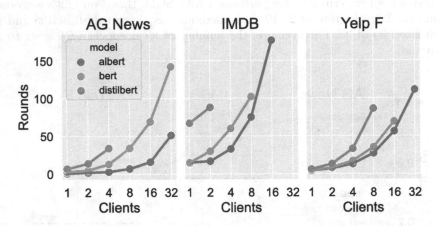

Fig. 2. Number of training rounds required to reach 90% of the non-federated baseline test accuracy. Omittions occur when the target is not reached in 100 (Yelp F) or 200 rounds (AG News, IMDB). Lower is better.

Examining the number of rounds necessary to achieve 90% of the non-federated baseline accuracy (Fig. 2) yields a similar observation. While all models perform worse with more clients, ALBERT and BERT mostly reach the target accuracy within the allocated number of rounds until 32 clients are used. DistilBERT on the other is unable to reach the target accuracy at 16 clients for Yelp F, and as low as 4 clients for IMDB.

4.3   Dynamics of Fine-Tuning

The test accuracy during fine-tuning (Fig. 3) allows a more complete understanding of how well FEDAVG scales for language model fine-tuning. While some scenarios (e.g. Yelp F. with BERT) show a gradual degradation with the number of clients, other configurations are more adversely affected by the increasing number of clients. In some instances the accuracy stays constant over a large period, sometimes even at the random classifier baseline for the whole (DistilBERT on IMDB) or part (DistilBERT on AG News) of the experiment when the number of clients is high.

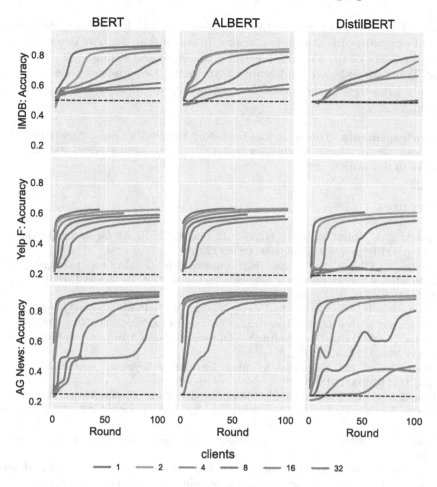

Fig. 3. Test accuracy (higher is better) over communication rounds for our scenarios. The random classifier baseline is shown as a dashed line.

5 Discussion

In this paper, we have evaluated the performance of Transformer-based language models fine-tuned in a federated setting. While BERT and ALBERT seem to learn each task quickly (Fig. 3), DistilBERT has a much slower learning progression in the federated setup. A possible explanation is the process of distillation during pre-training, but further research is needed to fully understand the cause. We demonstrated that BERT and ALBERT scale well up to 32 clients (Fig. 1), but found a substantial drop in performance compared in DistilBERT compared to its own baseline. Furthermore, DistilBERT requires more rounds to achieve the same performance. Investigating these issues in training DistilBERT with FL may be a promising direction for future research. Conversely, these results indicate that FL can be sensitive to the number of clients, highlighting the importance of evaluating FL at different scales. In conclusion, we have demonstrated the applicability of the federated learning paradigm and evaluated it on a number of Transformer-based models up to 32 clients. Our findings show that the relatively large sizes of these models are generally not prohibitive for federated learning.

Acknowledgements. This work was funded by VINNOVA (grant 2019-05156). We would also like to thank AI Sweden and CGit for providing additional compute resources to this work.

References

1. Adhikari, A., Ram, A., Tang, R., Lin, J.: DocBERT: BERT for document classification. arXiv preprint arXiv:1904.08398 (2019)
2. Devlin, J., Chang, M.W., Lee, K., Toutanova, K.: BERT: pre-training of deep bidirectional transformers for language understanding. In: NAACL-HLT (1), pp. 4171–4186 (2019)
3. Ge, S., Wu, F., Wu, C., Qi, T., Huang, Y., Xie, X.: FedNER: medical named entity recognition with federated learning. arXiv preprint arXiv:2003.09288 (2020)
4. Guha, N., Talwalkar, A., Smith, V.: One-shot federated learning. arXiv preprint arXiv:1902.11175 (2019)
5. Hard, A., et al.: Federated learning for mobile keyboard prediction. CoRR abs/1811.03604 (2018)
6. Hsu, T.M.H., Qi, H., Brown, M.: Measuring the effects of non-identical data distribution for federated visual classification. arXiv preprint arXiv:1909.06335 (2019)
7. Karimireddy, S.P., Kale, S., Mohri, M., Reddi, S.J., Stich, S.U., Suresh, A.T.: Scaffold: stochastic controlled averaging for on-device federated learning. arXiv preprint arXiv:1910.06378 (2019)
8. Konečný, J., McMahan, B., Ramage, D.: Federated optimization: distributed optimization beyond the datacenter. arXiv preprint arXiv:1511.03575 (2015)
9. Lan, Z., Chen, M., Goodman, S., Gimpel, K., Sharma, P., Soricut, R.: ALBERT: a lite BERT for self-supervised learning of language representations. In: International Conference on Learning Representations (2020)
10. Lee, J., et al.: BioBERT: a pre-trained biomedical language representation model for biomedical text mining. Bioinformatics **36**(4), 1234–1240 (2020)
11. Li, T., Sanjabi, M., Beirami, A., Smith, V.: Fair resource allocation in federated learning. arXiv preprint arXiv:1905.10497 (2019)
12. Lin, T., Kong, L., Stich, S.U., Jaggi, M.: Ensemble distillation for robust model fusion in federated learning. arXiv preprint arXiv:2006.07242 (2020)
13. Listo Zec, E., Mogren, O., Martinsson, J., Sütfeld, L.R., Gillblad, D.: Federated learning using a mixture of experts (2020)
14. Liu, D., Miller, T.: Federated pretraining and fine tuning of BERT using clinical notes from multiple silos (2020)
15. Maas, A.L., Daly, R.E., Pham, P.T., Huang, D., Ng, A.Y., Potts, C.: Learning word vectors for sentiment analysis. In: Proceedings of the 49th Annual Meeting of the Association for Computational Linguistics: Human Language Technologies, Portland, Oregon, USA, pp. 142–150. Association for Computational Linguistics, June 2011

16. McMahan, B., Moore, E., Ramage, D., Hampson, S., Arcas, B.A.: Communication-efficient learning of deep networks from decentralized data. In: Singh, A., Zhu, J. (eds.) Proceedings of the 20th International Conference on Artificial Intelligence and Statistics, vol. 54, , Fort Lauderdale, FL, USA, pp. 1273–1282. PMLR, 20–22 April 2017

17. McMahan, H.B., Ramage, D., Talwar, K., Zhang, L.: Learning differentially private recurrent language models. arXiv preprint arXiv:1710.06963 (2017)

18. Mohri, M., Sivek, G., Suresh, A.T.: Agnostic federated learning. In: Chaudhuri, K., Salakhutdinov, R. (eds.) Proceedings of Machine Learning Research, Long Beach, California, USA, vol. 97, pp. 4615–4625. PMLR, 09–15 June 2019

19. Paszke, A., et al.: Pytorch: an imperative style, high-performance deep learning library. In: Wallach, H., Larochelle, H., Beygelzimer, A., d' Alché-Buc, F., Fox, E., Garnett, R. (eds.) Advances in Neural Information Processing Systems 32, pp. 8024–8035. Curran Associates, Inc. (2019)

20. Sanh, V., Debut, L., Chaumond, J., Wolf, T.: DistilBERT, a distilled version of BERT: smaller, faster, cheaper and lighter. In: EMC^2 Workshop at NeurIPS 2019 (2019)

21. Sun, C., Huang, L., Qiu, X.: Utilizing BERT for aspect-based sentiment analysis via constructing auxiliary sentence. arXiv preprint arXiv:1903.09588 (2019)

22. Sun, C., Qiu, X., Xu, Y., Huang, X.: How to fine-tune BERT for text classification? In: Sun, M., Huang, X., Ji, H., Liu, Z., Liu, Y. (eds.) CCL 2019. LNCS (LNAI), vol. 11856, pp. 194–206. Springer, Cham (2019). https://doi.org/10.1007/978-3-030-32381-3_16

23. Vaswani, A., et al.: Attention is all you need. In: Advances in Neural Information Processing Systems, pp. 5998–6008 (2017)

24. Wolf, T., et al.: HuggingFace's transformers: state-of-the-art natural language processing. arXiv abs/1910.03771 (2019)

25. Yang, W., et al.: End-to-end open-domain question answering with BERTserini. arXiv preprint arXiv:1902.01718 (2019)

26. Zeng, Z., Deng, Y., Li, X., Naumann, T., Luo, Y.: Natural language processing for EHR-based computational phenotyping. IEEE/ACM Trans. Comput. Biol. Bioinf. **16**(1), 139–153 (2018)

27. Zhang, X., Zhao, J., LeCun, Y.: Character-level convolutional networks for text classification. In: Cortes, C., Lawrence, N.D., Lee, D.D., Sugiyama, M., Garnett, R. (eds.) Advances in Neural Information Processing Systems 28, pp. 649–657. Curran Associates, Inc. (2015)

28. Zhao, Y., Li, M., Lai, L., Suda, N., Civin, D., Chandra, V.: Federated learning with non-IID data. arXiv preprint arXiv:1806.00582 (2018)

Overcoming the Knowledge Bottleneck Using Lifelong Learning by Social Agents

Sergei Nirenburg[✉], Marjorie McShane, and Jesse English

Language-Endowed Intelligent Agents Lab, Rensselaer Polytechnic Institute,
Troy, NY 12180, USA

Abstract. In this position paper we argue that the best way to overcome the notorious *knowledge bottleneck* in AI is using lifelong learning by social intelligent agents. Keys to this capability are deep language understanding, dialog interaction, sufficiently broad-coverage and fine-grain knowledge bases to bootstrap the learning process, and the agent's operation within a comprehensive cognitive architecture.

Keywords: Artificial intelligent agents · Computational cognitive modeling · Lifelong learning · Cognitive architecture · Reasoning

The dominant AI paradigm today, which involves sophisticated analogical reasoning using machine learning, geared toward modeling the structure and processes of the human brain, not the content that drives its functioning. As a result, the emphasis in applications involving language processing is on developing sophisticated methods for manipulating uninterpreted results of perception (such as textual strings). This approach has a core weakness: the inability of systems to carry out self-aware reasoning or realistically explain their behavior. Attaining human-level performance in artificial intelligent agents is predicated on modeling how the architectures and algorithms used in implementing such agents handle the knowledge supporting decision-making, especially when related to conscious, deliberate behavior. Sufficient amounts of different kinds of knowledge must be amassed to emulate the knowledge humans have at their disposal to support commonsense decision-making during a variety of perception interpretation, reasoning, and action-oriented tasks. The availability of such knowledge to the artificial intelligent agents is, thus, a core prerequisite for this program of work. The conceptual complexities and the slow pace of the knowledge acquisition efforts in the classical AI paradigm led most of the AI practitioners to the conclusion that the field is facing an insurmountable "knowledge bottleneck." So, the task of knowledge acquisition was deemed to be impossible to tackle directly. Hence the well-known paradigm shift toward empirical methods.

Still, if the goal is developing systems that claim to model conscious human functioning, ignoring the "knowledge bottleneck" is not an option. Systems that aspire to emulate human capabilities of understanding, reasoning and explanation must constructively address the issue of knowledge acquisition and maintenance, which is a prerequisite for sustaining the lifelong operation of knowledge-based reasoning systems. This objective

© Springer Nature Switzerland AG 2021
E. Métais et al. (Eds.): NLDB 2021, LNCS 12801, pp. 24–29, 2021.
https://doi.org/10.1007/978-3-030-80599-9_3

is one of the central directions of R&D in our work on developing language-endowed intelligent agent (LEIA) systems. In the most general terms, our approach to overcoming the knowledge bottleneck is to develop agents (LEIAs) that can learn knowledge automatically by understanding natural language texts and dialog utterances. This can only be facilitated by the availability of a language interpreter system that extracts -- and represents in a metalanguage anchored in a formal ontological model of the world – the semantic and pragmatic/discourse meanings of natural language utterances and text. Such a system, in turn, would require significant knowledge support.

Over the past several decades, our team has developed a comprehensive language interpreter, OntoSem (the latest version is described in some detail in [1]), whose supporting knowledge resources include the ontological world model of ~9,000 concepts (~165,000 RDF triples) and the English semantic lexicon with ~30,000 word senses. In our R&D on overcoming the knowledge bottleneck we use OntoSem and its knowledge resources to *bootstrap* the process of automatic knowledge acquisition through language understanding.

OntoSem differs from practically all extant semantic and pragmatic analyzers in several ways, detailed in [1]: (a) it pursues ontologically-grounded semantic and pragmatic interpretation of inputs; (b) it determines how deeply to analyze inputs based on *actionability* requirements, which requires integrating reasoning about action with reasoning about language processing [9]; (c) it tackles a comprehensive inventory of difficult language communication phenomena such as lexical and referential ambiguity, fragments, ellipsis, implicatures, production errors, and many more; and (d) it facilitates lifelong learning – of lexical units in the lexicon as well as concepts and concept properties in the ontology necessary to express the meanings of lexical units.

OntoSem is the language interpretation module of OntoAgent, a cognitive architecture that serves as a platform for developing LEIA systems [2, 3]. OntoAgent is implemented as a service-based environment that consists of (a) a network of processing services, (b) a content service (comprised of several non-toy knowledge bases), and (c) an infrastructure service that supports system functioning, system integration, and system development activities. [INCLUDE A GENERIC ONTOAGENT DIAGRAM] OntoAgent operates at a level of abstraction that supports interoperability across the various perception, reasoning, and action services by standardizing input and output signals generated by all the "in-house" services. These signals are interoperable *M*eaning *R*epresentations, called XMRs, in which X is a variable describing a particular type of meaning – e.g., *visual* meaning (VMR) or *text* meaning (TMR). XMRs are formulated using a uniform knowledge representation schema that is compatible with the representation of static knowledge resources stored in a LEIA's memory system. Atoms of XMRs are semantically interpreted by reference to their descriptions in the LEIA's ontological world model, which is an important part of its long-term semantic memory.

To-date, proof-of-concept OntoAgent-based systems have been built that demonstrate the learning (either through dialog or utterances gleaned from text corpora) of ontological concepts and their properties; lexicon entries [5, 6], complex events (scripts) [7, 8] and even elements of the agent's knowledge about other agents (their "theory of mind," goals, plans, personality characteristics, biases, etc. [24]. Work is ongoing on extending the coverage and the typology of entities that a LEIA can learn. Clearly, a lot remains to be done. Strategically, continuing to develop the bootstrapping approach (with an option for human acquirers to "touch up" the agent's bootstrapping resources whenever human resources permit) is the best path toward overcoming the knowledge acquisition bottleneck. Space limitations do not allow detailed descriptions of any of the above. In this position paper we, therefore, discuss programmatic matters and refer the reader to publications where detailed descriptions of relevant phenomena and processes can be found.

Learning in OntoAgent can operate in an "opportunistic" mode, in which learning processes are spawned as a consequence of the LEIA's having encountered lacunae or inconsistencies in its knowledge resources while performing their regular tasks in whatever domain they are implemented. This process aims to model the way humans continuously enrich their vocabularies and their understanding of the world while engaging in a variety of activities not overtly associated with learning. It is a never-ending process of continual honing of the understanding of meanings of lexical units that should be very familiar to anybody who has ever operated in a language environment other than that of one's mother tongue. At this point, we concentrate on language inputs but enhancing the "opportunistic learning" method by taking into account the results of interpretation of other perception modalities, such as visual scene recognition, is a natural extension.

In what follows we briefly describe two examples of opportunistic learning. Consider a class of situations in which agents encounter an unknown word or phrase during language understanding within an application. In such a situation the agents first carry out a minimum of coarse-grained learning of the meaning of this word with the objective of generating a minimally acceptable underspecified meaning representation of the input utterance. This stage of learning relies as supporting knowledge on standard lexicon entry templates, the results of syntactic parsing, and the semantic analysis of known portions of the clause (mainly through unidirectional application of selectional restrictions encoded in lexical entries of known words in the input). For example, if the agent doesn't know the word *tripe* in the input *Mary was eating tripe*, it will learn a new lexical entry for *tripe* and, using the information that a) in the input sentence *tripe* is the direct object of *eat*, and b) that direct objects typically link to the THEME case role of the concept underlying the meaning of the verb in the input, have the semantics of *tripe* tentatively – pending further downstream specification – interpreted as a member of the ontological subhierarchy rooted at the concept INGESTIBLE, which is the THEME of INGEST, the concept used to interpret the semantics of the most frequent sense of the verb *eat* (For detailed descriptions and examples of this process see [5, 6].)

Similarly, when an agent encounters an unknown use (lexical sense) of a known word or phrase, it *coerces* the known meaning using an inventory of template-conversion rules. For example, the utterance *Mary* **rulered** *a pencil to John* will be interpreted as (in plain English, for clarity), 'Mary transferred possession of a pencil to John using a ruler' [4]. If the resulting interpretations of such inputs are actionable, the agent need not (at least immediately) pursue deeper learning. If they are not actionable, then the agent can attempt to recover in various ways, such as learning by reading from a corpus [5, 6] or entering into a dialog with a human collaborator (if present).

Another mode of LEIA learning is deliberate, dedicated learning, meaning that learning is the specific goal that the LEIA is pursuing at the time. Deliberate learning can be realized as interactive learning by instruction, as individual learning by reading or as a combination of these two methods. (Deliberate learning can also take place without an immediate perceptual trigger – the agent can use its reasoning capabilities to derive new knowledge through the application of rules of reasoning over its stored knowledge. This approach to learning has been in pursued in AI throughout its existence. We do not address this "internal reasoning" mode of learning in this paper.) The expectation in deliberate learning is that the natural language inputs to the system are texts or dialog utterances that are to be interpreted as training instructions. While the dedicated learning mode can be used to learn ontological concepts and lexical units, an important application of this mode is to teach LEIAs how to perform a variety of tasks and how they should assess various states of the world in preparation to making their decisions about action. To-date, we have developed and demonstrated two proof-of-concept systems of deliberate learning by language-based instruction in interactive dialogs between agents and human team members: a) a LEIA integrated into a furniture-building robot that learned ontological scripts using language instruction by a human [7, 8], and b) a LEIA integrated into a self-driving vehicle application that was how to behave in a variety of situations, such as how to get to various places, how to react to unexpected road hazards (e.g., a downed tree), and how to behave in complex situations, such as at a four-way stop [10]. The latter application also incorporated the opportunistic learning mode.

Irrespective of a particular mode, all learning based on language communication is made possible by close integration of several capabilities of LEIAs: a) advanced, broad-coverage language understanding; b) reasoning about domain-oriented tasks; and c) a set of heuristic rules guiding the learning process as such and thus representing LEIA's expertise as learners. As already mentioned, all of the above capabilities are predicated on the availability of a shared knowledge environment that both bootstraps learning and is continuously expanded and honed as a result of learning.

OntoAgent R&D belongs to the area of cognitive systems (e.g., [11, 12]). A number of research teams develop architectures that pursue aims that are broadly similar to those of OntoAgent. Systems and architectures such as DIARC [13], Companions [14], Icarus [15], Rosie [16] and Arcadia [17] all have salient points of comparison. Fundamental comparison of these and other systems is not feasible in this space. Here we will briefly address just a few points related to the scope and integration of language processing into cognitive architectures.

Within the field of cognitive systems, a growing number of projects has been devoted several aspects of language understanding, a response to the fact that the knowledge-lean paradigm currently prevalent in NLP has not been addressing, or therefore serving, the needs of sophisticated agent systems. For example, Mohan et al. [18] added a language processing component to a Soar agent, Forbus et al. [19, 20] investigated learning by reading, Allen et al. [21] demonstrated learning information management tasks through dialog and capturing user's operations in a web browser, Scheutz et al. [22] demonstrated learning objects and events through vision and language, Lindes and Laird [23] integrated a language understanding module into their Rosie robot.

Several characteristics set OntoAgent-based systems apart from many other contributions in this area [1]. First, they integrate language processing with other perception modalities (such as interoception and simulated vision) as well as reasoning, action and the management of the agent's episodic, semantic and procedural memory. Second, and most importantly, the language processing component treats many more linguistic phenomena than others, and is capable of multiple types of ambiguity resolution that is seldom if ever addressed in other cognitive systems with language processing capabilities. Third, OntoAgent-based systems learn not only lexicon and ontology but also scripts, plans and elements of the "theory of mind" of other agents. One planned enhancement is to include learning entries in the opticon (which is the correlate in the vision interpretation task of the lexicon in language processing), that will support grounding the results of language interpretation with the of visually recognized objects and events on the basis of the ontology underlying both visual and language interpretation in OntoAgent. Integration of OntoAgent with an embodied robotic system is reported in (7, 8]. The integration of a simulated vision perception in an autonomous vehicle system with OntoAgent is reported in [10].

OntoAgent has more features relevant to learning than those space constraints allow us to present in this position paper. Thus, LEIAs also maintain a long-term episodic memory of the text and utterances they have processed with OntoSem. This allows the LEIAs in certain cases to use analogical reasoning to minimize their efforts by retrieving (and then optionally modifying) stored TMRs instead of generating them "from scratch" using OntoSem. The long-term episodic memory also serves as the repository of the LEIA's knowledge about instances (exemplars) of concepts in its ontology, which facilitates a variety of additional reasoning capabilities, such as inductive learning or the maintenance of specific memories about other agents.

Another topic that we can only mention in this paper is hybridization of OntoAgent. At present, OntoAgent-based systems already incorporate results of (imported) modules (for example, a syntactic parser and a vision perception system) implemented in the empirical machine learning paradigm. We are working on applying empirical methods for filtering inputs to the learning-by-reading module of OntoAgent and investigating integration of these paradigms for the decision-making tasks across all the architecture modules.

To recap, keys to overcoming the knowledge bottleneck in AI include starting with sufficiently broad and deep, high-quality bootstrapping knowledge bases (lexicon and ontology); endowing agents with broad and deep language understanding capabilities; working within a knowledge-centric agent environment; enabling agents to learn both independently and in collaboration with people; and strategically keeping human developers in the loop as knowledge engineers to enforce the high quality of the learned resources.

References

1. McShane, M., Nirenburg, S.: Linguistics for the Age of AI, p. 445. MIT Press, Cambridge (2021). https://direct.mit.edu/books/book/5042/Linguistics-for-the-Age-of-AI
2. Nirenburg, S., McShane, M., English, J.: Content-centric computational cognitive modeling. Adv. Cogn. Syst. (2020)
3. English, J., Nirenburg, S.: OntoAgent: implementing content-centric cognitive models. In: Proceedings of the Annual Conference on Advances in Cognitive Systems (2020)

4. McShane, B.: Understanding of complex and creative language use by artificial intelligent agents. Italian J. Comput. Linguist. (submitted)
5. English, J., Nirenburg, S.: Ontology learning from text using automatic ontological-semantic text annotation and the Web as the corpus. In: Papers from the AAAI 2007 Spring Symposium on Machine Reading (2007)
6. Nirenburg, S., Oates, T., English, J.: Learning by reading by learning to read. In: Proceedings of the International Conference on Semantic Computing (2007)
7. Nirenburg, S., Wood, P.: Toward human-style learning in robots. In: AAAI Fall Symposium on Natural Communication with Robots (2017)
8. Nirenburg, S., McShane, M., Wood, P., Scassellati, B., Magnin, O., Roncone, A.: Robots learning through language communication with users. In: Proceedings of NLDB 2018 (2018)
9. McShane, M., Nirenburg, S., English, J.: Multi-stage language understanding and actionability. Adv. Cogn. Syst. 6, 119–138 (2018)
10. Nirenburg, S., English, J., McShane, M.: Artificial intelligent agents go to school. In: 34th International Workshop on Qualitative Reasoning (submitted)
11. Lieto, A., Lebiere, C., Oltramari, A.: The knowledge level in cognitive architectures: Current limitations and possible developments. Cogn. Syst. Res. 48, 39–55 (2018)
12. Langley, P., Laird, J., Rogers, S.: Cognitive architectures: research issues and challenges. Cogn. Syst. Res. 10, 141–160 (2009)
13. Scheutz, M., Schermerhorn, P., Kramer, J., Anderson, D.: First steps toward natural human-like HRI. Auton. Robot. 22, 411–423 (2007)
14. Forbus, K., Klenk, M., Hinrichs, T.: Companion cognitive systems: design goals and lessons learned so far. IEEE Intell. Syst. 24, 36–46 (2009)
15. Choi, D., Langley, P.: Evolution of the ICARUS cognitive architecture. Cogn. Syst. Res. 48, 25–38 (2018)
16. Mohan, S., Laird, J.: Learning goal-oriented hierarchical tasks from situated interactive instruction. In: Proceedings of AAAI, pp. 387–394 (2014)
17. Bridewell, W., Bello, P.: A theory of attention for cognitive systems. Adv. Cogn. Syst. 4 (2016)
18. Mohan, S., Mininger, A.H., Laird, J.E.: Towards an indexical model of situated language comprehension for real-world cognitive agents. Adv. Cogn. Syst. 3, 163–182 (2013)
19. Forbus, K., Riesbeck, C., Birnbaum, L., Livingston, K., Sharma, A., Ureel, L.: Integrating natural language, knowledge representation and reasoning, and analogical processing to learn by reading. In: Proceedings of AAAI 2007 (2007)
20. Forbus, K., Lockwood, K., Sharma, A.: Steps towards a 2nd generation learning by reading system. In: AAAI Spring Symposium on Learning by Reading (2009)
21. Allen, J., et al.: PLOW: a collaborative task learning agent. In: Proceedings of the Twenty-Second Conference on Artificial Intelligence (AAAI 2007) (2007)
22. Scheutz, M., Krause, E., Oosterveld, B., Frasca, T., Platt, R.: Spoken instruction-based one-shot object and action learning in a cognitive robotic architecture. In: Proceedings of AAMAS 2017 (2017)
23. Lindes, P., Laird, J.: Toward integrating cognitive linguistics and cognitive language processing. In: Proceedings of ICCM 2016 (2016)
24. McShane, M., Nirenburg, S., Jarrell, B.: Modeling decision-making biases. Biol.-Inspired Cogn. Architect. (BICA) J. 3, 39–50 (2013)

Methodological Approaches

Word Embedding-Based Topic Similarity Measures

Silvia Terragni⬤, Elisabetta Fersini^(✉)⬤, and Enza Messina⬤

University of Milano-Bicocca, Milan, Italy
s.terragni4@campus.unimib.it,
{elisabetta.fersini,enza.messina}@unimib.it

Abstract. Topic models aim at discovering a set of hidden themes in a text corpus. A user might be interested in identifying the most similar topics of a given theme of interest. To accomplish this task, several similarity and distance metrics can be adopted. In this paper, we provide a comparison of the state-of-the-art topic similarity measures and propose novel metrics based on word embeddings. The proposed measures can overcome some limitations of the existing approaches, highlighting good capabilities in terms of several topic performance measures on benchmark datasets.

Keywords: Topic modeling · Topic similarity · Word embeddings

1 Introduction

Topic models [7,10,24] are a suite of probabilistic models that aim at extracting the main themes (or "topics") from a collection of documents. When a topic model automatically generates a set of topics underlying a given corpus, few of them could be similar while others could be different. For instance, a topic about technology, characterized by the words "card video monitor cable vga", is more similar to the topic "gif image format jpeg color" than one about animals ("cat animal dog cats tiger"). Methods for automatically determining the similarity between topics have several potential applications, such as the validation of the quality of the topic modeling output for determining potential overlaps between pairs of topics [2] and document retrieval based on topic proximity [10].

To estimate the similarity between topics, several metrics have been introduced in the state of the art. Most of them are based on word tokens and usually adopt a list of top-t terms to estimate if two topics are related. On the other hand, few approaches exploit the probability distribution of the words denoting the topics to compute the similarity between themes. These distribution-based measures suffer from the high dimensionality of the vocabulary, generating solutions that do not strongly correlate with human judgment [1]. On the contrary, approaches that focus only on the word tokens of a topic [5,26] ignore that two words could be lexicographically different but denoting a similar meaning. For instance, the words *cat* and *kitten* should not be considered totally dissimilar. A

© Springer Nature Switzerland AG 2021
E. Métais et al. (Eds.): NLDB 2021, LNCS 12801, pp. 33–45, 2021.
https://doi.org/10.1007/978-3-030-80599-9_4

preliminary investigation that partially addressed the above problems has been introduced in [1]. They represent the words of a topic as vectors in a semantic space constructed from an external source or from the corpus using Pointwise Mutual Information (PMI). However, this approach is computationally expensive, requiring to compute the probability of the co-occurrence for each pair of words in the corpus, and does not take into account the more recent advances in Word Embeddings [9,18,21], that have already proved their benefits in several NLP applications and topic modeling [3,20]. Moreover, this approach does not take into account that the topics extracted are actually ranked lists of words, where the rank provides useful insight. In particular, if two topics contain the same words but at different ranking positions, this aspect should be considered when evaluating the similarity of the generated solution.

We therefore propose new topic similarity metrics that exploit the nature of word embeddings and take into consideration topics as ranked lists of words. We demonstrate in the experimental evaluation that these metrics can discover semantically similar topics, also outperforming the state-of-the-art topic similarity metrics.

The paper is organized as follows. In Sect. 2, the main state-of-the-art topic similarity measures are described. In Sect. 3, we present the proposed metrics, which are based on Word Embeddings. In Sect. 4, the experimental investigation is detailed. In Sect. 5, we outline the conclusions and future work.

2   Topic Similarity/Distance Measures: State of the Art

The goal of topic modeling is to extract K topics from a document corpus, where each topic is represented as a multinomial distribution over the vocabulary, usually referred to as *word-topic distribution*. Researchers usually consider the top-t most probable words (from the word-topic distribution) to represent a topic. This top-t ranked list of words is usually called *topic descriptor* [4]. The word-topic distribution and topic descriptors are the two key elements that can be exploited to estimate the similarity between two themes. In what follows, we will review the most relevant topic similarity measures that have been proposed in the literature. The topic descriptor of a topic i will be referred to as t_i, represented by its top-t most likely words, i.e. $t_i = \{v_0, v_1, \ldots, v_{t-1}\}$, where v_k is a word of the vocabulary V. We will refer to the word distribution of a topic i as β_i, which is a multinomial distribution over the vocabulary V. In particular, $\beta_i(v)$ represents the probability of the word v in the topic i.

We will introduce in the following subsections the metrics already available in the state of the art, by roughly dividing them into metrics that are based on the counts of the shared word tokens and metrics that are based on the probability distributions.

2.1   Measures Based on Shared Word Tokens

A simple way to compute the topic similarity is based on the number of words that two topics share. These measures ignore that two words may be different in their lexicographic representation but semantically similar.

Average Jaccard Similarity (JS). The ratio of common words in two topics can be measured by using Jaccard Similarity [13].[1] The Jaccard Similarity (JS) between t_i and t_j is defined as follows:

$$JS(t_i, t_j) = \frac{|t_i \cap t_j|}{|t_i \cup t_j|} \tag{1}$$

This measure varies between 0 and 1, where 0 means that the topics are completely different, and 1 means that topics are similar to each other.

Rank-Biased Overlap (RBO). To consider the ranking of the words, one can use Rank-Biased Overlap (RBO) [27], exploited in Bianchi et al. [5] in the topic modeling context. It is based on a probabilistic model in which a user compares the overlap of two ranked lists (that in our case correspond to two topics) at incrementally increasing depth. The user can stop to examine the lists at a given rank position according to the probability p, enabling therefore the metric to be top-weighted and consequently giving more weight to the top words of a topic. The smaller p, the more top-weighted the metric is. When $p = 0$, only the top-ranked word is considered. The metric ranges from 0 (completely different topic descriptors) to 1 (equal topic descriptors).

RBO is based on the concept of *overlap at depth* h between two lists, which is the number of elements that the lists share when only the first h words are considered. For example, the overlap at depth 2 between the lists $l_1 = \{cat, animal, dog\}$ and $l_2 = \{animal, kitten, animals\}$ is 1. The average overlap is defined as the proportion of the overlap at depth h over h. Therefore, the RBO measure when evaluating two topics is computed as the expected value of the average overlap that the user observes when comparing two lists.

Average Pairwise Pointwise Mutual Information (PMI). In [1], the authors present a similarity metric based on Pointwise Mutual Information (PMI). The authors adapt the PMI coherence to measure topic similarity by computing the average pairwise PMI between the words belonging to two topics. More formally, the PMI between the topics i and j is defined as:

$$PMI(t_i, t_j) = \frac{1}{t^2} \sum_{u \in t_i} \sum_{v \in t_j} PMI(u, v) \tag{2}$$

where t is the number of words of each topic.

---

[1] This approach has been used in [26] to compute the distance between topics.

2.2   Measures Based on Probability Distributions

Instead of considering the top-words, we can consider the word-topic distribution to compute the distance between metrics. However, these metrics may be sensitive to the high dimensionality of the vocabulary [1].

Average Log Odds Ratio (LOR). In [11], the topic similarity is computed using the average log odds ratio (LOR) that is defined as follows:

$$LOR(\beta_i, \beta_j) = \sum_{v \in V} \mathbb{1}_{\mathbb{R}_{\neq 0}}(\beta_i(v)) \mathbb{1}_{\mathbb{R}_{\neq 0}}(\beta_j(v)) |\log(\beta_i(v) - \beta_j(v)| \qquad (3)$$

where $\mathbb{1}_A(x)$ is an indicator function defined as 1 if $x \in A$ and 0 otherwise. This metric computes the distance between the distributions associated with two topics, so it is a dissimilarity metric.

Kullback-Leibler Divergence (KL-DIV). A widely used measure to determine the similarity between two topics is the Kullback-Leibler Divergence [2,22, 25], which measures the distance from a given topic's distribution over words to another one. It is defined as follows:

$$KL - DIV(\beta_i, \beta_j) = \sum_{v \in V} \beta_i(V) \log \frac{\beta_i(v)}{\beta_j(v)} \qquad (4)$$

Notice that this metric is not symmetric and its domain ranges from 0 (when two distributions are identical) to infinity. In fact, this metric represents a dissimilarity score. Other metrics based on computing the distance between distributions include the Jensen Shannon Divergence and the cosine similarity [1].

3   Word Embedding-Based Similarity

To overcome the absence of semantics in the traditional similarity measures available in the state of the art, one can resort to the use of word embeddings to capture conceptual relationships between words. In the word embedding spaces, the vector representations of the words appearing in similar contexts tend to be close to each other [18]. We can therefore exploit the nature of word embeddings and define new metrics to estimate how much two topic descriptors are similar.

Word Embedding-Based Centroid Similarity (WECS). The most simple strategy, originally designed in [6] for a cross-lingual task, consists of computing the centroids of two topic descriptors t_i and t_j and then estimating their similarity. Let be $\overrightarrow{t_i}$ the vector centroid of the topic descriptor t_i computed as the average of word embeddings considering all the words belonging to the topic i.

The Word Embedding-based Centroid Similarity between two topics is estimated as $WECS(t_i, t_j) = sim(\overrightarrow{t_i}, \overrightarrow{t_j})$, where sim is a measure of similarity between vectors, i.e. cosine similarity.

Word Embedding-Based Pairwise Similarity (WEPS). An alternative to WECS consists of averaging the pairwise similarity between the embedding vectors of the words composing the topic descriptors. We define the similarity between two topic descriptors t_i and t_j as follows:

$$WEPS(t_i, t_j) = \frac{1}{t^2} \sum_{v \in t_i} \sum_{u \in t_j} sim(w_v, w_u) \tag{5}$$

where t represents the number of words of each topic, and w_v and w_u denote the word embeddings associated with words v and u respectively.

Word Embedding-Based Weighted Sum Similarity (WESS). A simple way to combine the probability distributions and the word embeddings is to compute the sum of the word embeddings of the words in the vocabulary, where the sum is weighted by the probability of each term in the topic. Then, we compute the similarity between the resulting word embeddings.

More formally, let be $b_i = \sum_{v \in V} \beta_i(v) \cdot w_v$ the weighted sum of the word embeddings of the vocabulary for the topic i. Therefore, the WESS for the topic i and j is defined as $sim(b_i, b_j)$.

Word Embedding-Based Ranked-Biased Overlap (WERBO). We can extend RBO and define a new metric of similarity that is top-weighted and makes use of word embeddings. Given the lists $l_1 = \{cat, animal, dog\}$ and $l_2 = \{animal, kitten, animals\}$, the words *cat* and *kitten* are similar, even though they are lexicographically different. It follows that their overlap at depth 2 should be higher than 1. We therefore generalize the concept of overlap to handle word embeddings instead of simple word tokens.

Algorithm 1. Calculate generalized overlap at depth h

Input: t_i, t_j topic descriptors composed of n words; h depth of the list, where $h \leq n$

1: **for** $u := 1, \ldots, h$ **do**
2:     **for** $v := 1, \ldots, h$ **do**
3:         $sim[w_u^i, w_v^j] := similarity(w_u^i, w_v^j)$
4:     **end for**
5: **end for**
6: $overlap := 0$
7: **while** sim is not empty **do**
8:     $max\_value := max(sim)$
9:     $w_u^i, w_v^j := get\_indices(max\_value)$
10:     remove all entries of w_u^i and w_v^j from sim
11:     $overlap := overlap + max\_value$
12: **end while**
13: **return** $overlap$

Algorithm 1 shows how to compute the generalized overlap between two topic descriptors t_i and t_j. First of all, we compute the similarity between all the pairs of word embedding vectors w_u^i and w_v^j belonging to the two topics i and j (line 1–5). The associative array sim (line 3) is indexed by the tuple (w_u^i, w_v^j) and contains all the computed similarities. Subsequently (line 7–12), we process the associative array sim to get the words that are the most similar, to then update the overlap variable. In particular, the algorithm searches for the tuple (w_u^i, w_v^j) that has the highest similarity in sim (line 8), removes from sim all the entries containing w_u^i or w_v^j (line 9–10) and finally updates the overlap by adding the highest similarity value corresponding to the tuple (w_u^i, w_v^j) (line 12). For example, let us compute the generalized overlap at depth 3 of the word lists $l_1 = \{cat, animal, dog\}$ and $l_2 = \{animal, kitten, animals\}$. The result will be $sim(animal, animal) + sim(cat, kitten) + sim(animals, dog)$, because $(animal, animal)$ are identical vectors and should be summed first, then $(cat, kitten)$ are the second most similar vectors, and finally $(animals, dog)$ are the remaining vectors and should be summed at last.

In the proposed algorithm, $similarity(w_u^i, w_v^j)$ is the angular similarity between the vectors associated with the word embeddings related to the words u and v respectively[2]. Notice that this approach is based on a greedy strategy that estimates the overlapping by considering first the most similar embeddings of the words available in the top-h list. We will then refer to this approach as **WERBO-M**. Instead of computing the similarity between each word embedding, an alternative metric can compute the centroid of the embeddings at depth h. In this way, the overlap at depth h is just defined as $similarity(\overrightarrow{t_i}, \overrightarrow{t_j}) \cdot h$, where $\overrightarrow{t_i}$ and $\overrightarrow{t_j}$ are the centroids of the topics t_i and t_j respectively. We will refer to this metric as **WERBO-C**.

Weighted Graph Modularity (WGM). We can rethink two topic descriptors in the form of a graph. Each word represents a node in the graph, while the edges denote the similarity between the words. Considering two topics composed of their own words (nodes), the intra-topic similarity connections should be higher than the extra-topic similarity connections with any other topic. We can express this idea by using the measure of modularity, which estimates the strength of division of a graph into modules (in our case, topics).

Let $G = (U, E)$ be a fully connected graph, where U is the words related to t_i and t_j and E are weighted edges denoting the similarity between pairs of word embeddings. In particular, an edge weight is defined as $A_{uv} = sim(w_v, w_u)$, where $(u, v) \in E$, $v, u \in U$ and $sim(\cdot, \cdot)$ is the angular similarity between two word embeddings. Given the graph G, originating from two topic descriptors t_i and t_j, the Weighted Graph Modularity (WGM) can be estimated as:

$$WGM(t_i, t_j) = \frac{1}{2m} \sum_{v, u \in U(G)} [A_{vu} - \frac{k_v k_u}{2m}]\mathbb{1}_{vu} \qquad (6)$$

---

[2] We use the angular similarity instead of the cosine because we require the overlap to range from 0 to 1.

where k_v and k_u denote the degrees of the nodes v and u respectively, m is the sum of all of the edge weights in the graph, and $\mathbb{1}_{vu}$ is an indicator function defined as 1 if v and u are words belonging to the same topic, 0 otherwise. Modularity ranges from $-1/2$ (non-modular topics) to 1 (fully separated topics). Therefore, it should be considered as a dissimilarity score.

4 Experimental Investigation

4.1 Experimental Setting

Compared measures. Before proceeding with the description of the validation strategy and the performance measures adopted for a comparative evaluation, we summarize the investigated measures. In particular, in Table 1 we provide details about all the metrics, reporting their main features:

- TD, which denotes if the metric considers the top-t words of the descriptors;
- PD, that reports if the metric considers the topic probability distribution;
- WE, which indicates if the metric overcomes the limitation of the discrete representation of words by using Word Embeddings;
- TW, that identify if the metric is top-weighted, i.e. the words at the top of the ranked list are more important than the words in the tail.

The implementations of the measures are integrated into the topic modeling framework OCTIS [23], available at https://github.com/mind-lab/octis.

Table 1. Summary of the characteristics of the metrics presented in this paper. The newly proposed metrics are reported in bold.

Similarity/Distance measure	TD	PD	WE	TW
Jaccard Similarity (JS) [26]	✓			
Rank-biased Overlap (RBO) [27]	✓			✓
Pointwise Mutual Information (PMI) [1]	✓			
Average Log Odds Ratio (LOR) [11]		✓		
Kullback-Leibler Divergence (KL-DIV) [22]		✓		
Word embedding-based Centroid Similarity (WECS)	✓		✓	
Word Embedding Pairwise Similarity (WEPS)	✓		✓	
Word Embedding-based Weighted Sum Similarity (WESS)		✓	✓	
Word Embedding-based RBO - Match (WERBO-M)	✓		✓	✓
Word Embedding-based RBO - Centroid (WERBO-C)	✓		✓	✓
Weighted Graph Modularity (WGM)	✓		✓	

Validation Strategy. To validate the proposed similarity measures, and compare them with the state-of-the-art ones, we selected the most widely adopted topic model to produce a set of topics to be evaluated. In particular, we trained

Latent Dirichlet Allocation (LDA) [8] on two benchmark datasets, i.e. BBC news [16] and 20 NewsGroups.[3], originating 50 different topics per dataset.[4] For the pre-processing, we removed the punctuation and the English stop-words[5], and we filtered out the less frequent words, obtaining a final vocabulary of 2000 terms.

Given the topics extracted by LDA, we disregarded those with a low value of topic coherence, measured by using Normalized Pointwise Mutual Information (NPMI) [17] on the dataset itself as a reference corpus. Then we randomly sampled 100 pairs of topics (for each dataset) that have been evaluated by three annotators, by considering the top-10 words. In particular, the annotators have rated if two topics were related to each other or not, using a value of 0 (not related topics) and 1 (similar topics). The final annotation of each pair of topics has been determined according to a majority voting strategy on the rates given by the three annotators.

For the metrics that are based on the topic descriptors, we considered the top-10 words of each topic. Regarding the metrics that are based on word embeddings, we used Gensim's[6] Word2Vec model to compute the embedding space on the corpus with the default hyperparameters. The co-occurrence probabilities for the estimation of PMI have been computed on the training dataset. For the metrics that represent dissimilarity scores, such as KL-DIV, the LOR and WGM metrics, we considered their inverse.

Performance Measures. We evaluated the capabilities of all the topic similarity metrics, both the ones available in the state of the art and the proposed ones, by measuring Precision@k, Recall@k and F1-Measure@k.

In particular, Precision@k (P@k) is defined as the fraction of the number of retrieved topics among the top-k retrieved topics that are relevant and the number of retrieved topics among the top-k retrieved topics. Recall@k (R@k) is defined as the fraction of the number of retrieved topics among the top-k retrieved topics that are relevant and the total number of relevant topics. F1-Measure@k (F1@k) is defined the harmonic mean between P@k and R@k, i.e. $F1@k = 2(P@k \cdot R@k)/(P@k + R@k)$.

4.2   Experimental Results

Table 2 shows the results for the BBC News dataset in terms of P@k, R@k and F1@k by varying k for 1 to 5. As a first remark, we can see that the metrics that are based on the shared word tokens only, i.e. the Jaccard Distance (JD) and Rank-biased Overlap (RBO), achieve the lowest performance. KL-DIV and LOR, which are based only on the topic-word probability distributions, outperform the baselines JD and RBO, but they are not able to outperform the proposed measures that consider the word embeddings similarities. The most

---

[3] http://people.csail.mit.edu/jrennie/20Newsgroups/.

[4] We trained LDA with the default hyperparameters of the Gensim library.

[5] We used the English stop-words list provided by MALLET: http://mallet.cs.umass.edu/.

[6] https://radimrehurek.com/gensim/.

Table 2. Precision@K, Recall@K and F1-Measure@k on the BBC News dataset.

	k	State-of-the-art metrics					Proposed metrics					
		JD	RBO	PMI	LOR	KL-DIV	WESS	WEPS	WECS	WERBO-M	WERBO-C	WGM
P@K	1	0.818	0.864	0.955	0.846	0.909	0.909	0.955	**1.000**	**1.000**	**1.000**	0.818
	2	0.727	0.705	**0.864**	0.769	0.750	0.795	0.841	0.841	**0.864**	**0.864**	0.795
	3	0.652	0.667	0.803	0.667	0.652	0.742	0.788	0.773	**0.818**	0.788	0.773
	4	0.557	0.557	0.705	0.596	0.602	0.682	0.705	0.693	**0.716**	**0.716**	0.693
	5	0.482	0.491	0.573	0.492	0.536	0.573	**0.582**	**0.582**	0.582	0.582	0.573
	Avg	0.647	0.657	0.706	0.674	0.690	0.740	0.774	0.778	**0.796**	0.790	0.730
R@K	1	0.348	0.364	0.417	0.423	0.402	0.409	0.417	**0.439**	0.439	**0.439**	0.379
	2	0.545	0.534	**0.663**	0.641	0.587	0.614	0.648	0.648	0.659	**0.663**	0.621
	3	0.697	0.712	0.871	0.776	0.716	0.803	0.856	0.833	**0.879**	0.845	0.833
	4	0.784	0.784	0.977	0.885	0.848	0.951	0.977	0.966	**0.989**	**0.989**	0.966
	5	0.867	0.879	0.989	0.910	0.932	0.985	**1.000**	**1.000**	**1.000**	**1.000**	0.989
	Avg	0.648	0.655	0.783	0.727	0.697	0.752	0.780	0.777	**0.793**	0.787	0.758
F1@K	1	0.456	0.479	0.539	0.521	0.517	0.524	0.539	**0.570**	**0.570**	**0.570**	0.480
	2	0.589	0.574	**0.708**	0.651	0.617	0.650	0.689	0.689	0.705	**0.708**	0.656
	3	0.644	0.660	0.798	0.675	0.645	0.734	0.783	0.765	**0.809**	0.777	0.765
	4	0.627	0.627	0.786	0.677	0.673	0.762	0.786	0.775	**0.797**	**0.797**	0.775
	5	0.595	0.605	0.698	0.610	0.654	0.697	**0.709**	**0.709**	0.709	0.709	0.698
	Avg	0.582	0.589	0.706	0.627	0.621	0.673	0.701	0.701	**0.718**	0.712	0.675

competitive metric with respect to the proposed ones is the PMI, which obtains comparative results to the word-embedding metrics for $k = 2$. These results suggest that considering a richer representation of topical words helps in retrieving semantically similar topics to a given target topic. In particular, WERBO-M and WERBO-C reach the highest scores in most of the cases. This means that not only the meaning of the words are important when evaluating the similarity of two topics, but also the position of each word in the topic matters. In fact, WERBO-M and WERBO-C outperform the metrics WEPS and WECD that do not take into consideration the rank of the words.

Table 3 reports the results on the 20NewsGroups dataset. Here, the obtained results are similar to the previous dataset. All the word embedding-based metrics outperform the state-of-the-art ones. In particular, WERBO-C outperforms the other metrics or obtain comparable results in most the cases. Even if WESS is the similarity metric that obtains the best performance on average, the results obtained by WERBO-C and WERBO-M are definitely comparable. Also on this dataset PMI seems to be the most competitive metric, however the word-embedding metrics metrics outperform it in most of the cases.

We report in Table 4 two examples of topics evaluated by the considered similarity/distance measures. The first example reports two topics, that clearly represent two distinct themes, likely *religion* and *technology*. In this case, all the proposed metrics can capture the diversity of the two topics as well as the measure of the state of the art. On the other hand, the second example reports two related topics about *technology*. We can easily notice that while all the

Table 3. Precision@K, Recall@K and F1-Measure@k on 20 NewsGroups.

	k	State-of-the-art metrics					Proposed metrics					
		JD	RBO	PMI	LOR	KL-DIV	WESS	WEPS	WECS	WERBO-M	WERBO-C	WGM
P@K	1	0.833	0.833	**1.000**	0.833	0.833	**1.000**	**1.000**	0.958	0.958	0.917	0.958
	2	0.646	0.667	0.813	0.792	0.792	**0.833**	0.813	**0.833**	0.813	**0.833**	**0.833**
	3	0.569	0.569	0.681	0.653	0.667	0.694	0.694	0.694	**0.708**	**0.708**	0.694
	4	0.458	0.458	0.583	0.563	0.583	0.583	0.583	0.583	**0.604**	**0.604**	0.583
	5	0.408	0.408	0.492	0.492	0.492	**0.500**	**0.500**	**0.500**	**0.500**	**0.500**	**0.500**
	Avg	0.583	0.587	0.714	0.666	0.673	**0.722**	0.718	0.714	0.717	0.713	0.714
R@K	1	0.424	0.424	**0.542**	0.375	0.396	**0.542**	**0.542**	0.500	0.500	0.459	0.500
	2	0.581	0.591	0.758	0.667	0.737	**0.779**	0.758	**0.779**	0.758	0.772	**0.779**
	3	0.705	0.701	0.869	0.793	0.848	0.890	0.890	0.890	**0.904**	**0.904**	0.890
	4	0.734	0.734	0.950	0.866	0.950	0.950	0.950	0.950	**0.974**	**0.974**	0.950
	5	0.807	0.807	0.974	0.946	0.974	**0.988**	**0.988**	**0.988**	**0.988**	**0.988**	**0.988**
	Avg	0.650	0.651	0.819	0.730	0.781	**0.830**	0.825	0.821	0.825	0.819	0.821
F1@K	1	0.522	0.522	**0.653**	0.487	0.501	**0.653**	**0.653**	0.612	0.612	0.570	0.612
	2	0.566	0.580	0.727	0.681	0.706	**0.748**	0.727	**0.748**	0.727	0.744	**0.748**
	3	0.587	0.585	0.709	0.670	0.692	0.725	0.725	0.725	**0.739**	**0.739**	0.725
	4	0.527	0.527	0.674	0.640	0.674	0.674	0.674	0.674	**0.696**	**0.696**	0.674
	5	0.510	0.510	0.610	0.607	0.610	**0.621**	**0.621**	**0.621**	**0.621**	**0.621**	**0.621**
	Avg	0.542	0.545	0.675	0.617	0.637	**0.684**	0.680	0.676	0.679	0.674	0.676

measures of the state of the art suggest that the two topics are completely different because of their low values (e.g. JS = 0.053 and KL-DIV = −4.415), the proposed metrics can capture their actual similarity.

Table 4. Qualitative comparison of the considered measures. Since KL-DIV, LOR and WGM represent dissimilarity scores, they are reported as their inverse.

Topic 1	Topic 2	Metrics	Topic 1	Topic 2	Metrics
god	ftp	JS = 0	tiff	window	JS = 0.053
christian	fax	RBO = 0	gif	application	RBO = 0.057
christianity	pub	PMI = −0.042	image	manager	PMI = 0.327
religion	graphics	LOR = −3.204	format	display	LOR = −2.110
faith	computer	KL-DIV = −4.36416	jpeg	color	KL-DIV = −4.415
christ	software	WESS = −0.145	formats	widget	WESS = 0.787
sin	version	WEPS = −0.0941	color	mouse	WEPS = 0.402
people	mail	WECS = −0.183	images	screen	WECS = 0.565
view	gov	WERBO-M = 0.472	complex	button	WERBO-M = 0.651
paul	mit	WERBO-C = 0.120	resolution	user	WERBO-C = 0.170
		WGM = −0.102			WGM = −0.015
Ground Truth = unrelated topics			Ground Truth = similar topics		

5   Conclusions

In this paper, we investigated and compared several topic similarity metrics. These measures are particularly useful for data analysis tasks [10,19], i.e. when a user may want to identify topics that are similar for the theme of interest. We proposed several metrics that exploit word embeddings and take into account the ranking of words in the topic descriptors. We experimentally proved that the proposed metrics outperform the state-of-the-art ones. We believe that these metrics should be considered in topic modeling visualization tools [11,12,15,22,23] for improving their performance and allow a user to obtain relevant results. As future work, different word embeddings methods could be investigated, also considering the word embeddings deriving from the state-of-the-art contextualized language models, e.g. BERT [14].

References

1. Aletras, N., Stevenson, M.: Measuring the similarity between automatically generated topics. In: Proceedings of the 14th Conference of the European Chapter of the Association for Computational Linguistics, pp. 22–27 (2014)
2. AlSumait, L., Barbará, D., Gentle, J., Domeniconi, C.: Topic significance ranking of LDA generative models. In: Buntine, W., Grobelnik, M., Mladenić, D., Shawe-Taylor, J. (eds.) ECML PKDD 2009. LNCS (LNAI), vol. 5781, pp. 67–82. Springer, Heidelberg (2009). https://doi.org/10.1007/978-3-642-04180-8_22
3. Batmanghelich, K., Saeedi, A., Narasimhan, K., Gershman, S.: Nonparametric spherical topic modeling with word embeddings. In: Proceedings of the Conference, vol. 2016, p. 537. Association for Computational Linguistics (2016)
4. Belford, M., Namee, B.M., Greene, D.: Ensemble topic modeling via matrix factorization. In: Proceedings of the 24th Irish Conference on Artificial Intelligence and Cognitive Science, AICS 2016, vol. 1751, pp. 21–32 (2016)
5. Bianchi, F., Terragni, S., Hovy, D.: Pre-training is a hot topic: contextualized document embeddings improve topic coherence. In: Proceedings of the Joint Conference of the 59th Annual Meeting of the Association for Computational Linguistics and the 11th International Joint Conference on Natural Language Processing (ACL-IJCNLP 2021). Association for Computational Linguistics (2021)
6. Bianchi, F., Terragni, S., Hovy, D., Nozza, D., Fersini, E.: Cross-lingual contextualized topic models with zero-shot learning. In: Proceedings of the 16th Conference of the European Chapter of the Association for Computational Linguistics, EACL 2021, pp. 1676–1683 (2021)
7. Blei, D.M.: Probabilistic topic models. Commun. ACM **55**(4), 77–84 (2012)
8. Blei, D.M., Ng, A.Y., Jordan, M.I.: Latent Dirichlet allocation. J. Mach. Learn. Res. **3**, 993–1022 (2003)
9. Bojanowski, P., Grave, E., Joulin, A., Mikolov, T.: Enriching word vectors with subword information. Trans. Assoc. Comput. Linguist. **5**, 135–146 (2017)
10. Boyd-Graber, J.L., Hu, Y., Mimno, D.M.: Applications of topic models. Found. Trends Inf. Retr. **11**(2–3), 143–296 (2017)

11. Chaney, A.J., Blei, D.M.: Visualizing topic models. In: Proceedings of the 6th International Conference on Weblogs and Social Media. The AAAI Press (2012)
12. Chuang, J., Manning, C.D., Heer, J.: Termite: visualization techniques for assessing textual topic models. In: International Working Conference on Advanced Visual Interfaces, AVI 2012, pp. 74–77. ACM (2012)
13. Deng, F., Siersdorfer, S., Zerr, S.: Efficient jaccard-based diversity analysis of large document collections. In: Proceedings of the 21st ACM International Conference on Information and Knowledge Management, pp. 1402–1411 (2012)
14. Devlin, J., Chang, M., Lee, K., Toutanova, K.: BERT: pre-training of deep bidirectional transformers for language understanding. In: Proceedings of the 2019 Conference of the North American Chapter of the Association for Computational Linguistics: Human Language Technologies, NAACL-HLT 2019, pp. 4171–4186 (2019)
15. Gardner, M.J., et al.: The topic browser: an interactive tool for browsing topic models. In: NIPS Workshop on Challenges of Data Visualization, vol. 2, p. 2 (2010)
16. Greene, D., Cunningham, P.: Practical solutions to the problem of diagonal dominance in kernel document clustering. In: Proceedings of the 23rd International Conference on Machine Learning (ICML 2006), pp. 377–384. ACM Press (2006)
17. Lau, J.H., Newman, D., Baldwin, T.: Machine reading tea leaves: automatically evaluating topic coherence and topic model quality. In: Proceedings of the 14th Conference of the European Chapter of the Association for Computational Linguistics, EACL 2014, pp. 530–539 (2014)
18. Mikolov, T., Sutskever, I., Chen, K., Corrado, G.S., Dean, J.: Distributed representations of words and phrases and their compositionality. In: Advances in Neural Information Processing Systems 26: 27th Annual Conference on Neural Information Processing Systems 2013, pp. 3111–3119 (2013)
19. Newman, D.J., Block, S.: Probabilistic topic decomposition of an eighteenth-century American newspaper. J. Assoc. Inf. Sci. Technol. **57**(6), 753–767 (2006)
20. Nguyen, D.Q., Billingsley, R., Du, L., Johnson, M.: Improving topic models with latent feature word representations. Trans. Assoc. Computat. Linguist. **3**, 299–313 (2015)
21. Pennington, J., Socher, R., Manning, C.D.: Glove: global vectors for word representation. In: Proceedings of the 2014 Conference on Empirical Methods in Natural Language Processing (EMNLP), pp. 1532–1543 (2014)
22. Sievert, C., Shirley, K.: LDAvis: a method for visualizing and interpreting topics. In: Proceedings of the Workshop on Interactive Language Learning, Visualization, and Interfaces, pp. 63–70 (2014)
23. Terragni, S., Fersini, E., Galuzzi, B.G., Tropeano, P., Candelieri, A.: OCTIS: comparing and optimizing topic models is simple! In: Proceedings of the 16th Conference of the European Chapter of the Association for Computational Linguistics: System Demonstrations, EACL 2021, pp. 263–270 (2021)
24. Terragni, S., Fersini, E., Messina, E.: Constrained relational topic models. Inf. Sci. **512**, 581–594 (2020)
25. Terragni, S., Nozza, D., Fersini, E., Messina, E.: Which matters most? Comparing the impact of concept and document relationships in topic models. In: Proceedings of the First Workshop on Insights from Negative Results in NLP, Insights 2020, pp. 32–40 (2020)

26. Tran, N.K., Zerr, S., Bischoff, K., Niederée, C., Krestel, R.: Topic cropping: leveraging latent topics for the analysis of small corpora. In: Aalberg, T., Papatheodorou, C., Dobreva, M., Tsakonas, G., Farrugia, C.J. (eds.) TPDL 2013. LNCS, vol. 8092, pp. 297–308. Springer, Heidelberg (2013). https://doi.org/10.1007/978-3-642-40501-3_30
27. Webber, W., Moffat, A., Zobel, J.: A similarity measure for indefinite rankings. ACM Trans. Inf. Syst. **28**(4), 20:1–20:38 (2010)

Mixture Variational Autoencoder of Boltzmann Machines for Text Processing

Bruno Guilherme Gomes[(✉)], Fabricio Murai, Olga Goussevskaia,
and Ana Paula Couto da Silva

Universidade Federal de Minas Gerais, Belo Horizonte, Brazil
{brunoguilherme,murai,olga,ana.coutosilva}@dcc.ufmg.br

Abstract. Variational autoencoders (VAEs) have been successfully used to learn good representations in unsupervised settings, especially for image data. More recently, mixture variational autoencoders (MVAEs) have been proposed to enhance the representation capabilities of VAEs by assuming that data can come from a mixture distribution. In this work, we adapt MVAEs for text processing by modeling each component's joint distribution of latent variables and document's bag-of-words as a graphical model known as the Boltzmann Machine, popular in natural language processing for performing well in a number of tasks. The proposed model, MVAE-BM, can learn text representations from unlabeled data without requiring pre-trained word embeddings. We evaluate the representations obtained by MVAE-BM on six corpora w.r.t. the perplexity metric and accuracy on binary and multi-class text classification. Despite its simplicity, our results show that MVAE-BM's performance is on par with or superior to that of modern deep learning techniques such as BERT and RoBERTa. Last, we show that the mapping to mixture components learned by the model lends itself naturally to document clustering.

1 Introduction

Digital libraries and online social networks are current examples of ecosystems where large volumes of textual data are generated by users at https://www.overleaf.com/project/601fe4b6632b9e6672a3d137 every instant. On average, it is estimated that 500 million tweets are posted daily on Twitter[1], while 600 articles are created every day on Wikipedia.[2]. Similar figures also hold for other digital platforms such as Amazon Review, Yahoo Answers and Yelp Reviews. This explains in part the ever increasing importance of analyzing user-generated patterns in large textual data sets for Natural Language Processing (NLP) research.

In the last two decades, probabilistic graphical models (PGMs) have underpinned many successful applications in NLP [12,15]. Many popular word

---

[1] http://www.tweetstats.com/.
[2] https://en.wikipedia.org/wiki/Wikipedia:Statistics.

© Springer Nature Switzerland AG 2021
E. Métais et al. (Eds.): NLDB 2021, LNCS 12801, pp. 46–56, 2021.
https://doi.org/10.1007/978-3-030-80599-9_5

embeddings methods, such as word2vec [13] and GloVe [16], are based on simple Bayesian networks, which are PGMs defined over directed acyclic graphs. The success of PGMs in NLP stems from their ability to use unlabeled samples effectively for learning complex patterns in the data by allowing to explicitly specify dependencies among variables. For some more complex PGMs, exact inference is intractable due to the calculation of high-dimensional integrals. In these cases, variational inference techniques for approximating conditional distributions have been proposed and successfully applied to address the computational complexity issues [8,10].

In computer vision, similar approximations have been used in non-deterministic neural network models for learning compact image representations in unsupervised settings. These models, called variational autoencoders [8], typically consist of two networks respectively called encoder and decoder. The role of the encoder is to obtain a compact representation – an encoding – of an input image through non-linear transformations. This encoding is combined with some noise, i.e., a random variable sampled from a Gaussian distribution, and passed onto the decoder, whose role is to recover the original images through more non-linear transformations. More recently, a mixture variational autoencoder (MVAE) was proposed to make better use of the latent representation space [7].

In this paper, we propose a novel framework based on MVAE for text processing. Each mixture component models the joint distribution of the latent variables and the bag-of-words vector that represents a document. This distribution is represented as the graphical model known as the Boltzmann Machine, popular in NLP for performing well in a number of tasks and for being efficiently trained with variational learning due to its simple structure [2]. Despite the current trends in deep learning, we show that a shallow network can be effectively used as an encoder.

Our model, named MVAE-BM[3], can learn text representations from unlabeled data without requiring pre-trained word embeddings. MVAE-BM takes as input the bag-of-word vector representing a document and outputs its latent representation. We evaluate the representations obtained by MVAE-BM using six corpora w.r.t. the perplexity metric and accuracy on text classification. In spite of its simplicity, our results demonstrate that MVAE-BM's performance is on par with or superior to that of sophisticated deep learning techniques such as BERT [4] and RoBERTa [9]. Last, we show that the association between text and mixture component learned by the model lends itself naturally to document clustering.

2   Related Work

The task of learning patterns in large textual data sets has received significant interest in the last two decades. Here we discuss the main fronts of research related to our work.

---

[3] https://github.com/brunoguilherme1/MVAE-BM/.

Probabilistic Graphical Models (PGMs): PGMs provide a declarative language for blueprinting prior knowledge and valuable relationships in complex datasets. They contributed to fundamental advances in NLP, such as Topic Modeling [20] and word embedding [13,16]. A simple, yet effective graphical model used for language modeling is the Boltzmann Machine. This technique represents texts as bags-of-words and aims to learn their latent representation [2]. While these models represent documents using vectors of binary latent variables (since they are based on the Restricted Boltzmann Machine), MVAE-BM employs dense continuous document representations that are both expressive and easy to train.

Variational Autoencoder (VAE): VAE is a generative model that can be seen as an improved version of a standard autoencoder. VAE models are able to learn meaningful representations from the data in an unsupervised fashion. Variational inference with the re-parameterization trick was initially proposed in [8] and thereafter VAE has been widely adopted as a generative model for images [7]. Our MVAE-BM builds its encoder networks based on the VAE strategy [8] for the estimation of the latent variables present in the Boltzmann Machine and the Gumbel-Softmax strategy [6] to efficiently estimate the latent indicator variable of the mixture model.

Recently, several studies have presented efficient ways of combining PGM and VAE to solve NLP problems, with similar outcomes to MVAE-BM. In [23] an approach is presented for text modeling with latent information explicitly modeled as a Dirichlet variable. [12] and [11] introduced a generic variational inference framework for generative and conditional models of text, as well as alternative neural approaches for topic modeling. More recently, [15] combined non-parametric distribution models with VAE for text modeling.

Even though a mixture model using VAE has already shown promising results [7], MVAE-BM differs from the techniques listed above because it uses two neural networks to encode its latent variables and, in this way, it provides an estimation of the Boltzmann Machine as well as its mixture.

3   The MVAE-BM Model: Mixture Variational Autoencoder of Boltzmann Machines

In this section, we present MVAE-BM, an unsupervised model for document representation, based on mixture variational autoencoders. We first briefly introduce how variational autoencoders are used to estimate latent representations.

3.1   Background on Variational Autoencoders

A variational autoencoder (VAE) is a generative model which combines the encoder-decoder architecture for unsupervised learning with variational inference. In a VAE, the latent variables are sampled from a distribution (typically Gaussian) whose parameters are computed by passing the input through the encoder. VAE modifies the autoencoder network by replacing the latent variable h of an input x with a learned posterior recognition model $p_\theta(h|x)$. Let

$X = \{\boldsymbol{x}^{(n)}\}_{n=1}^{N}$ be a dataset comprised of N i.i.d. samples from a random variable $\boldsymbol{x}$, and h be an unobserved continuous random variable, assuming that $\boldsymbol{x}$ is dependent on h. The marginal distribution of $\boldsymbol{x}$ is defined as:

$$p(\boldsymbol{x}; \theta) = \int p(h; \theta) p(\boldsymbol{x}|h; \theta) dh. \tag{1}$$

In practice, the integral in Eq. (1) is intractable [8]. Hence, VAE uses a recognition model $q_\phi(h|\boldsymbol{x})$ to approximate the true posterior $p_\theta(h|\boldsymbol{x})$. So, instead of maximizing the marginal likelihood directly, the objective function becomes the variation lower bound, a.k.a. the evidence lower bound (ELBO) of the marginal:

$$\mathcal{L}(\boldsymbol{x}; \theta, \phi) = \mathbb{E}_{q_\phi(h|\boldsymbol{x})}[\log p_\theta(\boldsymbol{x}|h)] - \mathrm{KL}(q_\phi(h|\boldsymbol{x})\|p_\theta(h)),$$

where $q_\phi(h|\boldsymbol{x})$ is the approximation distribution variational for the true posterior $p_\theta(h|\boldsymbol{x})$. In the VAE model, $q_\phi(h|\boldsymbol{x})$ is known as the recognition (encoder) model, and $p_\theta(\boldsymbol{x}|h)$, the decoder model. Both encoder and decoder models are implemented via neural architectures. As discussed in [8], optimizing the marginal log-likelihood is essentially equivalent to maximizing $\mathcal{L}(\boldsymbol{x}; \theta, \phi)$, i.e., the ELBO, which consists of two terms. The first term is the expected reconstruction error, indicating how well the model can reconstruct data, given a latent variable. The second term is the KL divergence between the approximate posterior and the prior, acting as a regularization term that forces the learned posterior to be as close to the prior as possible. The prior $p_\theta(h)$ and the variational posterior $q_\phi(h|\boldsymbol{x})$ are frequently chosen from conjugate distribution families, allowing the KL divergence to be calculated analytically [6,8].

3.2   Proposed Model

MVAE-BM is an unsupervised learning model where two vectors of hidden variables, $\boldsymbol{h} \in \mathbb{R}^H$ and $\boldsymbol{c} \in \mathbb{R}^K$, are used for representing documents. Let V be the vocabulary and $\boldsymbol{x} \in \mathbb{R}^{|V|}$ be the bag-of-words representation of a document. We consider the generative model $p(\boldsymbol{x}, \boldsymbol{h}, \boldsymbol{c}) = p_\pi(\boldsymbol{c}) p(\boldsymbol{h}) p_\Theta(\boldsymbol{x}|\boldsymbol{h}, \boldsymbol{c})$, in which the latent variable $\boldsymbol{h}$ is generated from a centered multivariate Gaussian $\mathcal{N}(\boldsymbol{0}, \mathbf{I})$, and the latent indicator $\boldsymbol{c}$ is generated from a categorical distribution Multinomial$(\boldsymbol{\pi})$. The latent indicator $\boldsymbol{c} = [c_1, c_2, \dots, c_K]$ satisfies the conditions $c_i \in \{0, 1\}$, $\sum_{i=1}^{K} c_i = 1$. Each $\boldsymbol{x}$ is associated with an unique sample of $\boldsymbol{h}$, and is generated from a single component in the mixture model $p_\Theta(\boldsymbol{x}|\boldsymbol{h}, \boldsymbol{c})$. The generative process is given by:

$$\boldsymbol{c} \sim \prod_{k=1}^{K} \pi_k^{c_k}, \tag{2}$$

$$\boldsymbol{h} \sim \mathcal{N}(\boldsymbol{0}, \mathbf{I}),$$

$$\boldsymbol{x}|\boldsymbol{h}, \boldsymbol{c} \sim \prod_{k=1}^{K} p_{\Theta(k)}(\boldsymbol{x}|\boldsymbol{h})^{c_k},$$

where K is the predefined number of components in the mixture, and each component $p_{\Theta^{(k)}}(x|h)$ is an energy function based on the Boltzmann machine [14] parameterized by $\Theta^{(k)}$. For $K = 1$, it reduces to a VAE. In a VAE, an encoder network is used for learning a function $q_\phi(h|x)$ that compresses documents' original representation into a low-dimensional continuous space. In a MVAE, an additional encoder network is needed to learn the function $q_\eta(c|x)$ that clusters documents into specific groups. We found that using a simple Multi-Layer Perceptron (MLP) with two hidden layers for each of MVAE-BM's encoders works well in practice. For the decoder model $p_\Theta(x|h, c) = \prod_{k=1}^{K} p_{\Theta^{(k)}}(x|h)^{c_k}$, MVAE-BM uses a simple softmax decoder to reconstruct the document by independently generating words given c and h.

To maximize the log-likelihood of a document x, we derive the ELBO of $\mathcal{L}(x; \Theta, \phi, \eta)$:

$$\mathbb{E}_{q_\phi(h|x)q_\eta(c|x)}\left[\sum_{k=1}^{K} c_k \log p_{\Theta^{(k)}}(x|h)\right] - \mathrm{KL}(q_\phi(h|x)\|p(h)) - \mathrm{KL}(q_\eta(c|x)\|p(c)).$$

(3)

The conditional probability over words in a document $p_{\Theta_k}(x|h)$ is modeled by the multinomial logistic regression energy with parameters $\Theta^{(k)} = (R^{(k)}, b^{(k)})$:

$$p_{\Theta^{(k)}}(x|h) = \frac{1}{Z} \exp(-E(x; h, \Theta^{(k)})),$$

$$E(x; h, \Theta^{(k)}) = -h^\top R^{(k)} x - (b^{(k)})^\top x,$$

where Z is the partition function, $R^{(k)} \in \mathbb{R}^{H \times |V|}$ is the semantic word embedding and $b^{(k)} \in \mathbb{R}^{|V|}$ is the bias term for the k-th mixture component. Figure 1 depicts the complete architecture for the recognition and generative models. A vector x representing a document passes through two neural networks (encoders) in parallel to obtain the latent representations c and h used by the mixture of Boltzmann machines.

The posterior approximation $q_\phi(h|x)$ is conditioned on the current document x. The inference network $q_\phi(h|x)$ is modeled as:

$$q_\phi(h|x) \sim \mathcal{N}(h|\mu(x), \mathrm{diag}(\sigma^2(x))),$$

$$l = g(f_{A_2}(g(f_{A_1}(x)))),$$

$$\mu = f_{A_3}(l),$$

$$\log \sigma = f_{A_4}(l),$$

where $f_{A_i}(.)$ is the function represented by a linear layer A_i, $i = 1, \ldots, 4$, and $g(.)$ is an activation function. For each document x, the neural network computes the parameters μ and σ that parameterize the distribution of the latent variable h. Since the prior $p(h)$ is a standard Gaussian, the KL-Divergence $\mathrm{KL}(q_\phi(h|x)\|p(h))$ can be computed analytically [8].

For $q_\eta(c|x)$ we use a Gumbel-softmax as a proxy for the true posterior. The Gumbel-softmax [6,10] is a continuous approximation for sampling from a categorical distribution. More specifically, the recognition $q_\eta(c|x)$ is given by:

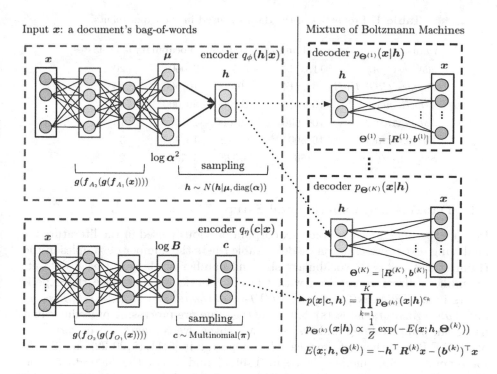

Fig. 1. MVAE-BM encoders $q_\phi(h|x)$ and $q_\eta(c|x)$ compress document x into latent representations h and c. Each of the K decoders is a Boltzmann machine that computes $p_{\Theta^{(k)}}(x|h)$ through the energy function $E(x; h, \Theta^{(k)})$. The mixture is controlled by the latent indicator vector c.

$$q_\eta(c_i = 1|x) \sim \frac{\exp((\log(B_i) + \epsilon_i)/\tau)}{\sum_{j=1}^{K} \exp((\log(B_j) + \epsilon_j)/\tau)},$$
$$\log(B) = g(f_{O_2}(g(f_{O_1}(x)))),$$

where $\epsilon_i \sim \text{Gumbel}(0, 1)$ and $f_{O_1}(.)$ and $f_{O_2}(.)$ represent linear layers. The approximation is accurate for a discrete distribution when the hyperparameter τ (known as 'temperature') goes to 0 and smooth for $\tau > 0$. Hence, using this approach, the KL-Divergence $\text{KL}(q_\eta(c|x)||p(c))$ can be easily evaluated [6].

Finally, to compute the expectation term in Eq. (3), we use the "reparameterization trick" proposed in [8] (for $q_\phi(h|x)$) and in [6] (for $q_\eta(c|x)$).

4 Experimental Results

In this section we describe the datasets used in our experiments, study MVAE-BM's hyperparameters and analyze MVAE-BM's performance on three different learning tasks: topic modeling, text classification, and document clustering.

Table 1. Properties of the datasets used in the experiments.

Dataset	Training set	Test set	Vocabulary	#Classes
20NewsGroups	11,314	7,531	2,000	20
Reuters (RCV1-v2)	794,414	10,000	10,000	90
Yelp Reviews	100,000	10,000	90,000	5
Yahoo Answers	100,000	10,000	20,000	10
TwitterHate	19,500	5,512	15,334	3
Subjectivity	9,756	3,323	5,563	2

4.1   Datasets and Experimental Setup

In our experiments, we leverage six corpora previously used in the literature for analyzing text representation models. Table 1 lists the number of samples in the training and test sets, vocabulary size and number of classes of each dataset. To make a direct comparison with the prior work, we reproduce the experiments in [20] (*20NewsGroups* and *RCV1-v2* datasets) and in [25] (*Yelp Reviews* and *Yahoo Answers* datasets), following the same pre-processing procedures and using the same training and test sets. Moreover, we compare the performance of MVAE-BM to the performance values reported in [5,11,15,20,22] and [23,25] for several baseline models, listed in Tables 2 and 3. For the *Subjectivity* and *TwitterHate* datasets, on the other hand, we created our own train-test splits, given that this information was not available from the related work.

Hyperparameter Configuration: For each dataset, the MVAE-BM's hyperparameters were chosen by grid search in the training set. The search was performed over the values 50, 200, 1,000, 2,000 for the number of neurons in each layer A_1, A_2, A_3, A_4 and O_1, respectively. Moreover, the search covered the values $1, 2, 4, 6, 8$ for the parameter O_2, which determines the number of components K in the mixture, and values $0.1, 0.5, 1$ for τ to obtain approximate categorical samples [6]. For the activation function **g**, we experimented with the *tanh* and *sigmoid* functions. The final hyperparameters can be found at.[4]

All of our experiments were executed on Google Colab[5]. Unlike more computationally expensive techniques, such as BERT and XLM-RoBERTa, MVAE-BM can be trained within a few minutes on platforms that provide public virtual machines. Its implementation, based on neural networks, is also suitable for parallelization via GPU/TPU.

4.2   Document Modeling

Here we evaluate the likelihood of documents left-out of the training set according to the model, using the perplexity metric. Perplexity measures how poorly

---

[4] https://github.com/brunoguilherme1/MVAE-BM/tree/main/hyperparameters.
[5] https://colab.research.google.com.

Table 2. Document modeling: perplexity values. (The latent dimension is indicated in parenthesis, and results not available in the original papers by dashes)

Model	20NewsGroups		RCV1	
	(50)	(200)	(50)	(200)
LDA	1,091	1,058	1,437	1,142
RSM	953	836	988	—
DocNade	836	—	742	—
GSM	787	829	717	602
fDARN	917	—	724	598
NVDLA	1,073	993	791	797
NVDM	836	852	563	550
NTM-R	775	763	—	—
NB-NTM	740	—	—	—
iTM-VAE-Prod	—	779	—	508
MVAE-BM	**730**	**740**	**550**	**504**

a probability model predicts a sample (lower is better), and is widely used with language models to measure their capacity to represent documents. Perplexity is defined as $\exp(-\frac{1}{D}\sum_{i=1}^{D}\frac{\log p(x_i)}{|x_i|})$, where D is the number of documents, and $|x_i|$ is the number of words in the document x_i. Following previous approaches, the variational lower bound (ELBO) is used to estimate $p(x_i)$ (which is actually an upper bound on perplexity [20]). A low perplexity indicates the model is good at predicting a given corpora.

Table 2 presents the perplexity metric of document modeling in *20News-Groups* and *RCV1-v2*, for latent variable dimensions 50 and 200 (shown as separate columns), for MVAE-BM and for 10 baselines: LDA [12], NVLDA [19], GSB [11], NVDM [12], NB-NTM [22], RSM [20], DocNADE [12], fDARN [20], SBN [12], NTM-R [5] and iTM-VAE-Prod [15]. These baselines represent a variety of techniques for topic modeling, some based on graphical models (LDA and RSM) and some based on *belief networks* and on deep networks (DocNADE, SBN, fDARN, NVDM).

MVAE-BM achieves the lowest perplexity values among all baselines in both datasets. Compared to the graphical models, MVAE-BM with a latent variable of dimension $H = 50$ in *RCV1-v2* performs even better than some baselines with 200 dimensions, which is likely due to the interaction between c and h, indicating that using c as an additional latent representation is more effective than increasing H.

4.3 Classification Based on Learned Representations

We now turn our attention to the task of text classification, using the representations learned by different models. In this *supervised* experiment the performance

Table 3. Document classification: models' accuracy (%). Baselines' results were transcribed from reference papers (dashes denote absent values).

Model	Yahoo Answers	Yelp Reviews	Subjectivity	TwitterHate
SCNN-VAE-Semi	65.0	52.0	—	—
CVAE	18.7	29.2	—	—
CVAE BoW	58.5	45.5	—	—
Dirichlet VAE	51.5	39.2	—	—
Dirichlet VAE BoW	59.0	46.3	—	—
BERT	67.6	52.5	87.7	78.2
RoBERTa	66.6	53.0	86.5	77.5
XLM-RoBERTa	69.2	52.5	76.2	74.2
DistilBERT	**70.1**	52.3	88.2	80.3
MVAE-BM	66.5	**55.3**	**89.2**	**82.3**

of MVAE-BM is compared against baselines based on VAE models: CNN-VAE [25] and Dirichlet-VAE [24] and on deep learning (Transformer) architectures: BERT [4], RoBERTa [9], XLM-RoBERTa [1], and DistilBERT [17].

The experiment consists of a document classification task on the test set of each dataset, performed by classifiers that were trained with the representations learned by MVAE-BM and the baseline models. We train a *logistic regression classifier* for the classification task. Since our main goal is to develop and evaluate text representations for classification tasks, we used the classifier standard implementation[6] without any optimizations.

Table 3 displays the classification accuracy obtained by each baseline and by MVAE-BM. For *Yelp Reviews* and *Yahoo Answers*, we transcribed the results of VAE and deep learning baselines from the original papers. For *Subjectivity* and *Twitter*, only the results for Transformers were found.

In *Yelp Reviews*, MVAE-BM has the highest accuracy. In Yahoo Answers, although the Transformer models and, in particular, DistilBERT, perform best, MVAE-BM outperforms the VAE baselines and its accuracy is on par with RoBERTa's. In *Subjectivity* and *TwitterHate* datasets, MVAE-BM achieves the highest accuracy among all the baselines, even though the deep learning models require significantly more computation power.

4.4 Document Clustering

In this section we evaluate how MVAE-BM performs at document clustering tasks. In general, automatic labeling can be done by applying any unsupervised method (e.g., K-means[7]) to the embeddings obtained for the documents. MVAE-BM, however, already includes a labeling of the data by means of the latent

---

[6] www.sklearn.com.

[7] https://github.com/UKPLab/sentence-transformers#clustering.

indicator vector $c = [c_1, c_2, \ldots, c_K]$, defined in Eq. (2). Since c is approximately a one-hot vector, it can be interpreted as a clustering of the input into K groups.

We aim to compare the quality of the clusters defined by MVAE-BM against those found by applying K-means to the text representations obtained using Transformer models.

Table 4 exhibits the results measured w.r.t. the Silhouette [18], the Calinski-Harabasz [21] and the Davies-Bouldin [3] clustering quality measures. We set MVAE-BM's and the baselines's hidden dimension h to 1024 and the number of clusters K in MVAE-BM and in K-means to the number of classes of each dataset (Table 1). The proposed model achieves the highest quality scores in almost all of the combinations (dataset, measure). In particular, in some cases (*Yahoo Answers*, *Yelp Reviews* and *TwitterHate*), the Silhouette score for MVAE-BM is one or two orders of magnitude higher than the baselines'.

Table 4. Clustering Score: SI (Silhouette), DB (Davies-Bouldin) and CA (Calinski-Harabasz). K-means used to cluster BERT variants' embeddings.

	TwitterHate			Subjectivity			20News			Yahoo Answers			Yelp Reviews		
	SI	DB	CA	SI	DB	CA	SI	DB	CA	SI	DB	CA	SI	DB	CA
BERT	0.03	3.23	378	0.005	5.4	110	0.11	6.3	623	0.01	10.34	654	0.05	11.69	781
DistilBERT	0.02	3.45	**367**	0.06	4.8	113	0.012	6.1	**589**	0.02	10.89	689	0.02	11.98	769
RoBERTa	0.01	3.43	378	0.05	5.2	112	0.01	6.5	650	0.04	10.45	623	0.018	11.67	720
XLM-RoBERTa	0.06	3.89	389	0.01	5.4	114	0.10	6.3	677	0.07	10.33	**656**	0.07	11.34	754
MVAE-BM	**0.23**	**3.12**	372	**0.15**	**4.1**	**98**	**0.23**	**6.0**	687	**0.33**	**10.01**	698	**0.46**	**11.23**	**712**

5   Conclusion

In this work, we presented MVAE-BM, a mixture of unsupervised latent models for language modeling. MVAE-BM is inspired by the Boltzmann machine and uses modern neural inference techniques to estimate the intractable latent distributions that appear in the model. In our experiments, we compared to more than 15 different baselines. In these tasks, our model outperformed all baselines in 5 of the 6 datasets used in this work. Apart from the performance gains, our model also has the advantage of learning text representations from unlabeled data without requiring pre-trained word embeddings. Those text representations can be applied with success in various learning tasks, including clustering.

References

1. Conneau, A., et al.: Unsupervised cross-lingual representation learning at scale. In: ACL (2020)
2. Dahl, G.E., Adams, R.P., Larochelle, H.: Training restricted Boltzmann machines on word observations. In: ICML (2012)
3. Davies, D.L., Bouldin, D.W.: A cluster separation measure. IEEE Trans. Pattern Anal. Mach. Intell. **PAMI**–1(2), 224–227 (1979). https://doi.org/10.1109/TPAMI.1979.4766909

4. Devlin, J., Chang, M., Lee, K., Toutanova, K.: BERT: pre-training of deep bidirectional transformers for language understanding. In: NAACL-HLT (2019)
5. Ding, R., Nallapati, R., Xiang, B.: Coherence-aware neural topic modeling. In: EMNLP (2018)
6. Jang, E., Gu, S., Poole, B.: Categorical reparameterization with Gumbel-softmax. In: ICLR (2017)
7. Jiang, S., Chen, Y., Yang, J., Zhang, C., Zhao, T.: Mixture variational autoencoders. Pattern Recognit. Lett. **128** (2019)
8. Kingma, D.P., Welling, M.: Auto-encoding variational Bayes. In: ICLR (2014)
9. Liu, Y., et al.: Roberta: a robustly optimized BERT pretraining approach. CoRR (2019). http://arxiv.org/abs/1907.11692
10. Maddison, C.J., Mnih, A., Teh, Y.W.: The concrete distribution: a continuous relaxation of discrete random variables. In: ICLR (2017)
11. Miao, Y., Grefenstette, E., Blunsom, P.: Discovering discrete latent topics with neural variational inference. In: ICML (2017)
12. Miao, Y., Yu, L., Blunsom, P.: Neural variational inference for text processing. In: ICML (2015)
13. Mikolov, T., Sutskever, I., Chen, K., Corrado, G.S., Dean, J.: Distributed representations of words and phrases and their compositionality. In: NeurIPS (2013)
14. Mnih, A., Gregor, K.: Neural variational inference and learning in belief networks. In: ICML (2014)
15. Ning, X., et al.: Nonparametric topic modeling with neural inference. Neurocomputing **399**, 296–306 (2020)
16. Pennington, J., Socher, R., Manning, C.: GloVe: global vectors for word representation. In: EMNLP (2014)
17. Reimers, N., Gurevych, I.: Making monolingual sentence embeddings multilingual using knowledge distillation. arXiv preprint arXiv:2004.09813 (2020)
18. Rousseeuw, P.J.: Silhouettes: a graphical aid to the interpretation and validation of cluster analysis. J. Comput. Appl. Math. **20**, 53–65 (1987)
19. Srivastava, A., Sutton, C.: Neural variational inference for topic models. In: NeurIPS (2016)
20. Srivastava, N., Salakhutdinov, R., Hinton, G.: Modeling documents with a deep Boltzmann machine. In: Conference on Uncertainty in Artificial Intelligence (2013)
21. Sugar, C.A., James, G.M.: Finding the number of clusters in a dataset. J. Am. Stat. Assoc. **98**(463), 750–763 (2003)
22. Wu, J., et al.: Neural mixed counting models for dispersed topic discovery. In: Annual Meeting of the Association for Computational Linguistics (2020)
23. Xiao, Y., Zhao, T., Wang, W.Y.: Dirichlet variational autoencoder for text modeling. CoRR (2018)
24. Xu, J., Durrett, G.: Spherical latent spaces for stable variational autoencoders. In: EMNLP (2018)
25. Yang, Z., Hu, Z., Salakhutdinov, R., Berg-Kirkpatrick, T.: Improved variational autoencoders for text modeling using dilated convolutions. In: ICML (2017)

A Modular Approach
for Romanian-English Speech Translation

Andrei-Marius Avram$^{(\boxtimes)}$, Vasile Păiş , and Dan Tufiş

Research Institute for Artificial Intelligence, Romanian Academy,
Bucharest, Romania
{andrei.avram,vasile,tufis}@racai.ro

Abstract. Automatic speech to speech translation is known to be highly beneficial in enabling people to directly communicate with each other when they do not share a common language. This work presents a modular system for Romanian to English and English to Romanian speech translation created by integrating four families of components in a cascaded manner: (1) automatic speech recognition, (2) transcription correction, (3) machine translation and (4) text-to-speech. We further experimented with several models for each component and present several indicators of the system's performance. Modularity allows the system to be expanded with additional modules for each of the four components. The resulting system is currently deployed on RELATE and is available for public usage through the web interface of the platform.

Keywords: Speech translation · Romanian-English · Bidirectional · Cascaded system

1 Introduction

The recent significant advances in automatic speech recognition (ASR), machine translation (MT) and text-to-speech (TTS) have been mainly driven by the development of deep learning models, higher computational power and greater data availability. These advancements have also aroused interest in converging them and creating more efficient speech to speech translation (S2ST) systems, thus further breaking down communication barriers between people that do not speak the same language.

However, the S2ST problem is far from being solved and there are currently two methods in approaching it: (1) cascaded systems and (2) end-to-end models. Although cascaded systems still outperform end-to-end models [11], they usually propagate the error from one module to another, making the whole system brittle and hard to analyse. End-to-end S2ST models do not have this issue and there are recent developments that try to shrink the gap between the two [13]. Yet, their performance is limited by the lack of speech translation datasets in comparison to the rich resources that are available for each individual field: ASR, MT or TTS.

© Springer Nature Switzerland AG 2021
E. Métais et al. (Eds.): NLDB 2021, LNCS 12801, pp. 57–63, 2021.
https://doi.org/10.1007/978-3-030-80599-9_6

The Romanian resources available for S2ST are quite scarce, being far from enough for training a competitive end-to-end model. Thus, in the context of the ROBIN project[1], we opted to create a cascaded system for Romanian to English and English to Romanian speech translation, by combining other existing components in a modular framework, allowing a similar low-latency S2ST mechanism. Our main contribution is the creation of this open source framework which allows different modules to be easily integrated into the system. We further analysed the end-to-end latency of various combinations of models and found out that in some cases, the whole system can obtain a near real-time performance, with a response time of around one second for Romanian to English speech translation. The system was also integrated in the RELATE platform [17].

The rest of the paper is organised as follows. The next section presents a review of the cascaded and end-to-end models, and of the previous S2ST systems for Romanian-English speech translation. Section 3 presents the models used for each module and how they were integrated into the RELATE platform. Finally, the paper ends with the conclusion and possible directions for future work in the Sect. 4.

2   Related Work

Extensive research was put into combining different modules within cascaded S2ST and some of the early work on speech translation used an ASR followed by a MT module [14]. However, this kind of approach makes the MT access the errors produced by the ASR and in [19] the authors propose to integrate the acoustic and the translation modules into a transducer that can decode the translated text directly from the audio signal. In addition, because a cascaded system is not naturally capable of maintaining the paralinguistic information, [1] proposed a model that can find the F0-based prosody features in an unsupervised manner and transfer the intonation to the synthesized speech.

One of the earliest attempts to create an end-to-end speech translation system was proposed by [7]. The model obtained a worse performance than a cascaded system, but since then several methods have been applied in order to boost their accuracy from which we can enumerate pre-training, multitask learning and attention passing [6,10]. In [13] the authors showed that by combining multitask training with synthetic data, the model slightly underperforms a cascaded baseline for Spanish-English S2ST.

The bidirectional Romanian-English speech translation has been also attempted in [9] by using a cascaded approach composed of three modules (ASR, MT, TTS) with additional textual corrections like spell checking and diacritic restoration of the ASR transcript, or letter-to-sound conversion and syllabification of the MT translation. They used the Google ASR API for transcribing the audio signal, and an in-house developed MT and TTS. However, their approach

---

[1] http://aimas.cs.pub.ro/robin/en/.

was highly coupled and the system was not able to easily adapt to new modules as they became available with improved performance.

3   System Overview

To approach the problem of S2ST, we used a cascaded system composed of four modules: ASR, textual correction (TC), MT and TTS. Each module contains one or more configurable models for both Romanian and English. This type of architecture allows us to easily integrate new modules and models into the platform and also to select a specific pipeline with respect to a potential problem demands. The overall architecture is depicted in Fig. 1. From the four modules, the TC can be skipped and it is marked with a dotted arrow.

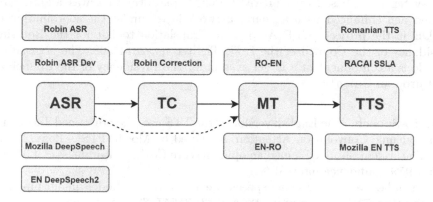

Fig. 1. The proposed S2ST cascaded architecture with the four modules. The Romanian models are depicted in the upper part and the English models in the lower part.

3.1   Modules Description

Automatic Speech Recognition. The models used for the Romanian speech recognition use deep neural networks and their architecture were based on Deep-Speech2 [2]. The models were developed in the ROBIN project in order to improve the transcription latency [3] of the system of that period. They also obtained a satisfying word error rate (WER) of 9.91% on a customized test set when combined with a language model. We provide two variants for this module: (1) the base version - `RobinASR` - presented in [4] and a development version that we continue to improve - `RobinASR Dev`.

For transcribing the English speech, we used the latest version of the speech-to-text model offered by Mozilla[2] that was based on DeepSpeech [12] - `Mozilla DeepSpeech` -, and also a DeepSpeech2 model[3] that was trained on

---

[2] https://github.com/mozilla/DeepSpeech.
[3] https://github.com/SeanNaren/deepspeech.pytorch.

LibriSpeech [15] - EN DeepSpeech2. Although the DeepSpeech2 architecture provided a deeper neural network with more parameters for each layer, Mozilla DeepSpeech turned out to be better and outperformed EN DeepSpeech2 with almost 3% on the LibriSpeech clean test set, obtaining a WER of 7.06%.

Transcription Correction. We currently offer only a version for Romanian textual correction that consists of (1) capitalizing the first letter of words from the transcription that are present on a known named entity list and (2) replacing the words with words from a vocabulary. In addition, we also employ a hyphen restoration for the RobinASR variant based on bi-gram and uni-gram statistics.

Machine Translation. Both Romanian to English and English to Romanian MT systems are based on eTRANSLATION platform that was additionally trained and enhanced with a neural network layer, under the coordination of TILDE in the project "CEF Automated Translation toolkit for the Rotating Presidency of the Council of the EU", TENtec no. 28144308. The translation module is a component of a larger system for the Presidency of the Council of the European Union[4].

Text-to-Speech. The English version of the TTS uses the pretrained Tacotron2 with Dynamic Convolution Attention [5] offered by MozillaTTS[5] - Mozilla EN TTS. The system obtained a median opinion score (MOS) naturalness of 4.310.06 with a 95% confidence interval.

To synthesize the Romanian speech, we integrate two models in our pipeline: (1) Romanian TTS developed in [18] and (2) RACAI SSLA developed in [8], that are based on Hidden Markov Models (HMM) to compute the most probable sequence of spectrograms. However, they offer a trade-off between speed and speech quality, with the Romanian TTS being the version that is slower, but with a higher quality of synthesis (3.150.73 MOS with a 95% confidence interval) and the RACAI SSLA being the version that is faster but with a lower quality of the produced speech.

3.2 RELATE Integration

All the modules are implemented as server processes, exposing their specific functionality as HTTP-based APIs. This allows for hosting the modules on different computing nodes and then integrate them via API calls into a single, unified framework. Furthermore, in order to allow easy interaction with the aggregated pipeline for speech to speech translation, we integrated it in the RELATE platform. We followed the approach described in [16] that allowed us to develop a platform component invoking each module as needed.

---

[4] https://ro.presidencymt.eu/#/text.
[5] https://github.com/mozilla/TTS.

The user is offered the possibility to either record in real-time using a micro-phone and have it translated or start by uploading an already existing sound file. Furthermore, the user is in complete control of the processing chain, being able to select for each step the desired module. However, default settings are pre-loaded, thus the user may use the framework without having to consider individual modules. Only modules available according to the user's choice of pri-mary language are presented in the interface, as depicted in Fig. 2 for English to Romanian S2ST.

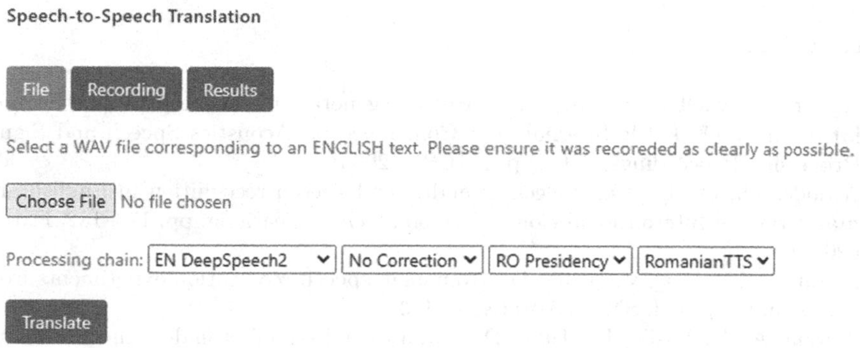

Speech-to-Speech Translation

| File | Recording | Results |

Select a WAV file corresponding to an ENGLISH text. Please ensure it was recoreded as clearly as possible.

| Choose File | No file chosen

Processing chain: | EN DeepSpeech2 ⌄ | | No Correction ⌄ | | RO Presidency ⌄ | | RomanianTTS ⌄ |

| Translate |

Fig. 2. Web interface for choosing pipeline modules.

When a translation process is started, the platform will call all the selected modules in order and aggregate the results. When this process is done, the user is presented with the final synthesized sound, obtained from the selected TTS module, and intermediate texts, obtained from the ASR and translation modules. The average response time of the whole cascaded system is around 1 s for Romanian to English and around 5 s for English to Romanian[6], while audio waves with less than 10 s are given as inputs.

4   Conclusions

This paper presented our work on creating a modular system for bidirectional Romanian-English speech translation that is composed of four modules that are put in a cascaded manner. Each module comes with a series of configurable models that allows a higher flexibility in choosing a specific processing pipeline. Furthermore, our architecture can be easily scaled by integrating new modules and models into the cascaded system. The whole system and its components were made publicly available for use on the RELATE platform[7].

---

[6] This slow down in latency is mostly caused by the Romanian TTS models that are based on HMMs.

[7] RO → EN: https://relate.racai.ro/index.php?path=translate/speech_ro_en EN → RO: https://relate.racai.ro/index.php?path=translate/speech_en_ro.

One direction for possible future work is to develop and integrate a neural based TTS for the Romanian language in order to reduce the latency of the current component, without compromising the speech synthesis quality. Another possible work is to make the source speech and target speech sound more alike by transferring the intonation from the source speech to the target speech.

Acknowledgement. This work was realized in the context of the ROBIN project, a 38 months grant of the Ministry of Research and Innovation PCCDI-UEFISCDI, project code PN-III-P1-1.2-PCCDI-2017-734 within PNCDI III.

References

1. Aguero, P., Adell, J., Bonafonte, A.: Prosody generation for speech-to-speech translation. In: 2006 IEEE International Conference on Acoustics Speech and Signal Processing Proceedings. vol. 1, p. I. IEEE (2006)
2. Amodei, D., et al.: Deep speech 2: end-to-end speech recognition in English and mandarin. In: International Conference on Machine Learning, pp. 173–182. PMLR (2016)
3. Avram, A.M., Păiş, V., TufiŞ, D.: Romanian speech recognition experiments from the robin project. ISSN 1843–911X, p. 103
4. Avram, A.M., Vasile, P., Tufis, D.: Towards a Romanian end-to-end automatic speech recognition based on deepspeech2. Proc. Rom. Acad. Ser. A. **21**, 395–402 (2020)
5. Battenberg, E., et al.: Location-relative attention mechanisms for robust long-form speech synthesis. In: ICASSP 2020–2020 IEEE International Conference on Acoustics, Speech and Signal Processing (ICASSP), pp. 6194–6198. IEEE (2020)
6. Bérard, A., Besacier, L., Kocabiyikoglu, A.C., Pietquin, O.: End-to-end automatic speech translation of audiobooks. In: 2018 IEEE International Conference on Acoustics, Speech and Signal Processing (ICASSP), pp. 6224–6228. IEEE (2018)
7. Bérard, A., Pietquin, O., Servan, C., Besacier, L.: Listen and translate: a proof of concept for end-to-end speech-to-text translation. arXiv preprint arXiv:1612.01744 (2016)
8. Boros, T., Dumitrescu, S.D., Pais, V.: Tools and resources for Romanian text-to-speech and speech-to-text applications. arXiv preprint arXiv:1802.05583 (2018)
9. Boroş, T., Tufiş, D.: Romanian-English speech translation. Proc. Roman. Acad. Ser. A **15**(1), 68–75 (2014)
10. Duong, L., Anastasopoulos, A., Chiang, D., Bird, S., Cohn, T.: An attentional model for speech translation without transcription. In: Proceedings of the 2016 Conference of the North American Chapter of the Association for Computational Linguistics: Human Language Technologies, pp. 949–959. Association for Computational Linguistics, San Diego (2016). https://doi.org/10.18653/v1/N16-1109. https://www.aclweb.org/anthology/N16-1109
11. Federico, M., et al. (eds.): Proceedings of the 17th International Conference on Spoken Language Translation. Association for Computational Linguistics, Online (2020). https://www.aclweb.org/anthology/2020.iwslt-1.0
12. Hannun, A., et al.: Deep speech: scaling up end-to-end speech recognition. arXiv preprint arXiv:1412.5567 (2014)
13. Jia, Y., et al.: Direct speech-to-speech translation with a sequence-to-sequence model. Proc. Interspeech **2019**, 1123–1127 (2019)

14. Ney, H.: Speech translation: coupling of recognition and translation. In: 1999 IEEE International Conference on Acoustics, Speech, and Signal Processing. Proceedings. ICASSP99 (Cat. No. 99CH36258), vol. 1, pp. 517–520. IEEE (1999)
15. Panayotov, V., Chen, G., Povey, D., Khudanpur, S.: LibriSpeech: an ASR corpus based on public domain audio books. In: 2015 IEEE International Conference on Acoustics, Speech and Signal Processing (ICASSP), pp. 5206–5210. IEEE (2015)
16. Păis, V., Tufiș, D., Ion, R.: Integration of Romanian NLP tools into the relate platform. In: International Conference on Linguistic Resources and Tools for Natural Language Processing (2019)
17. Păis, V., Tufiș, D., Ion, R.: A processing platform relating data and tools for Romanian language. In: Proceedings of The 12th Language Resources and Evaluation Conference, pp. 81–88. European Language Resources Association, Marseille (2020). https://lrec2020.lrec-conf.org/media/proceedings/Workshops/Books/IWLTP2020book.pdf
18. Stan, A., Yamagishi, J., King, S., Aylett, M.: The Romanian speech synthesis (RSS) corpus: building a high quality hmm-based speech synthesis system using a high sampling rate. Speech Commun. **53**(3), 442–450 (2011)
19. Vidal, E.: Finite-state speech-to-speech translation. In: 1997 IEEE International Conference on Acoustics, Speech, and Signal Processing, vol. 1, pp. 111–114. IEEE (1997)

NumER: A Fine-Grained Numeral Entity Recognition Dataset

Thanakrit Julavanich[1]([envelope]) [iD] and Akiko Aizawa[1,2] [iD]

[1] The University of Tokyo, Bunkyo, Tokyo 113-8654, Japan
thanakrit@g.ecc.u-tokyo.ac.jp
[2] National Institute of Informatics, Chiyoda, Tokyo 101-8430, Japan
aizawa@nii.ac.jp

Abstract. Named entity recognition (NER) is essential and widely used in natural language processing tasks such as question answering, entity linking, and text summarization. However, most current NER models and datasets focus more on words than on numerals. Numerals in documents can also carry useful and in-depth features beyond simply being described as cardinal or ordinal; for example, numerals can indicate age, length, or capacity. To better understand documents, it is necessary to analyze not only textual words but also numeral information. This paper describes NumER, a fine-grained **Num**eral **E**ntity **R**ecognition dataset comprising 5,447 numerals of 8 entity types over 2,481 sentences. The documents consist of news, Wikipedia articles, questions, and instructions. To demonstrate the use of this dataset, we train a numeral BERT model to detect and categorize numerals in documents. Our baseline model achieves an F1-score of 95% and hence demonstrating that the model can capture the semantic meaning of the numeral tokens.

Keywords: Named entity recognition · Numeral classification · Numeral understanding · Natural language understanding

1 Introduction

Named entity recognition (NER) is an NLP subtask that identifies and locates an entity in unstructured text, then classifies that entity into a predefined category. NER is an essential part of many NLP tasks and applications such as question answering, entity linking, and text summarization [17]. Most NER models and datasets are designed to focus on word entities—that is, the entity token consists of alphabetical characters—such as those denoting people, locations, and organizations. However, there is only a limited set of categories available for numerals, in which the token consists of numerical characters. For example, the CONLL-2003 corpus has no numeral entity type [13], and the OntoNotes 5 corpus has the types Percent, Money, Quantity, Ordinal, and Cardinal [15].

This work was supported by JST, AIP Trilateral AI Research, Grant Number JPMJCR20G9 and by NEDO, SIP-2 Program "Big-data and AI-enabled Cyberspace Technologies", Japan.

© Springer Nature Switzerland AG 2021
E. Métais et al. (Eds.): NLDB 2021, LNCS 12801, pp. 64–75, 2021.
https://doi.org/10.1007/978-3-030-80599-9_7

Equally as important as word tokens, numeral tokens also contain relevant information. Moreover, we need to categorize numeral entities in more detail than the current NER datasets can provide. In a real-life scenario, for example, in a biographical document, the text might include numerals describing the age, birth year, weight, and height of the subject of the biography. As shown in Fig. 1, an NLP application such as a question-answering task could include an inquiry using a monetary numeral entity, e.g. "50" in "who ordered more than 50 USD worth of meat today", or population numeral entity, e.g. "50,000" in "where is the nearest stadium with a capacity of more than 50,000 people." Understanding these numerals may help the model to better determine the correct part of the article or the right property in the knowledge base. For example, when the model recognizes the abovementioned monetary token "50", it can focus on the monetary property, e.g. the order's payment amount.

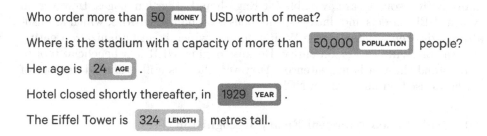

Fig. 1. Examples of numeral entities.

Previous studies have performed some research about numeral entities. For example, Min et al. [8] presented a numeral classification method using a rule-based approach. However, the focused classes were both semantic category, e.g., Money and Date, and syntactic categories, e.g., Number and Floatnumber. As a result, there may be a conflict of category taxonomy in this scheme. For example, the token "22.23" in the context "22.23 USD" can be categorized as both Money and Floatnumber. Another related work is the NTCIR-14 FinNum Task [2]. The authors published a dataset for fine-grained numeral entity recognition in social media data from StockTwits. This work focused on the financial domain with finance-related entity types such as buy price, sell price, and stop loss.

To better understand fine-grained numeral information, we created a numeral taxonomy by classifying the answer's property from the existing datasets for question answering over tabular data and text-to-SQL semantic parsing. Because each question focuses on a small domain, the unit and object token can be easily omitted. Therefore, it provides a more difficult task for the model to classify. We sort numerals into eight categories: Age, Population, Year, Date/Month, Length/Height, Money, Weight/Volume, and Generic.

This paper focuses on numerals in unstructured text, rather than primarily targeting word tokens like typical NER datasets. This work's main objective is to create an entity recognition dataset focused on numeral tokens. Moreover, we aim to ensure the model's ability to comprehend and capture the semantics of numerals before applying it to the downstream tasks. This work provides three main contributions. First, we annotate the numerals in the chosen text corpora with this taxonomy and construct a dataset for experiments. Second, we present the dataset focused on the cases where there is no unit token. Third, we conduct comprehensive investigations to compare the performance of different classification and entity recognition models. Our annotated dataset is published at https://github.com/Alab-NII/ValER.

2   Related Work

There are many approaches used to build NER systems, such as creating hand-crafted rules [10] or using a machine learning model [17]. Furthermore, today, many NER resources are available for English and other languages. However, in typical NER models and datasets, the focus is on named entities, which are usually word tokens such as "The White House", which is an Organization entity, or "Taylor Swift," a Person entity. Because of this, NER is a beneficial tool to understand the words in sentences. However, there is still a limited number of works focused on numerals in NER.

2.1   Rule-Based Numeral Entity Recognition

There have been several attempts to recognize or classify numeral tokens. Microsoft Recognizers Text[1] is an off-the-shelf tool for recognizing the numerals, units, and date/time expressed in documents. This library detects the unit token and matches pre-defined regular expression patterns to identify the numeral's type. For example, if the model detects the "USD" token, it can recognize the nearby numeral as belonging to the currency category. According to the numerals' units, it can detect numerals of the following four types, including Age, Currency, Dimension, and Temperature. However, if a sentence does not contain a unit token, recognition is difficult.

2.2   Machine Learning-Based Numeral Entity Recognition

For the machine learning approach, the capability of the model is dependent on the dataset used for training, especially the target entity types. At present, the available entity types of numeral tokens are still limited. For example, in the OntoNotes dataset [15], there are seven entity types for numerals: Date, Time, Percent, Money, Quantity, Ordinal, and Cardinal. Although this design provides some understanding of numeral tokens, there is still room to extend this structure, especially regarding the Quantity class. Entity class extension can help understand measurements such as length, duration, volume, or number of items.

---

[1] https://github.com/Microsoft/Recognizers-Text.

There have also been several datasets for numeral classification in a specific domain. For example, the NTCIR-14 FinNum task [2] focuses on numeral classification in informal financial documents. The focused entity types are finance-based, e.g., Quote, Change, Buy price, and Sell price. Several works have been submitted to this shared task based on state-of-the-art and well-known language models and architectures, e.g., CNN with ELMo word embeddings [1], RoBERTa-based models [6], and multi-layer perceptrons with LSTM [16]. However, as these studies were tailored to specific domains, they cannot be effectively applied to more general cases.

3 The NumER Dataset

The NumER dataset is created using documents from several datasets. Each numeral is annotated and categorized into one of the eight aforementioned categories. In total, the dataset consists of 2,481 sentences with approximately 56,111 tokens. Each sentence contains up to 19 numerals.

3.1 Annotation Scheme and Taxonomy

For the deeper context extraction of numerals, the first challenge is to define the entity types. We begin by focusing on the questions with numerals in the SPIDER dataset [18]. We identify the numerals and look for those containing hidden information. In particular, we focus on numerals that do not have a token to describe what they are. For example, in the sentence "This year, I am 25", the numeral 25 denotes an age without any token as its description. This particular case can cause difficulty in recognition when a numeral's description or unit is lacking.

We propose eight classes in total for numeral classification. Below we summarize the annotation guidelines for the eight classes.

AGE is the age of anything such as people, animals, or plants, as well as buildings or places such as monuments, parks, or schools.
POPULATION is the number of inhabitants or capacity for inhabitants in a specific area—for example, a country's population, number of stadium seats, or number of enrolled students.
YEAR is a year in any format, e.g., in a 4-digit format such as 2021 or a 2-digit format such as 95.
DATE/MONTH is a specific month or date such as Sunday 24th, the 8th month, or 4th of July.
LENGTH/HEIGHT is a measured size in two-dimensional space or time duration such as travel distance, human height, and running duration.
MONEY is any numeral related to money, such as a purchase amount, salary, or account balance.
WEIGHT/VOLUME is the measured weight or volume of anything such as a pet's weight, parcel's weight, or bottle's volume.

GENERIC is a broad category. A numeral that does not fall into any other category is considered to be part of the Generic type—for example, postcode, ID, or phone number.

In our annotation, we define a numeral entity as a token that contains only the numeral in the sentence without any characters or tokens describing its unit. However, we allow the following symbols in the numeral token to be annotated: "." (for floating-point numbers only), "-" (except when used to denote a range of values), and "/," as these symbols can exist in between numbers to connect multiple numeral groups to one entity, e.g., 3.14, 2021-01-01, and 2021/01/01.

3.2 Data Collection

The NumER dataset consists of documents gathered from four primary sources as follows.

- SPIDER [18] and SParC [19]: These text-to-SQL datasets consist of questions in formal and informal writing. Questions using numerals were collected.
- Wikidata [14]: This is an open data knowledge graph hosted by the Wikimedia Foundation. We queried the data using numeral properties both selectively and randomly. Sentences from the corresponding Wikipedia summaries describing the values of the selected properties were extracted.
- Epicurious[2]: A cooking recipe dataset. We extracted numeral-including sentences from the recipe instructions.
- News Category[3]: A dataset including news headlines from 2012 to 2018 obtained from HuffPost. Numeral-including sentences were extracted from the headlines and their summaries.

Annotation Process. The dataset was annotated by Amazon Mechanical Turk (MTurk) workers. Every worker had to pass our qualification test to test their understanding of the taxonomy before working. The qualification test consisted of four questions about several numeral entities that could result in different decisions depending on whether a respondent fully understood the defined taxonomy. For example, following our taxonomy, the numeral "300" in the sentence "Find a theatre with a capacity above 300 seats" is a Population entity because 300 is the number of humans that can fit in such a theatre. From a different point of view, the annotator might think 300 is the *volume* of a theatre.

We assigned three different MTurk workers to annotate each numeral. We formulated the annotation task as a classification task to reduce the difficulty for both the workers and the implementation. We extracted numerals using a heuristic method. The token is considered to be a numeral when one comma and one period are replaced, and only digits are left. For the token that contains hyphen or slash symbols, if the token can be parsed as a date or time, we

---

[2] https://www.kaggle.com/hugodarwood/epirecipes.
[3] https://www.kaggle.com/rmisra/news-category-dataset.

considered it as a single token. Otherwise, we split the token using those symbols. Then, we provided the extracted numeral and source sentence to the worker and asked them to categorize the given numeral.

Annotation Agreement. After the annotation process was finished, 74% of numerals yielded consistent annotation results by receiving the same decision from all three annotators. 25% of numerals had a majority (two) decision from two annotators. Only 1% of numerals had a split decision resulting from differences in the annotators' decisions.

The Kappa score between every two annotators was 76.6%, 76.7%, and 76.8%, considered an "almost perfect" agreement [9].

Annotation Conflict. The annotation conflicts were solved manually after calculating the inter-annotator agreement. For the majority-decision entities, we considered the majority decision as the correct entity type by default. Some sentences included both a majority-decision entity and a split-decision entity. We manually conducted further investigation to choose the final annotations for the split-decision entities to resolve conflicts.

3.3   Data Analysis

We split our data into training, development, and test sets by 70%, 10%, and 20%, respectively. We randomly split the data while maintaining the ratio between train/dev/test set in each class as close as possible to the ideal ratio. The general statistics of the dataset are presented in Table 1. Table 2 shows the data distribution in our dataset. The total number of entities is 5,477. Each class contains at least 300 entities. Year is the majority class with 1,580 entities. In contrast, Population is the minority class with 324 entities in total.

Each of our sentences contains at least one numeral and up to 19 numerals. 30% of our sentences contain at least three numerals in the sentence. Furthermore, 63% of the numeral tokens in our dataset belong to multiple classes and thus require disambiguation. For example, the numeral "20" can belong to the Age class, as in "age of 20", and the Length/height class, as in "20 cm". These conditions can help provide a more complex situation for the model to tackle. In addition, more than half of our numeral entities are numerals without a unit token.

3.4   Comparison to Other NER Datasets

As our dataset is focused on numeral tokens, there are only a few other datasets with the same focus. For example, Mandhan et al. [12] focused on numerals in a clinical text, such as those denoting blood pressure, temperature, pulse, heart rate, and drug dosage. Another example is NTCIR-14 FinNum [2], which focuses on the financial domain with informal documents gathered from Twitter. In contrast to these works, our dataset is focused on more general sentences in daily life.

Table 1. NumER dataset statistics.

	Train	Dev	Test
Number of sentences	1,737	248	496
Number of token	39,573	5,457	11,081
Number of entities	3,825	537	1,115
Type-token ratio	14%	30%	23%

Table 2. Distribution of the numeral entities in the dataset.

Entity type	Train	Dev	Test	Total
Age	329	47	94	470
Population	230	34	60	324
Year	1,139	141	300	1,580
Date/Month	647	81	177	905
Length/Height	536	98	206	840
Money	275	48	81	404
Weight/Volume	245	32	63	340
Generic	424	56	134	614
Total	3,825	537	1,115	5,477

4   Experimental Setup

This section describes the models used for benchmarking with our dataset. We benchmark two categories of models, including an off-the-shelf model and a pre-trained model that was fine-tuned on our training set. In every model, we configure the tokenizer to tokenize our focused numeral as one token. We also create another training set with a data augmentation technique to provide additional training data and improve the results.

4.1   Model Settings

SpaCy 2.3.5. [4]The spaCy NER model uses deep convolutional neural network and transition-based named entity parsing. We use spaCy NER using a blank English model. We train the model from scratch with our training set for the maximum of 30 epochs with a learning rate of 0.001 until there is no improvement for three epochs.

BERT. [3] We used BERT base models with both cased and uncased context, and fine-tuned the model on the NumER training set for three epochs with a learning rate of $5 \cdot 10^{-5}$ and a batch size of 32.

---

[4] https://v2.spacy.io/.

BiLSTM-CRF. [5] We used the BiLSTM-CRF model with Glove embedding [11] and trained for 15 epochs maximum with a learning rate of 0.001, dropout of 0.5, and batch size of 20 until there was no improvement for three epochs.

4.2   Data Augmentation

In our collected data, Population, Money, and Weight/Volume types can be considered minor classes because of the limited number of training data. To deal with the data sparsity, we create additional training data using the contextual augmentation technique [7] using the BERT language model. We randomly replace tokens in the sentence, including both words and numerals, with other suitable words predicted using BERT based on the original word's surrounding context. Thus, we keep the original label sequence unchanged.

Using the generated sentences, we created a new augmented dataset including the original sentences. In the augmentation, we replace randomly one to five tokens per sentence. For each original sentence, we generate five new sentences. In Table 3 we show examples of two source sentences and their augmented sentences.

Table 3. Example of data augmentation including two original sentences and three examples of augmented sentences for each.

Original Sentence	How many players have a weight greater than 220 or height shorter than 75?
Augmented	How can players have head weight greater than 220 its height shorter than 85?
	How would players have any breadth greater than 60 or height shorter than 75?
	How many players has their weight increased than 220 or height smaller than 75?
Original Sentence	It was built in 1974 to a height of 123 m.
Augmented	Bridge was built in 1922 to a height spanning 123 m
	It was built about 1974 to a height over 123 m
	It was built since 1974 to a heights of 123 metre

Because this method relies on randomization in choosing the token to replace, there is a chance that the predicted token from the language model may change the context of the sentence. This change can affect the entity type of a numeral token when its neighbour token is changed. To ensure the numeral's validity and type in the augmented sentences, we performed a manual check for every sentence that changed in the tokens near the annotated numeral token.

As a result of the augmentation process, 6,146 sentences are generated by contextual augmentation in addition to the original 1,737 training sentences, yielding a total of 7,883 sentences in the augmented training set. The development and testing set consists of the original 248 and 496 sentences, respectively.

5   Results

The models are evaluated on our test set to obtain entity-level precision, recall, and F1-score per class. The results obtained using an off-the-shelf model and trained models are reported in Table 4. We determine that all trained/fine-tuned models work better than the off-the-shelf tools. Every model achieved 100% F1-score in the span detection task. Overall, using the BERT model, we can achieve an F1-score of 95.2%. Date/month had the best results while Population had the worst. Using BiLSTM-CRF, we can reach an F1-score of 88.5%.

The overall best-performing model is BERT-cased fine-tuned on the NumER augmented training set. And every model trained on the augmented dataset performed better than those trained on the non-augmented dataset. It is encouraging that the BERT-based models are also able to capture the context of the numeral tokens. Table 5 describe each class's score of the best models.

Table 4. Overall results of baseline models in the NumER dataset.

Architecture	Training set	Precision	Recall	F1-score
BERT-cased	Augmented	0.953	0.952	0.952
BERT-cased	Non-augmented	0.943	0.936	0.938
BERT-uncased	Augmented	0.948	0.946	0.947
BERT-uncased	Non-augmented	0.947	0.947	0.946
BiLSTM-CRF	Augmented	0.886	0.884	0.885
BiLSTM-CRF	Non-augmented	0.865	0.862	0.863
spaCy	Augmented	0.856	0.855	0.856
spaCy	Non-augmented	0.820	0.818	0.818

Table 5. The results of the BERT-uncased model without data augmentation (UC-NOAUG), the BERT-cased model with data augmentation (C-AUG), and the SpaCy model with data augmentation (SpaCy-AUG). (P = Precision; R = Recall; F1 = F1-Score)

Entity type	UC-NOAUG			C-AUG			SpaCy-AUG		
	P	R	F1	P	R	F1	P	R	F1
Age	0.979	0.979	0.979	0.978	0.947	0.962	0.950	0.809	0.874
Population	0.791	0.883	0.835	0.794	0.833	0.813	0.533	0.706	0.608
Year	1.000	0.980	0.990	1.000	0.983	0.992	0.952	0.986	0.979
Date/Month	0.967	1.000	0.983	1.000	1.000	1.000	0.975	0.975	0.975
Length/Height	0.965	0.951	0.958	0.943	0.976	0.959	0.833	0.816	0.825
Money	0.920	0.988	0.952	0.952	0.988	0.970	0.950	0.792	0.864
Weight/Volume	0.846	0.873	0.859	0.814	0.905	0.857	0.533	0.813	0.658
Generic	0.885	0.812	0.847	0.917	0.835	0.874	0.854	0.625	0.722
Total	0.947	0.947	0.946	0.953	0.952	0.952	0.856	0.855	0.856

The results indicate that our augmentation technique helps the models to perform slightly better than they do using only the original data. For the models trained with non-augmented data, the worst-performing class is Population with an F1-score of 73.4%. After using the augmented data for training, the F1-score improves by almost 8% in the BERT-uncased model, which is the most significant improvement per class.

6   Application to Text-to-SQL Task

To demonstrate the benefit of the NumER dataset, we experimented on text-to-SQL tasks by incorporating the information from NumER into an existing text-to-SQL model. Such a process usually requires schema linking, a process to match the candidate value token in the question to its associated column. The numeral entity type information can benefit the schema linking process in the model. Namely, given the numeral entity type information, the model can perform the schema linking even when there is no overlap token between the query and candidate column names/values. Note that this is difficult for existing models that are based on surface-level string matching.

6.1   Model

We modified the IRNET model [4], a text-to-SQL model trained on the SPIDER dataset. IRNET is based on encoder-decoder architecture with a memory augmented pointer network. The input embedding for the natural language schema encoders is concatenated with additional information from NumER. Furthermore, the schema linking process is enhanced to consider our numeral entity types. The three modified components are described below.

First, we modified the question token type embedding which describes the referred schema and SQL command-related component in each token, including table, column, aggregated function, comparative word, superlative word, and numeral. We extended the embedding by adding eight more features to represent each NumER entity type using one-hot encoding.

Second, the column type embedding, which is used to keep track of which column is mentioned in the input question, is modified. We extended the embedding with eight more features in the same way as to question token type using the information from our manually annotated type of each column.

Finally, we modified the schema linking process in the preprocessing step. We performed the original schema linking process first. Then the type of each numeral token is recognized using the NumER model. We map the detected numeral type to the column with the same type. If there are multiple candidate columns, the column in the table with the matching name in question tokens is selected. If the mapped column was not detected in the original process, we add the token indicating the mapped column name in front of the numeral.

6.2 Result

We trained our modified version of IRNET model on the SPIDER training data and evaluated it using the SPIDER development dataset on the "exact set match without values" setting. We achieved 58.4% accuracy, compared to the vanilla IRNET model with 53.2% accuracy. As a result, the model benefits from the NumER model's information and has a performance improvement of 5.2%.

7 Conclusion

In this work, we present NumER, a fine-grained numeral entity recognition dataset that successfully classified numerals in a more generic domain than that of the typical NER dataset. The data consisted of non-specific-domain sentences from several sources. The collected sentences were in the form of articles, question, titles, and instructions. We conducted experiments by training models on our dataset and benchmarked using well-known models.

According to the results, the models can successfully capture the semantics of the numeral token. This shows that (I) our method can be used to extract information from numerals in the sentence, and (II) the numeral classification in a more generic domain is also possible and not limited to just a specific domain. In the future, we can extend our proposed taxonomy to more classes or adapt to match well-known ontology. We will also apply our model to current NLP challenges involving numerals to improve the result of target tasks and extend our data's usefulness.

References

1. Azzi, A.A., Bouamor, H.: Fortia1@ the NTCIR-14 FinNum task: enriched sequence labeling for numeral classification. In: Proceedings of the 14th NTCIR Conference on Evaluation of Information Access Technologies, pp. 526–538 (2019)
2. Chen, C.C., Huang, H.H., Takamura, H., Chen, H.H.: Overview of the NTCIR-14 FinNum task: fine-grained numeral understanding in financial social media data. In: Proceedings of the 14th NTCIR Conference on Evaluation of Information Access Technologies, pp. 19–27 (2019)
3. Devlin, J., Chang, M.W., Lee, K., Toutanova, K.: BERT: pre-training of deep bidirectional transformers for language understanding. In: Proceedings of the 2019 Conference of the North American Chapter of the Association for Computational Linguistics: Human Language Technologies, vol. 1 (Long and Short Papers), pp. 4171–4186. Association for Computational Linguistics, Minneapolis (2019). https://doi.org/10.18653/v1/N19-1423
4. Guo, J., et al.: Towards complex text-to-SQL in cross-domain database with intermediate representation. In: Proceedings of the 57th Annual Meeting of the Association for Computational Linguistics (ACL), pp. 4524–4535. Association for Computational Linguistics (2019). https://doi.org/10.18653/v1/P19-1444
5. Huang, Z., Xu, W., Yu, K.: Bidirectional LSTM-CRF models for sequence tagging. CoRR abs/1508.01991 (2015)

6. Jiang, M.T.J., Chen, Y.K., Wu, S.H.: CYUT at the NTCIR-15 FinNum-2 task: tokenization and fine-tuning techniques for numeral attachment in financial tweets. In: Proceedings of the 15th NTCIR Conference on Evaluation of Information Access Technologies, pp. 92–96 (2020)
7. Kobayashi, S.: Contextual augmentation: data augmentation by words with paradigmatic relations. In: Proceedings of the 2018 Conference of the North American Chapter of the Association for Computational Linguistics: Human Language Technologies, vol. 2 (Short Papers), pp. 452–457. Association for Computational Linguistics, New Orleans (2018). https://doi.org/10.18653/v1/N18-2072
8. Min, K., MacDonell, S., Moon, Y.-J.: Heuristic and rule-based knowledge acquisition: classification of numeral strings in text. In: Hoffmann, A., Kang, B., Richards, D., Tsumoto, S. (eds.) PKAW 2006. LNCS (LNAI), vol. 4303, pp. 40–50. Springer, Heidelberg (2006). https://doi.org/10.1007/11961239_4
9. Munoz, S., Bangdiwala, S.: Interpretation of Kappa and b statistics measures of agreement. J. Appl. Stat. **24**, 105–112 (1997). https://doi.org/10.1080/02664769723918
10. Nadeau, D., Sekine, S.: A survey of named entity recognition and classification. Lingvisticæ Investigationes **30**(1), 3–26 (2007). https://doi.org/10.1075/li.30.1.03nad
11. Pennington, J., Socher, R., Manning, C.D.: GloVe: global vectors for word representation. In: Empirical Methods in Natural Language Processing (EMNLP), pp. 1532–1543 (2014). https://doi.org/10.3115/v1/D14-1162
12. R., S.P., Mandhan, S., Niwa, Y.: Numerical atribute extraction from clinical texts. CoRR abs/1602.00269 (2016). https://doi.org/10.13140/RG.2.1.4763.3365
13. Tjong Kim Sang, E.F., De Meulder, F.: Introduction to the CoNLL-2003 shared task: language-independent named entity recognition. In: Proceedings of the Seventh Conference on Natural Language Learning at HLT-NAACL 2003, pp. 142–147 (2003). https://www.aclweb.org/anthology/W03-0419
14. Vrandečić, D., Krötzsch, M.: Wikidata: a free collaborative knowledgebase. Commun. ACM **57**(10), 78–85 (2014). https://doi.org/10.1145/2629489
15. Weischedel, R., et al.: OntoNotes release 5.0 (2013). https://doi.org/10.35111/XMHB-2B84
16. Wu, Q., Wang, G., Zhu, Y., Liu, H., Karlsson, B.: DeepMRT at the NTCIR-14 finnum task: a hybrid neural model for numeral type classification in financial tweets. In: Proceedings of the 14th NTCIR Conference on Evaluation of Information Access Technologies, pp. 585–595 (2019)
17. Yadav, V., Bethard, S.: A survey on recent advances in named entity recognition from deep learning models. In: Proceedings of the 27th International Conference on Computational Linguistics, pp. 2145–2158. Association for Computational Linguistics, Santa Fe (2018). https://www.aclweb.org/anthology/C18-1182
18. Yu, T., et al.: Spider: a large-scale human-labeled dataset for complex and cross-domain semantic parsing and text-to-SQL task. In: Proceedings of the 2018 Conference on Empirical Methods in Natural Language Processing, pp. 3911–3921. Association for Computational Linguistics, Brussels (2018). https://doi.org/10.18653/v1/D18-1425
19. Yu, T., et al.: SParC: cross-domain semantic parsing in context. In: Proceedings of the 57th Annual Meeting of the Association for Computational Linguistics, pp. 4511–4523. Association for Computational Linguistics, Florence (2019). https://doi.org/10.18653/v1/P19-1443

Cross-Domain Transfer of Generative Explanations Using Text-to-Text Models

Karl Fredrik Erliksson[1,2]([⊠]), Anders Arpteg[2], Mihhail Matskin[1],
and Amir H. Payberah[1]

[1] KTH Royal Institute of Technology, Stockholm, Sweden
{kferl,misha,payberah}@kth.se
[2] Peltarion, Stockholm, Sweden
anders@peltarion.com

Abstract. Deep learning models based on the Transformers architecture have achieved impressive state-of-the-art results and even surpassed human-level performance across various natural language processing tasks. However, these models remain opaque and hard to explain due to their vast complexity and size. This limits adoption in highly-regulated domains like medicine and finance, and often there is a lack of trust from non-expert end-users. In this paper, we show that by teaching a model to generate explanations alongside its predictions on a large annotated dataset, we can transfer this capability to a low-resource task in another domain. Our proposed three-step training procedure improves explanation quality by up to 7% and avoids sacrificing classification performance on the downstream task, while at the same time reducing the need for human annotations.

Keywords: Explainable AI · Generative explanations · Transfer learning

1 Introduction

There is a growing consensus that many practical machine learning (ML) applications require explainability, especially when these applications are subject to critical auxiliary criteria that are difficult to formulate mathematically, e.g., nondiscrimination, safety, or fairness [11,30]. Moreover, regulations such as the General Data Protection Regulation (GDPR) [13] equip people with a "right to explanation" for algorithmic decisions that significantly affect them. At the same time, deep neural networks (NNs) have achieved and even surpassed human performance in many tasks in natural language processing (NLP) and computer vision [15,43], which has motivated a large body of research over the last few years focusing on making NN predictions more explainable.

Explainability in ML has traditionally been approached from two perspectives; either by building models that provide inherent transparency and explainability [5,21,26], or by creating post-hoc explanations for an opaque model that

© Springer Nature Switzerland AG 2021
E. Métais et al. (Eds.): NLDB 2021, LNCS 12801, pp. 76–89, 2021.
https://doi.org/10.1007/978-3-030-80599-9_8

has already been trained [29,37,39]. This work falls into the former category where we teach a model to generate explanations as part of the prediction process, conceptually similar to how humans would be asked to motivate their reasoning for a specific decision. The explanations are formed by natural language, and we cast this as a supervised sequence-to-sequence (seq2seq) problem where the model learns from ground-truth explanations annotated by humans [4,36,40]. Natural language explanations provide a series of benefits compared to other common approaches, such as attributions methods and formal language. They are more easily accessible to non-expert end-users owing to the familiar format [4], and are often simpler to evaluate and annotate by humans. Narang et al. [32] recently investigated this approach and proposed a model called WT5 that achieves new state-of-the-art performance on various NLP explainability benchmarks [10]. However, this requires large amounts of annotated explanations during training and for many real-world applications this becomes a bottleneck.

We propose a three-step training procedure to transfer the ability to generate extractive explanations from a large easily-available dataset to a low-resource downstream task with a lack of annotated ground-truth explanations, in a potentially different domain. First, in the pre-training (PT) step, we train an initial language model using unannotated data. Then, in the explainability pre-training (EP) step, we teach the model the semantic meaning of an *explainability keyword*. Finally, we use this keyword during the fine-tuning (FT) step and at inference time to instruct the model to generate explanations for specific predictions. To summarize our contributions:

- Narang et al. in [32] provide a brief qualitative discussion regarding explainability transfer for WT5. We extend this work and provide a more thorough quantitative evaluation, including two popular seq2seq models, T5 [35] and BART [27]. We find that T5 consistently outperforms BART for extractive explanation generation across all our experiments.
- Using our proposed three-step training procedure, we show that the ability to generate extractive explanations can be transferred between tasks in different domains, and that it can result in both improved performance and explanation quality on a low-resource downstream task with few annotated explanations.
- We provide evidence that only a small number of samples from the downstream tasks need to be annotated with human explanations to achieve a significant boost in explanation quality.

Through the experiments, we see an increase of 7% and 5% in TF1 score (explanation quality) for T5-Base and T5-Large, respectively, when EP is performed.[1]

2   Background

In this section, we provide a brief background to seq2seq modelling in NLP and define the main idea of generative explanations.

---

[1] Code available at https://github.com/Peltarion/explainability_transfer.

2.1   Sequence-to-Sequence Models

Consider an NLP model $f : \mathcal{X} \rightarrow \mathcal{Y}$ where the input $x = (x_1, x_2, ..., x_{N_{\text{in}}}) \in \mathcal{X}$ and the output $y = (y_1, y_2, ..., y_{N_{\text{out}}}) \in \mathcal{Y}$ are both ordered sequences of tokens. By $\tilde{x}$ and $\tilde{y}$, we denote the corresponding raw input and output text. The model f is trained by maximizing the the the conditional probability $p(y_1, ..., y_{N_{\text{out}}} | x_1, ..., x_{N_{\text{in}}}) = \prod_i^{N_{\text{out}}} p(y_i | x_1, ..., x_{N_{\text{in}}}, y_1, ..., y_{i-1})$. At prediction time, an output sequence can be generated autoregressively by iteratively sampling $y_i \sim p(y_i | x_1, ..., x_{N_{\text{in}}}, y_1, ..., y_{i-1})$ either greedily or by methods like beam search.

Raffel et al. [35] introduced the idea of unifying all NLP tasks into a general common framework by treating them as seq2seq problems, referred to as the *text-to-text* framework. As an example, a binary classification problem with output classes {True, False} is posed as a generative task where the model is trained to explicitly generate the sequence of tokens corresponding to the target output class. This should be seen in contrast to other common BERT-based architectures [9], where a small model head tailored for a specific task and its format is attached on top of an encoder block to produce a probability distribution over the output classes. The raw input is formatted as $\tilde{x} =$ "⟨task_prefix⟩: ⟨input_text⟩", where the prefix is used to let the model know what type of task it is, e.g., "sentiment" for sentiment analysis. The target output is given by $\tilde{y} =$ "⟨target⟩", which in the case of classification problems would simply be the class label. This enables an easy way of transferring knowledge from one task to the other, thanks to the unified format. If the model would output anything other than the expected output classes during evaluation, it is considered as incorrect.

The Text-to-Text Transfer Transformer (T5) [35] is a model based on the above approach that was pre-trained on the large Common Crawl dataset [7], and has been demonstrated to achieve state-of-the-art performance on various NLP downstream tasks [43]. Apart from T5, many other seq2seq models have been used for tasks such as machine translation and text summarization. A recent popular model is BART [27], which is architecturally similar to T5 but using a different language model pre-training objective and number of hidden states in the embedding and feed-forward layers.

2.2   Generative Explanations

One way to approach explainability in deep learning is by letting a model produce explanations similar to how humans would motivate their reasoning. One of the earlier works by Hendricks et al. [16] considered generating "because of" sentences for a computer vision classification task. The text-to-text framework enables a new way to teach NLP models to produce generative explanations in a supervised fashion. This idea was recently explored in [32], where an extension of T5, called WT5 (short for "Why T5?"), was proposed. In this case, we simply prepend ⟨task_prefix⟩ in $\tilde{x}$ with the optional keyword "explain" and append the target output $\tilde{y}$ with "explanation: ⟨explanation⟩", where we assume that golden-truth annotated explanations are available for the task. The new input-output format thus becomes

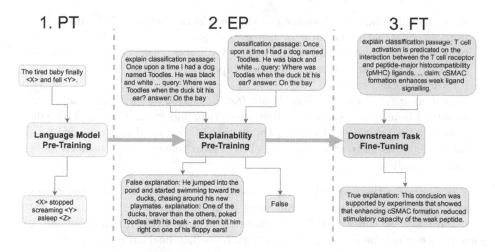

Fig. 1. The proposed three-step training procedure.

$$\tilde{x} = \text{“[explain] } \langle \texttt{task\_prefix} \rangle\text{: } \langle \texttt{input\_text} \rangle\text{”},$$

$$\tilde{y} = \text{“}\langle \texttt{target} \rangle \text{ [explanation: } \langle \texttt{explanation}_1 \rangle \text{] } \ldots \qquad (1)$$

$$\text{[explanation: } \langle \texttt{explanation}_M \rangle \text{]”},$$

where hard brackets denote optional explanation arguments and we allow for potentially multiple explanation sentences. An illustrative example of the input-output format is provided in Table 1. To simplify the annotation and evaluation process, it is helpful to consider the subset of *extractive explanations* that only consist of spans of tokens from the input text. This allows us to compute overlap statistics with respect to the ground truth to quantitatively measure the explanation quality [10].

3   Approach

The main focus of this work is to transfer explainability capabilities to a low-resource task in another domain with a potentially limited number of annotated explanations. Based on the procedure outlined in [32], we utilize seq2seq models to generatively produce natural language explanations alongside the original prediction task. To this end, we propose a three-step training procedure as illustrated in Fig. 1:

1. *Language model pre-training (PT)* is carried out in a self-supervised fashion on a large text corpus like C4 [35] (the yellow blocks in Fig. 1).[2]
2. *Explainability pre-training (EP)* is then performed on a large dataset with annotated explanations (the blue blocks in Fig. 1). Following the ideas in [32],

---

[2] Since all seq2seq models considered in this work have publicly released checkpoints from language model pre-training, this is used as starting point for step 2 in Fig. 1.

we teach the model the meaning of the "explain" keyword by uniformly at random constructing training instances with and without annotated explanations according to the format in Eq. (1). We hypothesize that this promotes a task-agnostic extractive explanation capability that can be extended also for various other tasks.

3. *Fine-tuning (FT) on the downstream task* is carried out with as many annotated explanations as are available (the green blocks in Fig. 1). At prediction time and during evaluation, the "explain" keyword is prepended to all instances, thus instructing the model to always generate explanations alongside its predictions.

Conceptually, there are no specific assumptions on the domain or semantics of the FT task, thus allowing the framework to be applicable broadly. To facilitate transferability, we consider FT tasks that can be cast into a similar input-output format as during the EP step, in this work text-classification problems.

4 Experiments

In this section, we first introduce the datasets, tasks, and evaluation metrics, and then evaluate our proposed approach for transferring generative explanation capabilities between tasks in potentially different domains. We do this in two different settings: (1) with all available annotated explanations, and (2) with limited annotated explanations during FT.

4.1 Datasets

We use three datasets in our experiments:

1. MultiRC [20][3]: a reading comprehension dataset consisting of multiple-choice questions for short paragraphs of text with annotated supporting evidence spans. We consider the binary classification of a given question and answer candidate pair.
2. FEVER [40] (see footnote 3): a large fact verification dataset extracted from Wikipedia that has been annotated by humans with supporting evidence spans. We consider claims that are either *supported* or *refuted*.
3. SciFact [42]: a small dataset where the task is to find abstracts from a corpus of research literature, and corresponding evidence sentences, that *support* or *refute* scientific and medical claims. We consider the subtask of text classification for a given claim-abstract pair and use the corresponding evidence sentences as ground-truth extractive explanations. Abstracts that do not contain any evidence for a claim are discarded, making the classification problem binary.

---

[3] We use the dataset versions distributed through the ERASER benchmark [10].

We use MultiRC and FEVER during the EP step and SciFact as the final downstream FT task, thus considering transfer from general English to the scientific and medical domain. To unify the input-output format and simplify transferability, all tasks are cast as binary classification problems where the output labels are {True, False}.

4.2 Evaluation

Consider the generic target output format for any of the introduced tasks,

$$\tilde{y} = y_{\text{label}} \ \texttt{explanation:} \ e_1 \ ... \ \texttt{explanation:} \ e_M, \tag{2}$$

where y_{label} is the target label, either True or False, and $\mathcal{E} = \{e_1, ..., e_M\}$ is the ground-truth explanation consisting of M sentences. The predicted output sequence $\hat{y}$ is assumed to follow the desired format and is split by the "explanation:" separator to form the predicted label $\hat{y}_{\text{label}}$ and explanation set $\hat{\mathcal{E}} = \{\hat{e}_1, ..., \hat{e}_{\hat{M}}\}$. If the model would output anything other than the desired format, this would be counted as part of the predicted label and thus resulting in both poor task performance and explanation quality.

We use four evaluation metrics in our experiments: F1 score for prediction task performance, as well as token-level F1 score (TF1), BLEU score [33], and ROUGE-L score [28] to measure extractive explanation quality. Each explanation sentence $e \in \mathcal{E}$ is tokenized and matched against all possible spans in the tokenized input text $\tilde{x}$. This forms a corresponding set of overlap tuples $\mathcal{S} = \{(e_{i_{\text{start}}}, e_{i_{\text{end}}}) \mid e \in \mathcal{E}\}$ of the start and end indices of the matched spans, and analogously $\hat{\mathcal{S}}$ from $\hat{\mathcal{E}}$. If an explanation does not exactly match any span, it is considered invalid and is discarded. TF1 is computed as the F1 score between $\hat{\mathcal{S}}$ and $\mathcal{S}$, averaged over all N samples in the dataset:

$$\text{TF1} = \frac{1}{N} \sum_{k=1}^{N} \frac{\text{P}_k \cdot \text{R}_k}{\text{P}_k + \text{R}_k}, \quad \text{P}_k = \frac{|\mathcal{S}^{(k)} \cap \hat{\mathcal{S}}^{(k)}|}{|\hat{\mathcal{S}}^{(k)}|}, \quad \text{R}_k = \frac{|\mathcal{S}^{(k)} \cap \hat{\mathcal{S}}^{(k)}|}{|\mathcal{S}^{(k)}|}. \tag{3}$$

The TF1 score significantly punishes generated outputs that deviate from the desired format, or if a generated explanation sentence is not exactly matching a span in the input text. To make the evaluation more nuanced, we also compute BLEU score and ROUGE-L score directly between the raw output text $\hat{y}$ and $\tilde{y}$. These metrics measure precision and recall-based overlap statistics, respectively, between shorter spans of different lengths and are not as binary as TF1. BLEU score has been previously used for abstractive explanation evaluation [4,32]. An illustrative example of the data post-processing procedure and the evaluation metrics are provided in Table 1.

Random Baseline. To put our results into a quantitative context, we construct a random baseline for each task. This is achieved by randomly sampling a predicted label according to the class weights in the training dataset. Additionally, we empirically estimate the probability mass function of the number of sentences

Table 1. Illustrative example of data post-processing and explanation quality evaluation metrics. Overlap spans are highlighted in gray.

Var.	Value
$\tilde{x}$	"explain classification passage: I had a dog named Toodles. He was black and white and had long floppy ears. He also had very short legs. Every Saturday we would go to the park and play Toodles' favorite game. query: What describes Toodles' legs? answer: Long"
$\tilde{y}$	"False explanation: I had a dog named Toodles. explanation: He also had very short legs."
$\hat{y}$	"False explanation: I had a dog called Toodles. explanation: He also had very short legs."
$\mathcal{E}$	{"I had a dog named Toodles.", "He also had very short legs."}
$\hat{\mathcal{E}}$	{"I had a dog called Toodles.", "He also had very short legs."}
$\mathcal{S}$	"explain classification passage: I had a dog named Toodles. He was black and white and had long floppy ears. He also had very short legs. Every Saturday we would go to the park and play Toodles' favorite game. query: What describes Toodles' legs? answer: Long"
$\hat{\mathcal{S}}$	"explain classification passage: I had a dog named Toodles. He was black and white and had long floppy ears. He also had very short legs. Every Saturday we would go to the park and play Toodles' favorite game. query: What describes Toodles' legs? answer: Long"

P: 100.00%   **R**: 50.00%   **TF1**: 66.67%   **BLEU**: 84.92%   **ROUGE-L**: 93.33%

M that constitute the extractive explanations in the training dataset. To form $\hat{\mathcal{E}}$, $\hat{M}$ is sampled independently from this distribution for each instance in the evaluation dataset, and the corresponding number of explanation sentences are then selected uniformly at random from the input text.

4.3   Model and Training Details

We consider two seq2seq models based on the Transformers architecture [41], namely T5 [35] and BART [27]. We analyze both the Base and the Large variants of T5 and the Large variant of BART. The experimental setup follows the training procedure outlined in Fig. 1, where MultiRC and FEVER are used during EP and SciFact is the FT task.

To teach the model to explain its predictions, EP instances are sampled with equal probability from a mixture of training samples with and without annotated explanations. Every time an explanation is added to the target output, the input text is prepended with the "explain" keyword as described in Sect. 2. This allows the model to learn the semantic meaning of the "explain" keyword, and the same format can be used during FT to generate explanations. We evaluate the model every 360 steps on the evaluation dataset and the checkpoint that achieves the lowest F1 score is used for further fine-tuning on the downstream task. After fine-tuning, average F1 and TF1 score is used as the final evaluation metric to select the best model checkpoint. For T5-Base and BART-Large, we repeat all experiments five times, and for T5-Large three times due to its large size and needed computational effort.[4]

Table 2. Validation set performance on SciFact with all annotated explanations.

Model	EP	F1	TF1	BLEU	ROUGE-L
T5-Large	None	84.0 ($\pm$2.9)	66.4 ($\pm$1.7)	71.9 ($\pm$1.3)	77.4 ($\pm$0.9)
	MultiRC	86.7 ($\pm$2.1)	**69.4** ($\pm$1.4)	73.2 ($\pm$1.4)	78.3 ($\pm$1.5)
	FEVER	88.4 ($\pm$1.6)	69.0 ($\pm$1.0)	74.3 ($\pm$2.2)	**79.2** ($\pm$0.5)
T5-Base	None	78.5 ($\pm$0.4)	64.6 ($\pm$0.7)	71.3 ($\pm$1.3)	75.8 ($\pm$0.5)
	MultiRC	81.9 ($\pm$1.4)	68.2 ($\pm$1.8)	**74.9** ($\pm$2.2)	78.4 ($\pm$2.1)
	FEVER	85.3 ($\pm$0.9)	69.2 ($\pm$0.5)	74.3 ($\pm$0.6)	78.8 ($\pm$0.3)
BART-Large	None	61.0 ($\pm$4.0)	37.7 ($\pm$6.6)	40.7 ($\pm$6.3)	57.7 ($\pm$7.0)
	MultiRC	85.8 ($\pm$2.2)	46.2 ($\pm$1.2)	42.9 ($\pm$1.8)	65.8 ($\pm$2.5)
	FEVER	**90.0** ($\pm$1.5)	45.0 ($\pm$0.6)	40.0 ($\pm$5.2)	64.5 ($\pm$3.5)
Random baseline	None	67.5 ($\pm$2.7)	19.1 ($\pm$1.8)	25.5 ($\pm$2.0)	32.4 ($\pm$1.7)

4.4   All Available Annotated Explanations for SciFact

Table 2 shows the results after FT with all available annotated explanations for SciFact. As a quantitative reference, we include a baseline for each model type when EP is not performed. These results are not directly comparable with [42], since we consider the subtask of label prediction and rationalization for the subset of refuted and supported claims. Overall, the T5-based models achieve significantly higher explanation quality compared to BART-Large, and we see consistent performance gains across all metrics when MultiRC or FEVER are used for EP. T5-Large achieves the highest TF1 score with a relative gain of 5%, closely followed by T5-Base that sees a relative gain of 7%. EP using FEVER has the highest positive impact on the prediction task performance (F1 score).

To provide a qualitative understanding of the generated explanations, three non-cherry picked examples for T5-Large with MultiRC during EP are shown in Table 3. The first claim is correctly refuted and the model generates all three sentences of the golden annotated explanation. The second claim is also correctly classified and in this case only one sentence constitutes both the predicted and golden explanation, which illustrates the flexibility in the generative approach. The last example is classified incorrectly, even though the model is extracting a majority of the actual golden explanation. This suggests two possible reasons; that the model is able to find the relevant part of the input but cannot infer the correct label from this, or that it generates a plausible explanation even though

---

[4] The hyperparameter settings for the different models and training phases are available in the public code repository.

Table 3. Non-cherry picked samples from the SciFact validation set for WT5-Large after MultiRC EP. Explanations in $\mathcal{S} \cap \hat{\mathcal{S}}$ are highlighted in green, $\hat{\mathcal{S}} \setminus \mathcal{S}$ in yellow, and $\mathcal{S} \setminus \hat{\mathcal{S}}$ in red (not present). The remaining part of the input text has been shortened.

Claim	Prediction
Taxation of sugar-sweetened beverages had no effect on the incidence rate of type II diabetes in India	False
BACKGROUND Taxing sugar-sweetened beverages (SSBs) has been proposed in high-income countries to reduce obesity and type 2 diabetes. ... The 20% SSB tax was anticipated to reduce overweight and obesity prevalence by 3.0% (95% CI 1.6%-5.9%) and type 2 diabetes incidence by 1.6% (95% CI 1.2%-1.9%) among various Indian subpopulations over the period 2014–2023, if SSB consumption continued to increase linearly in accordance with secular trends. However, acceleration in SSB consumption trends consistent with industry marketing models would be expected to increase the impact efficacy of taxation, averting 4.2% of prevalent overweight/obesity (95% CI 2.5–10.0%) and 2.5% (95% CI 1.0–2.8%) of incident type 2 diabetes from 2014–2023. ... CONCLUSION Sustained SSB taxation at a high tax rate could mitigate rising obesity and type 2 diabetes in India among both urban and rural subpopulations.	
Macrolides have no protective effect against myocardial infarction	True
CONTEXT Increasing evidence supports the hypothesis of a causal association between certain bacterial infections and increased risk of developing acute myocardial infarction. ... No effect was found for previous use of macrolides (primarily erythromycin), sulfonamides, penicillins, or cephalosporins. ...	
Stroke patients with prior use of direct oral anticoagulants have a lower risk of in-hospitality mortality than stroke patients with prior use of warfarin	**False**
Importance Although non-vitamin K antagonist oral anticoagulants (NOACs) are increasingly used to prevent thromboembolic disease, there are limited data on NOAC-related intracerebral hemorrhage (ICH). ... The unadjusted in-hospital mortality rates were 32.6% for warfarin, 26.5% for NOACs, and 22.5% for no OACs. Compared with patients without prior use of OACs, the risk of in-hospital mortality was higher among patients with prior use of warfarin (adjusted risk difference [ARD], 9.0% [97.5% CI, 7.9% to 10.1%]; adjusted odds ratio [AOR], 1.62 [97.5% CI, 1.53 to 1.71]) and higher among patients with prior use of NOACs (ARD, 3.3% [97.5% CI, 1.7% to 4.8%]; AOR, 1.21 [97.5% CI, 1.11-1.32]). Compared with patients with prior use of warfarin, patients with prior use of NOACs had a lower risk of in-hospital mortality (ARD, -5.7% [97.5% CI, -7.3% to -4.2%]; AOR, 0.75 [97.5% CI, 0.69 to 0.81]). ... Prior use of NOACs, compared with prior use of warfarin, was associated with lower risk of in-hospital mortality.	

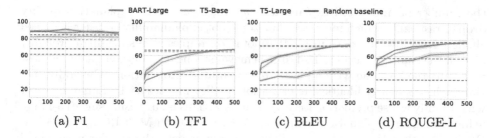

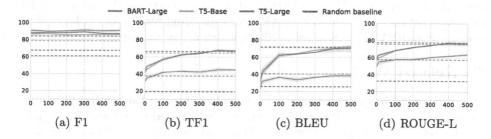

Fig. 2. Explainability transfer from MultiRC to SciFact. Evaluation metrics (a)–(d) with 95% confidence intervals as a function of number of annotated explanations during FT. Dashed lines correspond to the same values as EP *None* in Table 2. (Color figure online)

Fig. 3. Explainability transfer from FEVER to SciFact. (Color figure online)

this is actually not used in the label-prediction process. Since the training loss function encourages the same extractive explanations regardless of the label, there are no theoretical guarantees for explanation faithfulness. Wiegreffe et al. [45] investigate this phenomenon and provides some empirical evidence that there is indeed a robustness between the generated explanations and labels, but that further work in this area is needed.

4.5 Downstream Task with Limited Annotated Explanations

For most practical applications, annotated explanations on the target downstream task are scarce and costly to obtain. To evaluate the effectiveness of explainability transfer to alleviate these problems, we simulate scenarios with different number of available annotated explanations on SciFact. In all cases, EP is performed using all available explanations. Figure 2 depicts transfer from MultiRC to SciFact using $n_{exp} \in \{0, 10, 100, 200, 300, 400, 500\}$ out of 546 annotated training samples for SciFact.

For all models, there is an increase in prediction performance (F1 score) of performing EP, and it stays more or less constant regardless of n_{exp}. For the T5 models, we also see improved explanation quality across all metrics. This suggests that the EP procedure allows the model to be fine-tuned more effectively so that the WT5 explanation framework does not sacrifice task performance. As the number of annotated explanations approach zero, the explanation quality drops drastically, which indicates that zero-shot explainability transfer is indeed challenging. However, with just 200 annotated samples corresponding to roughly 35% of the training dataset, T5-Large achieves strong explanation quality almost matching the baseline with all available annotated explanations. Generally, T5-Base achieves nearly identical explanation quality metrics as T5-Large, however with slightly worse prediction task performance. This is surprising since T5-Large achieved higher explanation quality during EP on both MultiRC and FEVER. We believe that the small size of SciFact might benefit the smaller base model to more effectively transfer the explanation capability to the new task.

The explanation quality for BART is considerably lower than the T5 counterparts, meanwhile the prediction task performance is still competitive. BART is not as good at conforming to the strict extractive explanation format, which hurts the TF1 and BLEU score. ROUGE-L is also inferior but the gap to the T5 models is not as significant. We provide corresponding results for explainability transfer from FEVER to SciFact in Fig. 3, which follow the same general trends.

5   Related Work

Explainable ML has received a lot of research interest over the last few years and a comprehensive review of the field in general is provided in [14] and specifically in [8] related to applications for NLP. This work belongs to a class of methods that provide explainability by design and more specifically self-explaining systems, where the model itself produces an explanation as part of the prediction process. Attention-based models have mainly been considered for this purpose in NLP [6,23,44], much owing to the recent success of the Transformers architecture and the hope that this offers some inherent explainability "for free". However, the usefulness and validity of attention weights as explanations have been questioned [3,19,38].

Generative natural language explanations were studied in [4], who proposed an extended version of the SNLI dataset [2] with annotated abstractive explanations, and considered different seq2seq models for learning to generate such explanations. This work is based on [32], which approached the same problem by casting it into the T5 text-to-text framework [35]. Other previous works have also studied generative explanations for non-NLP tasks [12,16,22].

Another line of work for explainable NLP is based on rationalization pipelines that aim to produce extractive explanations by splitting the prediction process into two subsequent modules; a rationale extractor and a predictor [1,10,25,34]. The benefit of this approach is that it provides some faithfulness guarantees by construction since the predictor can only rely on the extracted rationales, however, potentially at the expense of prediction performance. The dilemma of faithful and plausible explanations was raised in [17] and was further studied in [45] for generative explanations. Both argue that self-explaining systems, although not guaranteedly faithful, can still be very useful in practice.

In the medical domain specifically, 1-dimensional CNNs with label-conditional attention have been explored for explainable ICD code prediction from discharge summaries [31]. The interpretability of Transformer attention weights in a medical context was analyzed and questioned in [18]. Recently, rationalization pipelines have been applied to medical and scientific text, for instance [42] and [24] utilize BERT-to-BERT models for SciFact and for classifying random clinical trials, respectively.

6   Conclusions

In this work, we have demonstrated that generating extractive explanations can be transferred from general English to tasks in the scientific and medical domain. Our proposed three-step training procedure with explainability pre-training improves explanation quality as well as prediction task performance on the downstream task. Furthermore, we see a large increase in explanation quality for only a small number of annotated explanations during fine-tuning, making it an attractive option for real-world use cases where annotations are limited and costly to obtain. An interesting direction for future work is to analyze the impact of specific weights of the classification and explanation objectives in the common loss function. We plan to shed further light on the faithfulness-plausibility dilemma by applying attribution methods (e.g., SHAP [29]) on top of the generated explanations. The practical usability of the generated explanations will also be further assessed by human evaluation studies. As an extension to cross-domain explainability transfer, the same approach can also be considered for explainability transfer across languages. We believe recent multilingual seq2seq models like mT5 [46] to be a promising candidate for this purpose.

References

1. Bastings, J., et al.: Interpretable neural predictions with differentiable binary variables. In: ACL (2019)
2. Bowman, S.R., et al.: A large annotated corpus for learning natural language inference. In: EMNLP (2015)
3. Brunner, G., et al.: On identifiability in transformers. In: ICLR (2019)
4. Camburu, O., et al.: e-SNLI: natural language inference with natural language explanations. In: NeurIPS (2018)
5. Chen, C., et al.: This looks like that: Deep learning for interpretable image recognition. In: NeurIPS (2019)
6. Clark, K., et al.: What does BERT look at? An analysis of BERT'S attention. In: ACL Blackbox NLP Workshop (2019)
7. Common Crawl. https://www.commoncrawl.org
8. Danilevsky, M., et al.: A survey of the state of explainable AI for natural language processing. In: AACL-IJCNLP (2020)
9. Devlin, J., et al.: BERT: pre-training of deep bidirectional transformers for language understanding. In: NAACL (2019)
10. DeYoung, J., et al.: ERASER: a benchmark to evaluate rationalized NLP models. In: ACL (2020)
11. Doshi-Velez, F., et al.: Towards a rigorous science of interpretable machine learning. arXiv preprint arXiv:1702.08608 (2017)

12. Ehsan, U., et al.: Rationalization: a neural machine translation approach to generating natural language explanations. In: AIES (2018)
13. EU: General Data Prodection Regulation (GDPR): Recital 71 (2018). https://www.privacy-regulation.eu/en/r71.htm
14. Guidotti, R., et al.: A survey of methods for explaining black box models. ACM Comput. Surv. (CSUR) **51**(5), 1–42 (2018)
15. He, K., et al.: Delving deep into rectifiers: surpassing human-level performance on ImageNet classification. In: ICCV (2015)
16. Hendricks, L.A., Akata, Z., Rohrbach, M., Donahue, J., Schiele, B., Darrell, T.: Generating visual explanations. In: Leibe, B., Matas, J., Sebe, N., Welling, M. (eds.) ECCV 2016. LNCS, vol. 9908, pp. 3–19. Springer, Cham (2016). https://doi.org/10.1007/978-3-319-46493-0_1
17. Jacovi, A., et al.: Towards faithfully interpretable NLP systems: how should we define and evaluate faithfulness? In: ACL (2020)
18. Jain, S., et al.: An analysis of attention over clinical notes for predictive tasks. In: Clinical NLP (2019)
19. Jain, S., et al.: Attention is not explanation. In: NAACL (2019)
20. Khashabi, D., et al.: Looking beyond the surface: a challenge set for reading comprehension over multiple sentences. In: NAACL (2018)
21. Kim, B., et al.: The Bayesian case model: a generative approach for case-based reasoning and prototype classification. In: NIPS (2014)
22. Kim, J., Rohrbach, A., Darrell, T., Canny, J., Akata, Z.: Textual explanations for self-driving vehicles. In: Ferrari, V., Hebert, M., Sminchisescu, C., Weiss, Y. (eds.) ECCV 2018. LNCS, vol. 11206, pp. 577–593. Springer, Cham (2018). https://doi.org/10.1007/978-3-030-01216-8_35
23. Kovaleva, O., et al.: Revealing the dark secrets of BERT. In: NeurIPS (2019)
24. Lehman, E., et al.: Inferring which medical treatments work from reports of clinical trials. In: NAACL (2019)
25. Lei, T., et al.: Rationalizing neural predictions. In: EMNLP (2016)
26. Letham, B., et al.: Interpretable classifiers using rules and Bayesian analysis: building a better stroke prediction model. Ann. Appl. Stat. **9**, 1350–1371 (2015)
27. Lewis, M., et al.: BART: denoising sequence-to-sequence pre-training for natural language generation, translation, and comprehension. In: ACL (2020)
28. Lin, C.: Rouge: a package for automatic evaluation of summaries. In: Text Summarization Branches Out (2004)
29. Lundberg, S., et al.: A unified approach to interpreting model predictions. In: NIPS (2017)
30. Miller, T.: Explanation in artificial intelligence: insights from the social sciences. Artif. Intell. **267**, 1–38 (2019)
31. Mullenbach, J., et al.: Explainable prediction of medical codes from clinical text. In: NAACL (2018)
32. Narang, S., et al.: WT5?! Training text-to-text models to explain their predictions. arXiv preprint arXiv:2004.14546 (2020)
33. Papineni, K., et al.: BLEU: a method for automatic evaluation of machine translation. In: ACL (2002)
34. Paranjape, B., et al.: An information bottleneck approach for controlling conciseness in rationale extraction. In: EMNLP (2020)
35. Raffel, C., et al.: Exploring the limits of transfer learning with a unified text-to-text transformer. JMLR (2020)
36. Rajani, N., et al.: Explain yourself! Leveraging language models for commonsense reasoning. In: ACL (2019)

37. Ribeiro, M., et al.: Why should i trust you? Explaining the predictions of any classifier. In: KDD (2016)
38. Serrano, S., et al.: Is attention interpretable? In: ACL (2019)
39. Sundararajan, M.: Axiomatic attribution for deep networks. In: ICML (2017)
40. Thorne, J., et al.: FEVER: a large-scale dataset for fact extraction and verification. In: NAACL (2018)
41. Vaswani, A., et al.: Attention is all you need. In: NIPS (2017)
42. Wadden, D., et al.: Fact or fiction: verifying scientific claims. In: EMNLP (2020)
43. Wang, A., et al.: Superglue: a stickier benchmark for general-purpose language understanding systems. In: NeurIPS (2019)
44. Wiegreffe, S., et al.: Attention is not not explanation. In: EMNLP-IJCNLP (2019)
45. Wiegreffe, S., et al.: Measuring association between labels and free-text rationales. arXiv preprint arXiv:2010.12762 (2020)
46. Xue, L., et al.: mT5: a massively multilingual pre-trained text-to-text transformer. arXiv preprint arXiv:2010.11934 (2020)

Semantic Relations

Semantic Relations

Virus Causes Flu: Identifying Causality in the Biomedical Domain Using an Ensemble Approach with Target-Specific Semantic Embeddings

Raksha Sharma[1]($\boxtimes$)(iD) and Girish Palshikar[2](iD)

[1] Indian Institute of Technology, Roorkee, India
`raksha.sharma@cs.iitr.ac.in`
[2] TCS Research, Tata Consultancy Services, Pune, India
`gk.palshikar@tcs.com`
`https://www.rakshasharma.com/`
`https://www.tcs.com/`

Abstract. Identification of Cause-Effect (CE) relation is crucial for creating a scientific knowledge-base and facilitate question-answering in the biomedical domain. An example sentence having CE relation in the biomedical domain (precisely Leukemia) is: *viability of THP-1 cells was inhibited by COR*. Here, *COR* is the cause argument, *viability of THP-1 cells* is the effect argument and *inhibited* is the trigger word creating a causal scenario. Notably CE relation has a temporal order between *cause* and *effect* arguments. In this paper, we harness this property and hypothesize that the temporal order of CE relation can be captured well by the Long Short Term Memory (LSTM) network with independently obtained semantic embeddings of words trained on the targeted disease data. These focused semantic embeddings of words overcome the labeled data requirement of the LSTM network. We extensively validate our hypothesis using three types of word embeddings, *viz., GloVe, PubMed*, and *target-specific* where the target (focus) is Leukemia. We obtain a statistically significant improvement in the performance with LSTM using GloVe and target-specific embeddings over other baseline models. Furthermore, we show that an ensemble of LSTM models gives a significant improvement ($\sim$3%) over the individual models as per the *t*-test. Our CE relation classification system's results generate a knowledge-base of 277478 CE relation mentions using a rule-based approach.

Keywords: Cause-effect relation extraction · Biomedical domain · Deep Neural Network (Long Short Term Memory) · Semantic embeddings

1 Introduction

The MEDLINE database is growing at the rate of 500,000 new citations each year. With such explosive growth, it is challenging to keep up to date with all

© Springer Nature Switzerland AG 2021
E. Métais et al. (Eds.): NLDB 2021, LNCS 12801, pp. 93–104, 2021.
https://doi.org/10.1007/978-3-030-80599-9_9

of the discoveries and theories in biomedical research. Thus, there is a need to provide automatic extraction of the user-oriented biomedical knowledge [1,4]. Cause-Effect (CE) relation is one such type of user-oriented biomedical knowledge. The semantic connection between a causal argument and its effect is referred to as a CE relation. For example, *virus causes flu* has a CE relation, where *virus* is the cause argument, and *flu* is the effect argument, and *causes* is the trigger argument creating causal relation. Moldovan et al. [18] reported that causal questions are answered with a very low precision score of 3.1%. It is crucial to answering causal questions with high precision in the biomedical domain as it is related to human life. Identifying CE relation from the biomedical data can produce a scientific knowledge-base that can facilitate answering user queries in the biomedical domain [8]. The following example illustrates the purpose of the identification of CE relation in the biomedical domain.

- Input: "Tumor cell killing was achieved by concerted action of necrosis apoptosis induction."

- Proposed Output: CE relation found with the following CE mentions:
  Causal Cue: *achieved by*
  Cause: *Concerted action of necrosis apoptosis induction*
  Effect: *Tumor cell killing*
- QA System based on the proposed output:
  Question: What is the effect of concerted action of necrosis apoptosis induction on tumor cells of Leukemia?
  Answer: Tumor cell killing

The correct answer to the question could help understanding the disease to the patient or diagnosing a terminal illness such as *Leukemia* to the doctors/patients. Utilizing cause-effect relations in the development of a question answering system leads to improved performance [8].[1] Another direct application domain is a scientific database dedicated to a disease. Record of arguments of CE relations for a disease *viz., cause, effect, the cue for causality* can form a scientific knowledge-base dedicated to the disease [22]. Such knowledge-bases can help scientists, doctors, and other users perform tasks such as diagnosis, exploring and validating hypotheses, understanding the state-of-the-art, and identifying opportunities for new research.

Various complex constructs are used to express causality in text. The simplest way of expressing CE relations in the text is by using generic *causative* verbs, such as *cause, lead, result.* Apart from this, different domains have their causative verbs, which are either new verbs specific to that domain (*e.g., over-express, up-regulate* in the biomedical domain) or generic verbs that have a special causative sense specific to that domain (*e.g., inhibit, express* in the biomedical domain). There are other complexities with the linguistic expression of CE relations in

---

[1] Causal questions are frequently used in general on Web. Naver Knowledge iN, http://kin.naver.com reported 130,000 causal questions from 950,000 sentence-sized database [18].

text. One is the negation of the apparent CE relation mention, *e.g.*, *However, the precise mechanisms by which BCR stimulation leads to accumulation of malignant cells remain incompletely understood.* Next is the use of discourse connectives like *after, while etc.,* to express causal linking between two arguments, *e.g.,* {*Cleaved caspase-3 was increased*}$_{Effect}$ *after* {*treatment of COR*}$_{Cause}$. The presence of linguistically complex constructs in the biomedical domain makes extraction of CE relation a more challenging task than in generic domains [22].

In this paper, we address a relatively novel problem: the identification of cause-effect relationships and their arguments in the biomedical domain for Leukemia. Leukemia is a group of cancers that begins in the bone marrow and results in high number of abnormal white blood cells (WBC), called *leukemia cells.* Leukemia is the most frequent type of cancer in children. In 2015, Leukemia was detected in 2.3 million people and resulted in 353, 500 deaths; the average five-year survival rate is 57% in the USA.[2] The exact causes for Leukemia are unknown, although some risk factors are known, including family history, smoking, and exposure to ionizing radiation or chemicals such as *benzene.* Table 1 shows some example sentences about Leukemia in which CE relation mentions, *viz.,* Cause (C), Effect (E), and Causal-Cue (CC) are present. Note that sometimes the CE relation mention does not include a causative verb, but a causal *cue phrase,* such as *due to, because, hence, therefore.*

The CE relations in leukemia are at widely different abstraction levels - from genetic, molecular, cellular, organ level, tissue level to patient-level as an entity. In the corpus, we can discern a finer structure to the CE relations, apart from the two standard arguments, cause and effect. For example, CE relations seem have associated with them additional optional information, such as *evidence* (see Table 1 (2)), or a *control condition i.e.,* a condition under which the causal relation holds (see Table 1 (3)). In the biomedical domain, a cause is often an *agent* (such as an organism, drug, compound), an *event*, an *action* or a *condition.* An *event* is any change in the physical state or property of one or more named entity instances. A *condition* is broadly any property or state of one or more named entity instances, which is sustained over reasonably long periods. An effect is often an event or a condition.

In this paper, we conceptualize that in a causal sentence, *cause* temporally precedes the *effect.* Long Short Term Memory (LSTM) network is a deep neural network having recurrent connections between the layers. It is tailored to process the text having a temporal order of words. Therefore the temporal order of CE relations can be captured well by the Long Short Term Memory network, which makes it a potential technique for the identification of CE relations in the biomedical domain. We present a CE relation identification system for the biomedical domain with the focus on Leukemia. First the system is formalized as a binary classification system with two classes, *viz., CE-Relation, Not-CE-Relation.* Next the sentences which are identified as CE Relation tag are used for the extraction of CE Relations arguments using a rule based system. Figure 1 shows the architecture of the proposed system. Stage-1 is the neural binary

---

[2] https://en.wikipedia.org/wiki/Leukemia.

Table 1. Examples of CE relations in leukemia.

(1) [Human T-cell leukemia virus type 1 (HTLV-1)]$_C$ [causes]$_{CC}$ a highly lethal [blood cancer]$_E$ or a chronic debilitating [disease of the spinal cord]$_E$
(1) The [co-expression of p96 (ABL/BCR)]$_C$ [enhanced]$_{CC}$ the [kinase activity]$_E$ and as a consequence, the [transformation potential of p185 (BCR/ABL)]$_E$
(3) While survival rates for ALL have improved, [central nervous system (CNS) relapse]$_C$ remains a significant [cause]$_{CC}$ of [treatment failure]$_E$ and [treatment-related morbidity]$_E$
(2) Using both [pharmacologic and genetic assays]$_E$, we show here that [inactivation of RIP1/RIP3]$_C$ [resulted]$_{CC}$ in [reduction of SOCS1 protein levels]$_E$ and [partial differentiation of AML cells]$_E$
(3) [Bone mass acquisition]$_E$ may be [compromised]$_{CC}$ in survivors of childhood acute lymphocytic leukemia due to various factors, including [adiposity]$_C$
(6) [cCMP-AM]$_C$ did not [induce]$_{CC}$ [apoptosis in K-562 cells, a human chronic myelogenous leukemia cell line,]$_E$ [due to]$_{CC}$ [rapid export via multidrug resistance-associated proteins]$_C$

classification model, which identifies whether a sentence has CE relation or not. Stage-2 performs extraction of CE relation constructs using a rule based system.

Though deep neural networks require a massive amount of labeled datasets for classification, the tagged data requirement is overcome by getting focused embeddings of words trained on a large unlabeled corpus specific to Leukemia. We compare our LSTM-based model with Multi-layer Perceptron (MLP) and Support Vector Machine (SVM) for CE relation identification using three types of word embeddings, *viz., GloVe, PubMed,* and *target-specific* where the target is Leukemia. Results (5) show that LSTM with target-specific embeddings outperformed all other reported models. Furthermore, we show that an ensemble of LSTM models trained using GloVe and target-specific embeddings gives a significant improvement ($\sim$3%) over the individual models.

The major contributions of the paper are as follows.

- We generate 2, 01, 066 embeddings specific to Leukemia using 60, 000 research papers on Leukemia from PubMed. We show the effectiveness of these focused (target-specific) embeddings over pre-trained embeddings for CE relation identification task.
- An ensemble of LSTMs trained using GloVe and target-specific embeddings produces an accuracy of **83.78**%, which is significantly greater over the individual models for the CE relation identification task.
- We generate a knowledge-base of 277478 CE relation mentions from the dataset of 60, 000 documents.

The rest of the paper is organized as follows. Section 2 discusses the related work. Section 3 describes the preparation of training data and the semantic embeddings used in the paper. Section 4 provides the experimental setup. Section 5 shows the results and Sect. 6 concludes the paper.

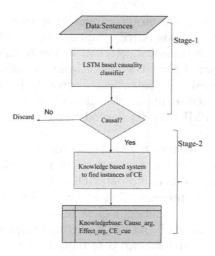

Fig. 1. Flowchart of the proposed system

2   Related Work

CE relation identification, in general, has been and continues to be well studied in the literature. Much of the work has attempted to discover CE relationships in the text by focusing on lexical and semantic constructs.

Kaplan et al. [11] wrote hand-coded rules considering causal scenario may vary from context to context. Joskowicz et al. [10] prepared a dedicated knowledge-base to build a causal analyzer for a Navy ship. Their objective was to understand a short narrative message about the Navy ship's equipment using CE relation. However, knowledge-based systems have low generalizability. In addition, building and maintaining a knowledge-based system is expensive for the targeted domain itself. Many researchers have used linguistic patterns to identify CE relations in the text without using any knowledge-base [7,12]. A few works used grammatical patterns to identify CE relations [8,13,21]. There are very few instances of combining grammatical patterns with machine learning to extract semantic relations, such as cause-effect [3,6]. In another work, cue phrases (cause triggering construct) with their probability were used to extract other lexical arguments of cause-effect relation [3]. Do [6] developed a minimally supervised approach based on focused distributional similarity and discourse connectives.

None of the work discussed so far has considered the complications of the biomedical domain. However, due to domain-specific vocabulary and constructs, conventional CE relation extraction methods are not suitable in the biomedical domain. Mihuailua et al. [16] defined an annotation scheme for enriching biomedical domain corpora with causality relations. Their scheme was used to annotate 851 causal relations to form BioCause, a collection of 19 open-access full-text biomedical journal articles. Mihuailua et al. [15] created several baselines and experimented with and compare various parameter settings for three algorithms, *i.e.*, Conditional Random Fields (CRF), Support Vector Machines (SVM) and Random Forests (RF) for causality detection in the biomedical domain. They also evaluated the impact of lexical, syntactic, and semantic features on each of the algorithms, and showed that semantics improves the performance in all cases. Sharma et al. [22] proposed an approach that deploys the linguistic cue indicating CE constructs and PMI between dependency relations for identification of CE relation in a sentence.

Knowledge-based and pattern-based approaches have severe coverage issues. They can only consider those instances for which knowledge or pattern can be derived by observing the training data. This paper presents a deep-neural-network-based supervised approach, that is, LSTM for CE relation identification with target-specific word embeddings as input. The use of target-specific semantic embeddings of words facilitates capturing complex CE relations while reducing the need for excessive labeled data requirements.

3   Training Data and Embeddings

Leukemia is a highly researched disease in the biomedical domain, having more than $3,02,926$ scientific documents on PubMed and more than $3,09,492$ on Nature. As CE relations can be expressed using various semantic constructs, we use distributed representation of a sentence capturing various characteristics of the text in terms of embeddings and then use them for training classification models. The training dataset and the embeddings used in the paper are described below.

3.1   Training Data

We extracted a set of 2500 sentences from Leukemia-related papers in PubMed and labeled them for the training of models. Two competent annotators were consulted to assign binary labels: *CE Relation (1)* and *Not CE Relation (0)*. The *Cohens k* between the annotators is 0.97 [2]. We used majority voting to determine the actual label. Section 5 reports the 5-fold cross-validation accuracy on this dataset.

3.2   Generic GloVe Embeddings

The Global Vector model [20] referred to as GloVe combines word2vec with the ideas drawn from matrix factorization methods, such as LSA [5]. We used *pre-trained* GloVe word embeddings of size 300. We refer to them as *generic* embeddings as they are trained on the Wikipedia 2014 dataset.[3]

---

[3] Download: https://nlp.stanford.edu/projects/glove/.

3.3 Target-Specific Word Embeddings

To obtain target-specific word embeddings where the target (focus) is Leukemia, we parsed and downloaded $60,000$ abstracts containing the term *Leukemia* from PubMed using the *Entrez* package of Biopython. The corpus has $5,12,061$ sentences and $1,22,29,561$ words. Embeddings of size 300 are learned from the corpus using the word2vec package [17]. Default parameters were used to train the model. The same dataset of $60,000$ documents is used to prepare the knowledge-base containing CE relation arguments.

3.4 Domain-Specific PubMed Embeddings

This is a set of pre-trained embeddings in the biomedical domain. The embeddings are trained on abstracts from PubMed without focusing on any particular disease.[4] Essentially, these semantic embeddings are trained using a domain-specific corpus, that is, the biomedical domain, but not specifically dedicated to the target for which classifier has to be trained, unlike our *target-specific word embeddings*.

Table 2. Statistics for the word embeddings

Embeddings	Vocab-Size	Words-Found
Generic-GloVe	400000	4995
Domain-specific	1999860	6323
Target-specific	201066	6678
Training Data	6740	6740

In addition to the above-described embeddings, we have observed the performance of LSTM with embeddings learned from the training dataset by LSTM's embeddings layer. Table 2 shows the statistics related to the embeddings used in this paper. Column 3 of Table 2 presents the number of words from the training dataset (Sect. 3.1) whose embeddings are found in the embeddings set.

4   Experimental Setup

This paper hypothesizes that the temporal order between cause-expression and effect-expression can be captured well by the Long Short Term Memory (LSTM) network. While Support Vector Machine (SVM) [15] and Multilayer Perceptron (MLP) are not tailored to process the sequential structure of words, hence not much suitable for CE relation identification. CE relation is a contextual property; LSTM generates hidden features representing the context. Hence it is a

---

[4] Available for download: http://evexdb.org/pmresources/vec-space-models/.

favorable architecture for CE relation identification. Lilleberg et al. [14] validated that word embeddings bring extra semantic features that help in text classification. Therefore, the use of independently trained semantic embeddings of words in place of words overcomes the labeled data requirement of LSTM. We provide a comparison among LSTM, SVM, and MLP using three types of semantic embeddings, viz., GloVe, PubMed, and target-specific where the target is Leukemia.

To train an SVM based classifier, we have used the publicly available Python-based Scikit-learn package [19]. Though results are reported with linear kernels due to their superior performance, we experimented with other polynomial kernels. Yin and Jin, [23] speculated that the sum of word embeddings is meaningful and can represent the document. For example, the sum of word embeddings of *Germany* and *capital* is close to the embedding of *Berlin* [17]. We adhered to the same convention to produce embeddings of sentences to train SVM-based classifiers with embeddings (Eq. 1).

Sentence (S) is having t_i token with v_i embedding:

$$S(t_1 : v_1; t_2 : v_2; ...; t_n : v_n),$$

v_i is an m dimension vector:

$$v_i = (v_{i1}, v_{i2}, ..., v_{im}),$$

Sentence embedding S of m dimension:

$$S = (\sum_{i=1}^{n} v_{i1}, \sum_{i=1}^{n} v_{i2}, ..., \sum_{i=1}^{n} v_{im}) \tag{1}$$

To implement MLP and LSTM, we used Keras functional API. The embedding layer of the LSTM network is initialized with the size of the embeddings. The middle layer is an LSTM layer, which is initialized with 256 activation units. The output layer is a dense layer having *sigmoid* as the activation function. MLP has the same settings, except the middle layer is a dense layer with 256 activation units.

Table 3. t-test ($\alpha = 0.05$) results for the systems having significant difference in accuracy.

	t-value	P-value
LSTM-GloVe *vs* SVM-BoW	1.88	0.04
LSTM-Target-specific *vs* SVM-BoW	2.18	0.0
Ensemble *vs* SVM-BoW	5.39	0.00
Ensemble *vs* LSTM-Generic-GloVe	5.19	0.00
Ensemble *vs* LSTM-Target-specific	4.08	0.00

Knowledge-Base Generation: The instances classified as CE relation by our system become the input to a rule-based system for extraction of CE relation mentions. We use the rule-based system proposed by Sharma et al. [22] for this purpose. It is specifically trained in an unsupervised manner to extract CE relation mentions from the bio-medical text. It is based on the principle that a known causal verb can be used to extract CE arguments, and known CE arguments can be used to discover unknown causative verbs (hence *co-discovery*). Point-wise mutual information (PMI) is used to measure the level of (linguistic) associations between a causative verb and its argument.

5 Results

We implemented 12 Systems to validate our hypothesis extensively. Figure 2 shows the 5-fold cross-validation accuracy concerning each system. **BoW** is the Bag-of-words model with SVM. **Train-MLP** and **Train-LSTM** are models trained on embeddings obtained from training data only with MLP and LSTM settings, respectively. Out of the remaining nine, three systems employ SVM, three use MLP, and three use LSTM, where each system in the collections individually trained using **PubMed** (Sect. 3.4), **GloVe** (Sect. 3.2), and **target-specific embeddings** (Sect. 3.3), respectively (Fig. 2).

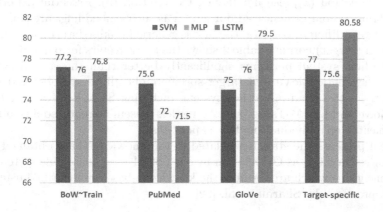

Fig. 2. 5-fold cross-validation accuracy in % for CE relation identification.

Mihuailua et al. [15] used Bag-of-words with SVM for causality identification. Figure 2 shows that SVM-BoW produces a 5-fold cross-validation accuracy of 77.2%. On the other hand, the performance of SVM with PubMed, GloVe, and target-specific embeddings is not significantly different from that of SVM-BoW. SVM is not able to incorporate the additional semantic information and contextual information provided by embeddings. The difference in the vocabulary of diseases (*e.g.*, Leukemia and Glioma have many dedicated words) makes PubMed embeddings (Sect. 3.4) inadequate for finding CE relations in Leukemia. On the other hand, LSTM with GloVe outperforms SVM-BoW by a significant margin. These generic word embeddings bring in additional favorable information that is not available in the training data. On the other hand, target-specific embeddings that are obtained from the data focusing on the targeted disease (Leukemia) performed the best with LSTM.

The performance of LSTM-Train is inferior to that of SVM-BoW as the data is not sufficiently large for LSTM. Use of pre-trained embeddings *viz.*, generic-GloVe, and target-specific embeddings reduce the labeled data requirement of LSTM. On the other hand, the performance of MLP is not significantly different from SVM. Both the algorithms are unable to capture the context formed by the sequence of words. CE relation has a long term dependency, *cause, effect* and *causality cue* mentions can be any words apart in the sentence. Long Short Term Memory (LSTM) network solves this problem by using gates to control the memorizing process [9].

Ensemble: We observed that an ensemble of LSTM models trained using generic *GloVe* and *target-specific* embeddings produced an accuracy of **83.78%**, which is significantly greater than the accuracy delivered by the individual classifier for CE relation identification. The classification probability value assigned by the individual classifier is averaged to obtain the ensemble classification probability. If the averaged probability is more than 0.5 for any instance, it is classified as having CE relation (1), else not having CE relation (0). Essentially, both the systems bring in complementary information as their embeddings are trained on two completely different corpora; the first is generic (Wikipedia) corpus, another is Leukemia (target) corpus. Table 3 shows the t-test results for pairs of systems where the first system performs significantly better than the second system. LSTM with generic-GloVe and target-specific embeddings are observed to be significantly better than any other system, including SVM-BoW as per t-test. LSTM-Golve and LSTM-Target-specific models' ensemble reported a significant improvement over individual models as per t-test.

Table 4 presents the statistics related to the knowledge-base obtained from the instances classified as CE relation by our LSTM-based ensemble system. CE relation mentions which are forming the knowledge-base are identified using the approach proposed by Sharma et al. [22].

Table 4. Statistics of Knowledge-base

CE_{Cue}	CE_{Cause}	CE_{Effect}
98778	87235	91465

6   Conclusion

Cause-Effect (CE) relation in a scientific text is an instance of knowledge required to be identified to answer causal questions. In this paper, we present that the long-term dependency between *cause* and *effect* expressions in a sentence can be captured well by the LSTM network for CE relation identification. The use of target-specific embeddings, which are learned from a corpus focused on the targeted disease, overcomes the labeled data requirement of LSTM. In addition, embeddings learned from a generic corpus (Wikipedia), *i.e.*, GloVe provides complementary information to the model. Results show that LSTM with target-specific embeddings and GloVe produce 80.5% and 79.5% accuracy, respectively, which is significantly better than models trained using Support Vector Machine and Multilayer Perceptron. Furthermore, an ensemble of the LSTM models trained using GloVe and target-specific embeddings produced an accuracy of 83.7%, which is significantly greater than the accuracy delivered by the individual classifier for CE relation identification. Furthermore, our CE relation classification system's results generate a knowledge-base of 277478 CE relation mentions using a rule-based approach.

References

1. Ananiadou, S., Mcnaught, J.: Text mining for biology and biomedicine. Citeseer (2006)
2. Berry, K.J., Mielke, P.W., Jr.: A generalization of cohen's kappa agreement measure to interval measurement and multiple raters. Educ. Psychol. Meas. **48**(4), 921–933 (1988)
3. Chang, D.-S., Choi, K.-S.: Causal relation extraction using cue phrase and lexical pair probabilities. In: Su, K.-Y., Tsujii, J., Lee, J.-H., Kwong, O.Y. (eds.) IJCNLP 2004. LNCS (LNAI), vol. 3248, pp. 61–70. Springer, Heidelberg (2005). https://doi.org/10.1007/978-3-540-30211-7_7
4. Cohen, K.B., Hunter, L.: Getting started in text mining. PLoS Comput. Biol. **4**(1), e20 (2008)
5. Deerwester, S., Dumais, S.T., Furnas, G.W., Landauer, T.K., Harshman, R.: Indexing by latent semantic analysis. J. Am. Soc. Inf. Sci. **41**(6), 391 (1990)
6. Do, Q.X., Chan, Y.S., Roth, D.: Minimally supervised event causality identification. In: Proceedings of the Conference on Empirical Methods in Natural Language Processing, pp. 294–303. Association for Computational Linguistics (2011)
7. Garcia, D.: COATIS, an NLP system to locate expressions of actions connected by causality links. In: Plaza, E., Benjamins, R. (eds.) EKAW 1997. LNCS, vol. 1319, pp. 347–352. Springer, Heidelberg (1997). https://doi.org/10.1007/BFb0026799
8. Girju, R.: Automatic detection of causal relations for question answering. In: Proceedings of the ACL 2003 Workshop on Multilingual Summarization and Question Answering-Volume 12, pp. 76–83. Association for Computational Linguistics (2003)
9. Hochreiter, S., Schmidhuber, J.: Long short-term memory. Neural Comput. **9**(8), 1735–1780 (1997)
10. Joskowicz, L., Ksiezyck, T., Grishman, R.: Deep domain models for discourse analysis. In: AI Systems in Government Conference, 1989, Proceedings of the Annual, pp. 195–200. IEEE (1989)
11. Kaplan, R.M., Berry-Rogghe, G.: Knowledge-based acquisition of causal relationships in text. Knowl. Acquisition **3**(3), 317–337 (1991)
12. Khoo, C.S., Kornfilt, J., Oddy, R.N., Myaeng, S.H.: Automatic extraction of cause-effect information from newspaper text without knowledge-based inferencing. Literary Linguist. Comput. **13**(4), 177–186 (1998)

13. Kim, H.D., et al.: Incatomi: integrative causal topic miner between textual and non-textual time series data. In: Proceedings of the 21st ACM International Conference on Information and Knowledge Management, pp. 2689–2691. ACM (2012)
14. Lilleberg, J., Zhu, Y., Zhang, Y.: Support vector machines and word2vec for text classification with semantic features. In: 2015 IEEE 14th International Conference on Cognitive Informatics & Cognitive Computing (ICCI* CC), pp. 136–140. IEEE (2015)
15. MIHĂILĂ, C., Ananiadou, S.: Recognising discourse causality triggers in the biomedical domain. J. Bioinform. Comput. Biol. 11(06), 1343008 (2013)
16. Mihăilă, C., Ohta, T., Pyysalo, S., Ananiadou, S.: Biocause: annotating and analysing causality in the biomedical domain. BMC Bioinform. 14(1), 2 (2013)
17. Mikolov, T., Chen, K., Corrado, G., Dean, J.: Efficient estimation of word representations in vector space. arXiv preprint arXiv:1301.3781 (2013)
18. Moldovan, D., Paşca, M., Harabagiu, S., Surdeanu, M.: Performance issues and error analysis in an open-domain question answering system. ACM Trans. Inf. Syst. (TOIS) 21(2), 133–154 (2003)
19. Pedregosa, F., et al.: Scikit-learn: Machine learning in python. J. Mach. Learn. Res. 12(Oct), 2825–2830 (2011)
20. Pennington, J., Socher, R., Manning, C.D.: Glove: global vectors for word representation. EMNLP 14, 1532–43 (2014)
21. Radinsky, K., Davidovich, S., Markovitch, S.: Learning causality from textual data. In: Proceedings of Learning by Reading for Intelligent Question Answering Conference (2011)
22. Sharma, R., Palshikar, G., Pawar, S.: An unsupervised approach for cause-effect relation extraction from biomedical text. In: Silberztein, M., Atigui, F., Kornyshova, E., Métais, E., Meziane, F. (eds.) NLDB 2018. LNCS, vol. 10859, pp. 419–427. Springer, Cham (2018). https://doi.org/10.1007/978-3-319-91947-8_43
23. Yin, Y., Jin, Z.: Document sentiment classification based on the word embedding. In: 4th International Conference on Mechatronics, Materials, Chemistry and Computer Engineering (2015)

Multilevel Entity-Informed Business Relation Extraction

Hadjer Khaldi[1,2(✉)] ⓘ, Farah Benamara[2], Amine Abdaoui[1],
Nathalie Aussenac-Gilles[2], and EunBee Kang[1]

[1] Geotrend, Toulouse, France
hadjer@geotrend.fr
[2] IRIT-CNRS, Toulouse, France

Abstract. This paper describes a business relation extraction system that combines contextualized language models with multiple levels of entity knowledge. Our contributions are three-folds: (1) a novel characterization of business relations, (2) the first large English dataset of more than $10k$ relation instances manually annotated according to this characterization, and (3) multiple neural architectures based on BERT, newly augmented with three complementary levels of knowledge about entities: generalization over entity type, pre-trained entity embeddings learned from two external knowledge graphs, and an entity-knowledge-aware attention mechanism. Our results show an improvement over many strong knowledge-agnostic and knowledge-enhanced state of the art models for relation extraction.

Keywords: Business relation extraction · Language model · Entity knowledge

1 Introduction

Binary relation extraction (RE) is a subtask of information extraction that aims at discovering semantic relations between two entity mentions in unstructured natural language texts [35]. In a dynamic business world, analyzing huge amount of textual content by business professionals to extract strategic information have become an arduous task, which makes automatic extraction of business relations between organizations (e.g., *startups, companies, non-profit organizations*, etc.) an essential tool for identifying links between specific market stakeholders and discovering new threats or opportunities [22]. For example, from the sentence *"[United Technologies Corporation]₁ defeats [Rolls-Royce]₂'s claim of patent infringement by jet engines."* extracted from the web, a RE system can identify the business relation LAWSUIT (*1,2*).

According to Zhao et al. [37], business relations involving organizations can be either Inner-Organizational (Inner-ORG) linking a company and its components (e.g. company-manager), or Inter-Organizational (Inter-ORG) for relations involving different companies (e.g. company-partner). In this paper, we

© Springer Nature Switzerland AG 2021
E. Métais et al. (Eds.): NLDB 2021, LNCS 12801, pp. 105–118, 2021.
https://doi.org/10.1007/978-3-030-80599-9_10

focus on binary Inter-ORG relations that may hold between two organizations. This is a domain-specific relation extraction task that is generally cast into a multiclass classification problem, where each class corresponds to a specific relation type [35]. Although domain-specific RE has already been explored (see for instance the biomedical [13] and food [27] domains), business RE has received much less attention in the literature. Current works in the field share three main limitations: (a) they rely on datasets that are either small (less than 1k instances) to train neural models or not freely available to the research community [4,33,37], (b) they generally consider only two relations (namely *Competition* and *Cooperation* [9,32]), and most importantly (c) the proposed models, either supervised [4,33] or semi-supervised [2,9,39], do not account for any prior knowledge about the organizations involved in a business relation.

In this paper, we aim to go one step further and overcome these limitations through three main contributions: **(1) a novel characterization of inter-organizational business relations** based on five relations that we believe are of particular importance for business professionals: INVESTMENT, COOPERATION, SALE-PURCHASE, COMPETITION, and LEGAL PROCEEDINGS, **(2) the first large English dataset of about 10k relation instances**[1] composed of sentences extracted from web documents and manually annotated according to this new characterization, **(3) a simple but effective multilevel entity-informed neural architecture for business relation extraction** built on top of BERT language model [5] without requiring its retraining (i.e. its original parameters and architecture are preserved). We consider for the first time three complementary levels of knowledge about entities: **(a)** generalization over entity type designed to force the classifier to reason at the entity type level rather than the entity mention, **(b)** pre-trained entity embeddings learned from external knowledge graphs, coming from Wikipedia2Vec [30], and exploring for the first time NASARI semantic vectors [3], and **(c)** an entity-knowledge-aware attention mechanism to determine the interactions between the relation representation and knowledge about entity pairs involved in the business relation as given by knowledge graphs. While each level alone has already been used for improving RE performances (see Sect. 2), as far as we know, no prior work conducted a systematic evaluation of the performances of RE *while combining knowledge from various levels*. When evaluated on our dataset, our models show an improvement (up to +2.4%) over many strong knowledge-agnostic and knowledge-enhanced state of the art models for RE. More importantly, our approach is able to better handle less frequent relations expressed in complex sentences.

2   Related Work

RE at the sentence level is an active research area in the Natural Language Processing (NLP) community [14,19]. Most studies target generic relations (e.g., hypernymy or cause-effect relationships) relying on popular manually annotated datasets such as SemEval-2010 Task 8 [7], ACE 2004 [17] and TACRED [36].

---

[1] https://github.com/Geotrend-research/business-relation-dataset.

Recent approaches are based on deep learning methods where both knowledge-agnostic and knowledge-informed models have been proposed (henceforth *Kag* and *Kin*, respectively). *Kag* RE models receive as input dense representations of words that can be either word embeddings, or position embeddings that encode the relative distance of each word from entity mentions in a sentence [6,10,38]. The use of pre-trained contextualized language models (PLM) has further improved the performances. See for instance R-BERT [28] and Shi et al. [23] who introduced entity masking into BERT to prevent overfitting. *Kin* RE on the other hand exploits factual knowledge about entities and words as given by external linguistic resources. For example, KnowBert [20] learns a knowledge-enhanced language model by incorporating knowledge from Wordnet and Wikipedia through a multitask end-to-end learning procedure that jointly learns language modeling and entity linking. Instead of modifying BERT language modeling objective and re-training its parameters (as done in KnowBert), other approaches align entity vectors to the original representations of the PLM (e.g., E-BERT [21]) or plug neural adapters outside the PLM to inject factual and linguistic knowledge (e.g., K-adapter [25]). Finally, other studies incorporate knowledge about entities via attention mechanisms [12,13].

While entity-enhanced models have shown to be quite effective for extracting generic and biomedical relations, their use in business RE has not been investigated yet. Most existing works make use of semi-supervised approaches relying on lexico-syntactic patterns that are often relation specific [2,9]. Supervised methods have also been recently proposed. For example, Yamamoto et al. [32] exploit generic information extraction systems to extract business relations from web news articles, while Collovini et al. [4] propose a specific framework based on Conditional Random Fields to extract relations between FinTech companies from Portuguese news texts.

In this paper, we propose the first *Kin* model for business RE based on simple neural architectures that require neither additional training to learn factual knowledge about entities nor alignment between each entity and its vector representation. Hence, knowledge about entities is viewed as external features to be injected into the relation classifier along with the sentence representation (as given by BERT). Compared to existing *Kin* models where sources of knowledge about entities (entity generalization, pre-trained entity embeddings (P-EE), entity-aware attention mechanism) have been considered independently, as far as we know, no prior work attempted to measure the impact of combining multiple levels of knowledge on the performances of RE. This paper, therefore, contributes to the field of generic RE with multilevel entity informed neural architectures but also to domain-specific RE with a new large dataset of five business relations.

3   Data and Annotation

Business relations are marginally present in knowledge bases (KB) such as DBpedia [1] where relations like *Subsidiary* and *Ownership_of* can be found [39]. Some business relations are nevertheless annotated in generic relation datasets with fairly low frequencies, such as *Employment/Membership/Subsidiary* in the ACE 2004 dataset [17]. Since there are no publicly available resources, we decided to compile our own business RE dataset. First, we define a characterization of Inter-ORG business relations according to which the dataset will be annotated. We start from a set of four relation types initially proposed by [37]: INVESTMENT, COOPERATION, SALE, and SUPPLY. Then, we combine the last two relations into SALE-PURCHASE, since we target non-oriented relations, i.e., $R(EO_1, EO_2) = R(EO_2, EO_1)$, EO_i being named entities of type organizations (henceforth ORG). Inspired by [9,32], we add COMPETITION and LEGAL PROCEEDINGS. Finally, the relation OTHERS accounts for the absence of a business relation between two ORG, referring to any other relation type between them.

Our dataset is new and is composed of sentences collected from the web[2] by requesting search engines API using a list of keywords related to various business activity fields such as *autonomous cars, 3D printing*, etc.[3] The sentences are selected according to two main criteria: **(i)** They must contain at least two entities of type ORG as predicted by both Spacy and StanfordNLP, two well known named entity taggers; and **(ii)** Sentences whose words are at least 95% of type ORG are discarded. Further details about relation type definitions, data acquisition, and data annotation rules are provided in the annotation guidelines (see the link in footnote 4). The collected sentences were manually annotated by nine non-domain-expert English speakers via the collaborative annotation platform *Isahit*[4]. The annotation was made in batches, each containing 2k instances. For each batch, 10% of the annotated data is re-annotated by experts. This helped to assess the quality of the annotations and improve annotation guidelines. Over 1k of re-annotated instances, the average Kappa between the annotators and the experts is 0.766 which is a strong agreement given the complexity of the task (many relations are implicitly expressed and the large context within the sentence (39 words on average) makes the annotation hard). Table 1 shows the total number of annotated relations as well as the distribution of instances in the train and test sets.

Table 1. Dataset statistics per relation type in the train and test sets.

	INVEST.	COMPET.	COOPERAT.	LEGAL.	SALE.	OTHERS	#Total
Train	281	1,675	627	50	248	5,647	**8,528**
Test	50	296	111	8	44	997	**1,506**
#All	**331**	**1,971**	**738**	**58**	**292**	**6,644**	**10,034**

---

[2] We consider textual contents from various sources and formats excluding those retrieved from social media, e-commerce, and code versioning websites.

[3] The set of keywords have been chosen by business intelligence experts.

[4] https://isahit.com/en/.

4   Multilevel Entity-Informed RE

We propose the model architecture shown in Fig. 1. It relies on BERT PLM as a sentence encoder to encode the input sentence tokens into contextualized representations, as it has shown to be a quite effective language encoder for RE (see Sect. 2). Following [24], we mark both the beginning and end of each entity involved in a relation by: $[E_{11}]$, $[E_{12}]$ for EO_1 and $[E_{21}]$, $[E_{22}]$ for EO_2. To deal with entity ambiguities (e.g., *Amazon* can refer to the river, the rainforest, as well as to the company), we link every EO_i to its unique disambiguated textual identifier in Wikipedia (Wikification) using BLINK [11][5], an open-source entity linker.

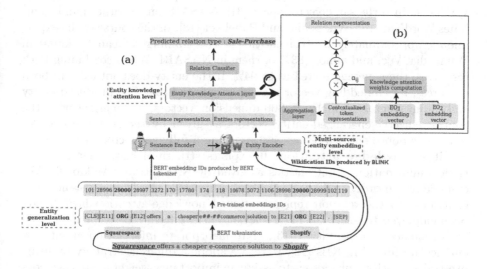

Fig. 1. (a) Our multilevel entity-informed model for business relation extraction and (b) a detailed description of our knowledge-attention mechanism.

We consider one simple and two complex aggregators to extract the most productive features from the contextualized representations of both sentence tokens and entity mentions as produced by the sentence encoder: **BizBERT**, the BERT PLM fine-tuned on our business dataset that uses the final hidden state of the classification token $[CLS]$, **BizBERT+CNN** a convolutional layer followed by a max-pooling and an activation function on top of BizBERT, and **BizBERT+BILSTM** that uses a BiLSTM layer instead on top of BizBERT.

The model can be augmented at multiple levels with knowledge about entities. We newly consider three main levels of additional knowledge:

- **Entity generalization.** We designed a generalization strategy that consists in replacing the target entities EO_i in the input sentence before giving it to

---

[5] https://github.com/facebookresearch/BLINK.

the sentence encoder by the generic tag ORG to prevent overfitting and help the model to reason at the entity type level rather than the entity mention itself which may be infrequent in the corpus or over-represented. For example, the entity pair *(Google, Microsoft)* can be very frequent for the relation COMPETITOR but rare for COOPERATION. This strategy, initially proposed for generic and clinical relations [23,26] is used for the first time for business RE.

- **Multi-sources entity embeddings.** The disambiguated IDs provided by the BLINK Wikification process, are used by the entity encoder to query two complementary external sources of knowledge about entities: Wikipedia2Vec and for the first time in RE, NASARI. Wikipedia2Vec implements the extended version of the skip-gram model to map words and entities from Wikipedia into the same vector space [31]. NASARI on the other hand, combines WordNet [16], Wikipedia, and BabelNet [18]. In the course of the experiments, approximately 92% of entities in the training set can be found in Wikipedia2Vec, and almost 83% of them in NASARI. When combining both resources, the coverage increases to 94%. If the entity does not exist in both resources, its embedding vector is randomly initialized. The produced entity embedding vectors, which are 300-dimension vectors, are merged with the contextualized generalized-entity vectors as given by the sentence encoder into a one dense entities representation using a fully connected layer.

- **Entity-Knowledge attention.** It exploits structural knowledge and statistical information about entities as given by NASARI and Wikipedia2Vec embeddings in order to focus on the most important words in a sentence that contribute to the relation representation. Knowledge-attention has already been employed to select the most relevant entities from KBs to be integrated with sentence representation [13], or to incorporate information about how entities are linked in KBs [12]. Here, we adopt a different strategy by using pre-trained entity embeddings to assign an importance weight a_{ij} to each contextualized token representation of an input sentence as presented in Fig. 1(b).

The final multilevel entity-informed sentence representation is fed into a relation classifier. We consider two configurations: *monotask learning* and *multitask learning*. The first one is a multi-class learning problem where the classifier has to predict the relation type that links a pair of entities (EO_i, EO_j) in a given sentence among the six relations that we consider (including OTHERS). The second configuration is designed to deal with data imbalance (cf. Table 1), following recent studies that show that jointly learning common characteristics shared across multiple tasks can have a strong impact on RE performances [29,34]. To this end, we jointly train two classifiers using multitask objectives. The first one performs *relation identification* to detect whether a business relation holds between a given entity pair or not (i.e., business vs. non-business). It is trained on a more balanced dataset (business (37%) vs. non-business (63%)) to optimize a binary cross-entropy loss. The second classifier performs *relation classification* and learns how to predict the relation type between two EO_i (this is a 6-class classification task) with a multi-class cross-entropy loss.

5   Experimental Settings and Baselines

We experiment with different models $\mathcal{M_E}$ while varying the aggregation layer $\mathcal{M}$ (BizBERT, BizBERT+CNN, BizBERT+BILSTM) and the entity knowledge levels $\mathcal{E}$ (t, wiki, nas, att) among entity type generalization (t), multi-source entity embeddings from either Wikipedia $(wiki)$ or NASARI (nas), and entity-attention (att).

In our experiments, the sentence encoder relies on the `bert-base-cased` model implemented in the HuggingFace library[6]. The sentence encoder always outputs a sentence representation of dimension 768, either using the BERT's [CLS] final embedding, a CNN with a kernel size set to 5 applied to all the contextualized embeddings, or a BiLSTM with hidden units set to 768 applied to the same contextualized embeddings. All the models $\mathcal{M_E}$ are trained either in a mono-task or a multitask configuration. BERT is fine-tuned on our business dataset for 5 *epochs* using the Adam optimizer with an initial learning rate of 2^{-5} and a batch size of 16.

Our multilevel entity-informed models have been evaluated on the test set[7] and compared to the best performing *Kag* and *Kin* state of the art models for RE, as follows.

- **CNN**^{Kag} [35]. This model is based on a convolutional neural network that uses FastText [15] pre-trained word embedding vectors of 300-dimension, three 1D convolutional layers, each one using 100 filters and a stride of 1, and different window sizes (3, 4 and 5 respectively) with a ReLU activation function. Each layer is followed by a max-pooling layer. The output layer is composed of a fully connected layer followed by a softmax classifier. The results reported here were obtained using a dropout of 50% and optimized using the Adam optimizer [8] with a learning rate of 10^{-3}.
- **Attention-BiLSTM**^{Kag} [38]. It adopts a BiLSTM model with an attention mechanism that attends over all hidden states and generates attention coefficients relying on FastText embeddings as input representation. During experiments, best results have been obtained using 100 hidden units, an embedding dropout rate of 70%, a final layer dropout rate of 70%, and an Adam optimizer learning rate of 1.
- **R-BERT**^{Kag} [28]. This is an adaptation of BERT for RE that takes into account entities representation in the relation instance representation. The model relies on the `bert-base-cased` model for English that is fine-tuned on our dataset for 5 *epochs*. R-BERT^{Kag} has been trained with the same hyper-parameters used to train our models.

---

[6] https://huggingface.co/bert-base-cased.
[7] All the hyperparameters were tuned on a validation set (10% of the train set).

– **KnowBert**[Kin] [20]. We also compare with KnowBert, one of the best *Kin* systems for RE[8]. KnwoBert comes up with three models either pre-trained with Wikipedia (**KnowBert-Wiki**), WordNet (**KnowBert-WordNet**), or with both resources (**KnowBert-W+W**). KnowBert-Wiki entity embeddings are learned using a skip-gram model directly from Wikipedia descriptions without using any explicit graph structure between nodes. Entity embeddings are then incorporated into BERT using knowledge-attention and re-contextualization mechanism. Embeddings in KnowBert-WordNet are learned from both Wordnet synset glosses and a knowledge graph constructed from word-word and lemma-lemma relations. KnowBert models are fine-tuned on our dataset for 5 *epochs* using the same hyper-parameters proposed in the original paper.

6   Results and Discussions

6.1   Baseline Results

Table 2 presents the results of state of the art *Kag* and *Kin* baselines in terms of macro-averaged F-score (F1), precision (P), and recall (R); best scores are in bold[9]. Among the four *Kag* models, R-BERT achieves the best scores. The results are however lower when compared to KnowBERT which confirms that injecting knowledge about entities is crucial for effective RE. KnowBERT-Wiki being the best baseline in terms of F1-score, we, therefore, consider this model as a strong baseline to compare with.

Table 2. Results of Knowledge-agnostic (*Kag*) and knowledge-informed (*Kin*) baselines.

Model[Kag]	P	R	F1	Model[Kin]	P	R	F1
CNN [35]	63.5	58.7	59.7	KnowBERT-Wiki [20]	**65.3**	71.9	**68.2**
Att.-BiLSTM [38]	59.4	54.3	56.3	KnowBERT-Wordnet [20]	63.6	71.5	67.0
R-BERT [28]	63.6	67.4	65.2	KnowBERT-W+W[Kin] [20]	64.2	**72.7**	67.5

6.2   Results of the Proposed Architectures (Monotask and Multitask)

Due to the high number of $\mathcal{M}_{\mathcal{E}}$ configurations (3 combinations for $\mathcal{M}$ and 16 for $\mathcal{E}$, leading to a total of 48 different models), we only present the best performing

---

[8] Among existing entity-informed models (cf. Sect. 2), at the time of performing these experiments, and as far as we know, only KnowBert and ERNIE were actually available to the research community. In this paper, we compare with Knowbert as it achieved the best results on the TACRED dataset (71.50% on F1-score) when compared to ERNIE (67.97%) [25].

[9] We also experimented with Entity-Attention-BiLSTM following [10] but the results were not conclusive.

ones. Table 3 summarizes our results. Due to space limitation and to better compare the contributions of level of knowledge, we present the entity type and P-EE sources (t, $wiki$, nas) horizontally, and the attention one (att) vertically along with the classifier setting (monotaks vs. multitask).

Table 3. Results of the monotask and multitask experiments. Best scores are underlined while bold ones are those that outperform the best baseline.

Model	monotask			monotask$_{att}$			multitask			multitask$_{att}$		
	P	R	F1	P	R	F1	P	R	F1	P	R	F1
BizBERT$_{wiki}$	64.3	67.9	65.7	63.2	70.8	66.6	**65.5**	70.5	67.6	63.8	71.9	67.4
BizBERT$_{wiki+t}$	**68.5**	71.9	**70.1**	**67.8**	**73.9**	**70.6**	67.2	70.9	**68.9**	**67.2**	71.2	**69.1**
BizBERT$_{nas}$	64.7	68.6	66.1	62.7	71.2	66.4	**65.6**	70.8	67.8	64.3	71.1	67.3
BizBERT$_{nas+t}$	**66.8**	70.6	**68.5**	**68.1**	**72.5**	**70.1**	**69.8**	69.8	**69.7**	**68.0**	71.7	**69.7**
BizBERT$_{nas+wiki+t}$	**67.9**	71.4	**69.5**	**67.8**	73.4	**70.4**	68.1	70.3	**69.1**	**67.5**	71.6	**69.4**
BizBERT+CNN$_{wiki}$	63.6	70.6	66.7	**65.5**	71.6	68.0	64.4	71.3	67.5	**65.4**	71.4	68.0
BizBERT+CNN$_{wiki+t}$	64.7	70.6	67.2	**66.2**	70.8	68.1	**66.3**	72.6	**69.1**	66.1	**72.9**	**69.0**
BizBERT+CNN$_{nas}$	61.6	71.3	65.6	64.9	71.2	67.7	63.1	71.8	66.9	**65.5**	72.1	**68.4**
BizBERT+CNN$_{nas+t}$	**68.1**	72.5	**69.9**	65.3	71.0	67.7	**68.1**	71.3	**69.3**	65.0	72.2	68.0
BizBERT+BILSTM$_{wiki}$	62.5	70.8	65.9	64.3	71.7	67.4	63.2	69.6	65.9	64.3	71.6	67.4
BizBERT+BILSTM$_{wiki+t}$	64.9	**72.0**	67.9	64.4	70.2	67.1	**67.6**	**73.5**	**70.1**	64.3	71.0	67.1
BizBERT+BILSTM$_{nas}$	63.3	71.0	66.5	64.1	71.2	67.0	64.3	68.2	65.8	64.3	71.1	67.1
BizBERT+BILSTM$_{nas+t}$	64.0	**72.0**	67.3	63.7	71.1	67.0	**65.5**	**72.5**	**68.4**	**67.0**	**72.5**	**69.3**

In the monotask configuration, we can observe that BizBERT results are better than BizBERT+CNN and BizBERT+BILSTM and that the sentence features obtained via BizBERT+BILSTM is the least productive. From the observed results, two other interesting findings can be drawn. First, models with only one level of entity knowledge do not outperform the KnowBERT baseline (e.g., $F1 = 67.7\%$ for BizBERT$_t$, $F1 = 66.7\%$ for BizBERT+CNN$_{wiki}$ and $F1 = 66.1\%$ for BizBERT$_{nas}$). Second, P-EE from NASARI are more productive than those from Wikipedia2Vec. See for example BizBERT$_{wiki} = 65.7\%$ vs. BizBERT$_{nas} = 66.1\%$ and BizBERT+BILSTM$_{wiki} = 65.9\%$ vs. BizBERT+BILSTM$_{nas} = 66.5\%$. This shows that even with NASARI low coverage rate when performing entity linking (83% vs. 92% for Wikipedia2vec), the relation classifier could capture important knowledge about entities and that P-EE built from multiple sources (BabelNet, WordNet synsets, Wikipedia pages) are of better quality than those built from Wikipedia alone.

When multiple levels of knowledge are injected into the model, most results increase outperforming the baseline. In particular, combining P-EE with generalization over entity type has been very productive, achieving 1.9% in terms of F1-score over the baseline when using $wiki + t$ with BizBERT. BizBERT$_{wiki+t}$ also outperforms its single level counterparts (i.e., BizBERT$_t$ and BizBERT$_{wiki}$) by 2.4% and 4.4% respectively. We observe the same tendency when training the models with $nas + t$ vs. nas and t alone. When relying on $wiki + nas + t$, the results are better than those obtained for $wiki + nas$, but still lower when compared to $wiki + t$. This can be explained by the weak converge of NASARI for the entities present in the test set. Finally, when the knowledge-attention layer is activated, almost all the models gained in terms of F1 score, yielding to the highest improvement (about 2.4%) over the baseline for BizBERT$_{wiki+t+att}$, our best model. This demonstrates that knowledge-attention is an important mechanism for RE when coupled with other levels of knowledge about entities regardless of the aggregation layer used. Overall, these results show that directly injecting knowledge about entities as external features to the relation classifier without neither PLM re-training nor architecture update is a simple and effective solution for RE. More importantly, multiple levels of knowledge are needed, the best level being Wikipedia P-EE when coupled with entity type and knowledge-attention.

The results of the multitask setting show the same general conclusions already drawn from the monotask experiments: multilevel knowledge about entities is better than injecting a single level alone. However, we notice that BizBERT scores are lower when compared to the monotask configurations while those of the BizBERT+CNN and BizBERT+BILSTM increased. Indeed, the BizBERT+BILSTM model with $nas + t$ beats the baseline with the highest difference in this multitask configuration (1.9% in terms of F1-score), which is still lower than the best performing model (i.e. BizBERT$_{wiki+t+att}$ in monotask setting). This shows that learning to classify business relations (monotask setting) is more effective than learning simultaneously both relation identification and relation classification (multitask setting). This implies that discriminating business from non-business relations is a much more complex task than discriminating between business relations, making the relation identification task harder. Two reasons behind that could be: (a) the dataset imbalance between business relation types and OTHERS relation type, and (b) the variability of relation patterns that could be included in the relation type OTHERS which make learning features about this class difficult.

6.3   Error Analysis

The F-scores per class achieved by BizBERT$_{wiki+t+att}$, our best performing model, are: INVESTMENT 67.9%, SALE-PURCHASE 41.3%, COMPETITION 77.6%, COOPERATION 67.8%, and LEGAL PROCEEDINGS 82.4%. When compared to KnowBERT-Wiki, the best baseline, our model gets better scores for COOPERATION (+5.5%), INVESTMENT (+3.6%), SALE-PURCHASE (+0.8%), and LEGAL PROCEEDINGS (+4.6%) whereas it fails to account for COMPETITION (−0,4%). It is interesting to note that our model is more effective than the baseline when it comes to classifying relations with few instances. This observation is more visible in complex sentences that contain more than 4 entities.

A closer look at the confusion matrices shows that both models do not perform well when differentiating between business relations and non-business relations (OTHERS). The multitask setting we developed did not help mitigating this, since it gave less effective results than the monotask one. This is more salient for SALE-PURCHASE and COOPERATION where 38% and 19% of instances respectively were predicted as OTHERS by our model. This is because OTHERS instances do not have common characteristics like the five business relations we consider, as it may represent any other relation that may exist between two ORG (e.g. attending the same event, etc.).

A manual analysis of misclassified relations reveals two other sources of error. The first one concerns sentences containing more than one relation between different entity pairs, as in (1). In this example, only the relation linking the two EO in bold has to be identified. Our best model predicts INVESTMENT (EO_1, EO_3), whereas the ground-truth annotation is OTHERS (EO_1, EO_3). Note that an INVESTMENT relation actually exists between EO_1 and EO_2. The second source of error arises from relations expressed metaphorically or indirectly, as in (2), where the expression *has issued Autonomous Vehicle Testing Permits* triggers a COMPETITION relation between *Volkswagen* and *Delphi Automotive*. However, the model predicts OTHERS.

(1) In 2001, [**Enel**]$_1$ acquired [Infostrada]$_2$, previously property of [**Vodafone**]$_3$: the cost of the operation was 7.21285 billion euro.

(2) Wheego and Valeo now join the likes of Google, Tesla, GM Cruise and Ford on the list of companies the Californian DMV has issued Autonomous Vehicle Testing Permits to, as well as [**Volkswagen**]$_1$, Mercedes Benz, [**Delphi Automotive**]$_2$ and Bosch.

7 Conclusion

This paper presented (a) the first large business dataset annotated according to a new characterization composed of five business relations, and (b) simple but effective multilevel entity informed neural architectures to extract those relations from web documents. We conducted for the first time a systemic evaluation of the contribution of different levels of knowledge, experimenting with entity type generalization, pre-trained embeddings from Wikipedia2vec and NASARI, and entity-knowledge-attention both in a monotask and multitask settings. Our results show that multiple levels of knowledge are needed for effective RE, beating very competitive knowledge-agnostic and knowledge-informed state of the art models. Our approach only requires entity knowledge as input alongside with the sentence representation provided by BERT pre-trained language model without any additional trained layer or parameters re-training. It is therefore generic and can be easily applied to extract other types of relations between named entities thanks to different sources of knowledge.

In the future, we plan to extend our model to handle implicitly expressed relations as well as to account for inner-organizational business relations.

References

1. Auer, S., Bizer, C., Kobilarov, G., Lehmann, J., Cyganiak, R., Ives, Z.: DBpedia: a nucleus for a web of open data. In: Aberer, K., et al. (eds.) ASWC/ISWC -2007. LNCS, vol. 4825, pp. 722–735. Springer, Heidelberg (2007). https://doi.org/10.1007/978-3-540-76298-0_52

2. Braun, D., Faber, A., Hernandez-Mendez, A., Matthes, F.: Automatic relation extraction for building smart city ecosystems using dependency parsing. In: Proceedings of NL4AI@ AI* IA, pp. 29–39. CEUR-WS.org (2018)

3. Camacho-Collados, J., Pilehvar, M.T., Navigli, R.: Nasari: integrating explicit knowledge and corpus statistics for a multilingual representation of concepts and entities. Artif. Intell. **240**, 36–64 (2016)

4. Collovini, S., Gonçalves, P.N., Cavalheiro, G., Santos, J., Vieira, R.: Relation extraction for competitive intelligence. In: Quaresma, P., Vieira, R., Aluísio, S., Moniz, H., Batista, F., Gonçalves, T. (eds.) PROPOR 2020. LNCS (LNAI), vol. 12037, pp. 249–258. Springer, Cham (2020). https://doi.org/10.1007/978-3-030-41505-1_24

5. Devlin, J., Chang, M.W., Lee, K., Toutanova, K.: Bert: pre-training of deep bidirectional transformers for language understanding. In: Proceedings of NAACL-HLT, no. 1 (2019)

6. Gupta, P., Rajaram, S., Schütze, H., Runkler, T.: Neural relation extraction within and across sentence boundaries. In: Proceedings of the AAAI Conference on Artificial Intelligence, pp. 6513–6520 (2019)

7. Hendrickx, I., et al.: SemEval-2010 task 8: Multi-way classification of semantic relations between pairs of nominals. In: Proceedings of the 5th International Workshop on Semantic Evaluation, pp. 33–38. ACL (2010)

8. Kingma, D.P., Ba, J.: Adam: a method for stochastic optimization. arXiv preprint arXiv:1412.6980 (2014)

9. Lau, R., Zhang, W.: Semi-supervised statistical inference for business entities extraction and business relations discovery. In: Proceedings of SIGIR Workshop, pp. 41–46 (2011)

10. Lee, J., Seo, S., Choi, Y.S.: Semantic relation classification via bidirectional LSTM networks with entity-aware attention using latent entity typing. Symmetry **11**(6), 785 (2019)

11. Li, B.Z., Min, S., Iyer, S., Mehdad, Y., Yih, W.T.: Efficient one-pass end-to-end entity linking for questions. In: Proceedings of EMNLP, pp. 6433–6441 (2020)

12. Li, J., Huang, G., Chen, J., Wang, Y.: Dual CNN for relation extraction with knowledge-based attention and word embeddings. Comput. Intell. Neurosci. **2019**, 1–10 (2019)

13. Li, Z., Lian, Y., Ma, X., Zhang, X., Li, C.: Bio-semantic relation extraction with attention-based external knowledge reinforcement. BMC Bioinform **21**, 1–18 (2020)

14. Martinez-Rodriguez, J.L., Hogan, A., Lopez-Arevalo, I.: Information extraction meets the semantic web: a survey. In: Semantic Web, pp. 1–81 (2020)

15. Mikolov, T., Grave, É., Bojanowski, P., Puhrsch, C., Joulin, A.: Advances in pre-training distributed word representations. In: Proceedings of LREC (2018)

16. Miller, G.A., Beckwith, R., Fellbaum, C., Gross, D., Miller, K.J.: Introduction to wordnet: an on-line lexical database. Int. J. Lexicography **3**(4), 235–244 (1990)

17. Mitchell, A., Strassel, S., Huang, S., Zakhary, R.: Ace 2004 Multilingual Training Corpus, p. 1. Linguistic Data Consortium, Philadelphia pp (2005)

18. Navigli, R., Ponzetto, S.P.: Babelnet: the automatic construction, evaluation and application of a wide-coverage multilingual semantic network. Artif. Intell. **193**, 217–250 (2012)
19. Pawar, S., Palshikar, G.K., Bhattacharyya, P.: Relation extraction: a survey. arXiv preprint arXiv:1712.05191 (2017)
20. Peters, M.E., et al.: Knowledge enhanced contextual word representations. In: Proceedings of EMNLP-IJCNLP, pp. 43–54 (2019)
21. Poerner, N., Waltinger, U., Schütze, H.: E-BERT: efficient-yet-effective entity embeddings for BERT. In: EMNLP, pp. 803–818. ACL (2020)
22. Sewlal, R.: Effectiveness of the web as a competitive intelligence tool. South African J. Inf. Manage. **6**(1), 1–16 (2004)
23. Shi, P., Lin, J.: Simple bert models for relation extraction and semantic role labeling. arXiv preprint arXiv:1904.05255 (2019)
24. Soares, L.B., FitzGerald, N., Ling, J., Kwiatkowski, T.: Matching the blanks: distributional similarity for relation learning. In: Proceedings of ACL, pp. 2895–2905 (2019)
25. Wang, R., et al.: K-adapter: Infusing knowledge into pre-trained models with adapters. arXiv preprint arXiv:2002.01808 (2020)
26. Wei, Q., et al.: Relation extraction from clinical narratives using pre-trained language models. In: AMIA Annual Symposium Proceedings, vol. 2019, p. 1236. American Medical Informatics Association (2019)
27. Wiegand, M., Roth, B., Lasarcyk, E., Köser, S., Klakow, D.: A gold standard for relation extraction in the food domain. In: Proceedings of LREC (2012)
28. Wu, S., He, Y.: Enriching pre-trained language model with entity information for relation classification. In: Proceedings of ACM CIKM 2019, pp. 2361–2364 (2019)
29. Yadav, S., Ramesh, S., Saha, S., Ekbal, A.: Relation extraction from biomedical and clinical text: unified multitask learning framework. IEEE/ACM Trans. Comput. Biol. Bioinform. (2020)
30. Yamada, I., et al.: Wikipedia2Vec: an efficient toolkit for learning and visualizing the embeddings of words and entities from Wikipedia. In: Proceedings of EMNLP: System Demonstrations, pp. 23–30 (2020)
31. Yamada, I., Shindo, H., Takeda, H., Takefuji, Y.: Joint learning of the embedding of words and entities for named entity disambiguation. In: Proceedings of The 20th SIGNLL CoNLL, pp. 250–259 (2016)
32. Yamamoto, A., Miyamura, Y., Nakata, K., Okamoto, M.: Company relation extraction from web news articles for analyzing industry structure. In: 2017 IEEE ICSC, pp. 89–92 (2017)
33. Yan, C., Fu, X., Wu, W., Lu, S., Wu, J.: Neural network based relation extraction of enterprises in credit risk management. In: 2019 IEEE BigComp, pp. 1–6 (2019)
34. Ye, W., Li, B., Xie, R., Sheng, Z., Chen, L., Zhang, S.: Exploiting entity BIO tag embeddings and multi-task learning for relation extraction with imbalanced data. In: Proceedings of ACL, pp. 1351–1360 (2019)
35. Zeng, D., Liu, K., Lai, S., Zhou, G., Zhao, J.: Relation classification via convolutional deep neural network. In: Proceedings of COLING: Technical Papers, pp. 2335–2344. ACL, Dublin City University (2014)
36. Zhang, Y., Zhong, V., Chen, D., Angeli, G., Manning, C.D.: Position-aware attention and supervised data improve slot filling. In: Proceedings of EMNLP, pp. 35–45. ACL (2017)
37. Zhao, J., Jin, P., Liu, Y.: Business relations in the web: semantics and a case study. J. Softw. **5**(8), 826–833 (2010)

118   H. Khaldi et al.

38. Zhou, P., et al.: Attention-based bidirectional long short-term memory networks for relation classification. In: Proceedings of ACL (Volume 2: Short Papers), pp. 207–212. ACL (2016)
39. Zuo, Z., Loster, M., Krestel, R., Naumann, F.: Uncovering business relationships: Context-sensitive relationship extraction for difficult relationship types. In: Proceedings of LWDA (2017)

The Importance of Character-Level Information in an Event Detection Model

Emanuela Boros[1]([⊠]) [iD], Romaric Besançon[2] [iD], Olivier Ferret[2] [iD],
and Brigitte Grau[3]

[1] University of La Rochelle, L3i, 17000 La Rochelle, France
emanuela.boros@univ-lr.fr
[2] Université Paris-Saclay, CEA, List, 91120 Palaiseau, France
{romaric.besancon,olivier.ferret}@cea.fr
[3] Université Paris-Saclay, CNRS, LIMSI, ENSIIE, 91405 Orsay, France
brigitte.grau@limsi.fr

Abstract. This paper tackles the task of event detection that aims at identifying and categorizing event mentions in texts. One of the difficulties of this task is the problem of event mentions corresponding to misspelled, custom, or out-of-vocabulary words. To analyze the impact of character-level features, we propose to integrate character embeddings, that can capture morphological and shape information about words, to a convolutional model for event detection. More precisely, we evaluate two strategies for performing such integration and show that a late fusion approach outperforms both an early fusion approach and models integrating character or subword information such as ELMo or BERT.

Keywords: Information extraction · Events · Word embeddings

1 Introduction

In this article, we concentrate more specifically on event detection, which implies identifying instances of specified types of events in a text. The notion of event in our work is classically defined as something that happens and covers a wide spectrum, from terrorist attacks to births or nominations. The instances of these events in texts, which are called *event mentions* or *event triggers*, are annotated as words or phrases that evoke a reference type of events. The most successful approaches developed for achieving this task are currently based on neural models, which have been intensively studied to overcome fundamental limitations, specifically the complex choice of features [2,9,25–27,29,36]. All these proposed models based on Convolutional Neural Networks (CNNs), Recurrent Neural Networks (RNNs), or even Graph Neural Networks (GNNs) rely on word embeddings, a general distributed word representation that is produced by training a

This work was partly supported by the European Union's Horizon 2020 research and innovation program under grants 770299 (NewsEye) and 825153 (Embeddia).

© Springer Nature Switzerland AG 2021
E. Métais et al. (Eds.): NLDB 2021, LNCS 12801, pp. 119–131, 2021.
https://doi.org/10.1007/978-3-030-80599-9_11

deep learning model on a large unlabeled dataset. Consequently, word embeddings replace the hard matches of words in the feature-based approaches with the soft matches of continuous word vectors. Hence, compared to previous rule-based or machine learning-based approaches, neural models are supposed to be less sensitive to the problem of unseen triggers since the distributed representations of words they exploit can account for the similarity between words.

However, this capacity may vary depending on the reasons why a trigger was not seen during the training of a model. We illustrate these different cases on the ACE 2005 dataset[1], a standard corpus used for evaluating event detection. An unseen trigger may be a morphological variant of a trigger already seen in the training set. For instance, *torturing* is not present in the training data but is a variant of *torture* and can be considered as a trigger for the same type of events, namely *Life.Injure*. Moreover, *torturing* is likely to be present among general pre-trained word embeddings and if so, a neural event extraction model is likely to successfully detect this trigger. The situation may be different when a trigger is absent from the training data because it corresponds to a misspelled version of a reference trigger. For instance, *aquitted* is part of the ACE 2005 test dataset for referring to a *Justice.Sentence* event while only *acquitted*, the correct form for that word, is present in the training data. In that case, we cannot assume that the unseen word is part of general word embeddings and as a consequence, has little chance to be detected as a trigger for a *Justice.Sentence* event.

From a more general perspective, the problem of missing word embeddings when using pre-trained models in the context of event extraction is not marginal. The ACE 2005 dataset, for instance, covers the most common events of national and international news (from a variety of sources selected from broadcast news programs, newspapers, news reports, internet sources, or transcribed audio) and thus, it contains different types of discourse, professional or noisy discussions prone to the presence of mistakes in spelling and custom words. As a result, in this dataset, 14.8% of the words are not part of the pre-trained embeddings provided by Google, trained with *word2vec* on *Google News* [23], 1.5% for the *GloVe* embeddings [30], and 4.5% for the fastText embeddings [12]. Different strategies were proposed and implemented for dealing with the issue of missing words in neural language models. For static word embeddings, fastText relies on a representation of words based on n-grams of characters. For contextual models, ELMo [31] exploits a character-based representation built with a CNN while BERT [3] adopts a mixed strategy based on subwords, called wordpieces, where a word is split into subwords when it is not part of a predefined and restricted vocabulary [13,15,21]. However, while BERT seems to be an interesting option for a large number of tasks in Natural Language Processing, its ability to handle noisy inputs is still an open question [32] or at least requires the addition of complementary methods [24]. This limitation may result from the dependence of BERT on a vocabulary. The alternative is to rely, as ELMo for instance, on a character model in which all words, including words with abnormal character combinations and misspellings, are processed similarly.

---

[1] https://catalog.ldc.upenn.edu/ldc2006t06.

Some researchers studied the application of CNNs to characters. For anti-spam filtering, the use of character-level n-grams was already experimented out of the context of deep learning models [14]. In [5], character-level embeddings were automatically learned and joined with pre-trained word embeddings in a CNN-based model for Part-of-Speech tagging. This architecture was also used for improving the performance of a Named Entity Recognition (NER) system in [4]. While character models have been used with success in several contexts for tackling the absence of pre-trained embeddings for all words, the use of CNNs to learn directly from characters was also investigated, without the need for any pre-trained embeddings [37]. Notably, the authors use a relatively deep network and apply it to sentiment analysis and text classification tasks. The application of character-level convolutions to language modeling was explored in [15] by using the output of a character-level CNN as the input to a Long Short Term Memory (LSTM) network at each time step. The same model is easily applied to various languages. However, the choice of CNN-based or LSTM-based character-level word embeddings did not affect the performance significantly [16, 22].

Our contributions in this article are more particularly focused on the integration of character-level features in event detection models for addressing the issue of unknown words. More specifically, we show that an event detection model exploiting a character-based representation is complementary to a word-oriented model and that their combination according to a late fusion approach outperforms an early fusion strategy.

2 Related Work

The current state-of-the-art systems for event extraction involve neural network models to improve event extraction. [27] and [2] deal with the event detection problem with models based on CNN. [28] improve the previous CNN models of [27] for event detection, slightly modifying the way CNNs are applied to sentences by taking into account the possibility to have non-consecutive n-grams as basic features instead of continuous n-grams. Both models use word embeddings for representing windows of text that are trained as the other parameters of the neural network.

The authors of [25] predict at the same time event triggers and their arguments in a joint framework with Bidirectional RNNs (Bi-RNNs) and a CNN and systematically investigate the usage of memory vectors/matrices to store the prediction information during the labeling of sentence features. Additionally, the authors augment their system with discrete local features inherited froms [17].

A GNN is advocated in [29] based on dependency trees to perform event detection with a pooling method that relies on entity mentions aggregating the convolution vectors. The authors of [20] consider also that arguments provide significant clues to this task and adopt a supervised attention mechanism to exploit argument information explicitly for event detection, while also using events from FrameNet.

Table 1. Statistics about unknown words in the ACE 2005 dataset.

	All words	Trigger words
Train	14,021	931
Test	3,553	219
Unknown words in test data	930 (26.2%)	66 (30.1%)
Unknown words with a known similar word	825	54

Further, some researchers have proposed other hybrid neural network models with different types of pre-set word embeddings that combine different neural networks to make use of each other's abilities. A hybrid neural network (a CNN and an RNN) [9] was developed to capture both sequence and chunk information from specific contexts and use them to train an event detector for multiple languages without any handcrafted features.

Some authors went beyond sentence-level sequential modeling, considering that these methods suffer from low efficiency in capturing very long-range dependencies. An approach that goes beyond sentence level [8] was proposed by using a document representation using an RNN model that can automatically extract cross-sentence clues.

Recently, different approaches that include external resources and features at a subword representation level have been proposed. For example, Generative Adversarial Networks (GANs) framework has been applied in event extraction [10,36]. Besides, reinforcement learning is used in [36] for creating an end-to-end entity and event extraction framework. An approach based on the BERT pre-trained model [35] attempts an automatic generation of labeled data by editing prototypes and filtering out the labeled samples through argument replacement by ranking their quality.

The problem of ambiguous indicators for particular types of events, i.e., the same word can express completely different events, such as *fired*, that can correspond to an *Attack* type of event or can express the dismissal of an employee from a job, is approached in [19] by using an RNN and cross-lingual attention to model the confidence of the features provided by other languages.

3   Motivation

Learning word representations from a corpus (word embeddings) allows us to derive a flexible similarity between words that takes into account a form of synonymy or relatedness between the words into the model. A drawback of this kind of representation is that unknown words (i.e. words unseen in the training corpus) are not well represented: they are generally associated with a random embedding even if these words are morphologically close to known words. Existing embeddings trained on very large collections of text, such as *word2vec* embeddings, which have proven their efficiency as initial embeddings for event

Table 2. Examples of unknown words focused on triggers.

Event type	Unknown/Closest trigger words
Start-Org	*creating/creation, opening/open, forging/forming, formed/form*
End-Org	*crumbled/crumbling, dismantling/dismantle, dissolved/dissolving*
Transport	*fleeing/flying, deployment/deployed, evacuating/evacuated*
Attack	*intifada/Intifada, smash/smashed, hacked/attacked, wiped/wipe*
End-Position	*retirement/retire, steps/step, previously/previous, formerly/former*

extraction, do not take into account these morphological similarities: no lemmatization or stemming or even case normalization is performed.

We present in Table 1 some statistics about unknown words on the dataset we will use for training and testing our proposed approach, the ACE 2005 dataset, using the standard training/validation/test split [11]. We report the size of the vocabulary for the whole dataset, the size of the vocabulary for the trigger words, the number of words in the test dataset not seen in the training dataset, and among those, the number of words for which a similar word (measured by a Levenshtein ratio of less than 0.3) can be found.

We see that there is an important number of words, even among the trigger words, that cannot be exploited by the models because they are not seen in the training corpus. Also, most of these words (more than 88%) have similar words in the training corpus which could be used to approximate their representation, as they are likely to be semantically close words. To illustrate this, we show in Table 2 examples of unknown words focused on triggers. The examples are pairs of words used as triggers in the test set associated with their closest trigger words (for an event of the same type) as seen in the training set (with a distance of Levenshtein ratio less than 0.3). We can see in this table that most of the pairs correspond to derivational morphology links. These semantic links are lost with the standard embedding models.

The integration of a character-based embedding model should be able to help in dealing with such cases by allowing to bridge the gap between the unknown words and representations of known words used for training the system. The same problem occurs with infrequent words, that could be better represented if they are processed at a character level.

4   Approach

Our approach lies in the standard supervised framework of event detection where the task is modeled as a word classification task: considering a sentence, we want to predict for each word of the sentence if it is an event trigger and associate it with its event type. The input of the system is therefore a target word in the context of a sentence and the output an event type or NONE for non-trigger words. To study the influence of character-based features, we rely on the CNN model proposed by [27] as a core model. This core architecture is used in the two

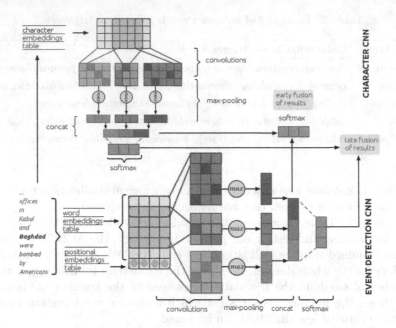

Fig. 1. Word + Character CNN.

components of our overall system: the Word model and the Character model. These two components are combined using either an early fusion approach or a late fusion approach, as illustrated by Fig. 1.

4.1 Word and Character CNN Models

In the Word CNN model, the context of a target word is formed by its surrounding words in the sentence, which constitutes the input of the convolution layer. To consider a limited-sized context, longer sentences are trimmed and shorter ones are zero-padded. We consider a context window for every trigger candidate where each token is associated with a word embedding and a relative position to the trigger candidate embedding. The word and position embeddings are concatenated and passed through the convolution layer. The concatenated output of convolutional filter maps forms, after a max-over-time pooling operation on each one, the representation of the input (target token and context) that finally goes through a softmax classification layer. The Character CNN model is very close to the Word CNN model, with two main differences: words are replaced by characters and there is no position embedding associated with each character.

4.2   Integration of Word and Character Models

Early Fusion. The first type of integration is the early fusion model, in which the two representations of the input sequence produced by the Word and Character CNNs (i.e., the concatenation of the output vectors of their filters) are concatenated before the fully-connected softmax classification layer. Using this type of integration allows joint learning of the parameters of the two models in the training phase.

Late Fusion. The late fusion integration of the Word and Character CNNs relies on the combination of the decisions of the two models, which are trained separately and therefore learn different characteristics of the candidate trigger. Indeed, the word-level CNN combines word and position embeddings that can capture syntactic and semantic information, and of course, the relative positions of words to the candidate trigger. The character-level CNN learns more local features from character n-grams and can capture morphological information. The late fusion focuses on the individual strength of these two models by applying the following rule: we always keep the Character CNN label, except if a trigger was detected by the Word CNN but not by the Character CNN. This strategy is motivated by the fact that the Word CNN model has good coverage whereas the Character model is more focused on precision.

5   Experiments and Results

5.1   Dataset

The evaluation is conducted on the annotated ACE 2005 corpus. We use the same split as previous studies with this dataset [25, 27]: 40 news articles (672 sentences) for the test set, 30 other documents (863 sentences) for the development set, and the remaining 529 documents (14,849 sentences) for the training set. Following the same line of work, we consider that a trigger is correct if its event type, subtype, and offsets match those of a reference trigger. We use Precision (P), Recall (R), and F-measure (F1) to evaluate the overall performance.

5.2   Hyperparameters

For the Word CNN, we consider a sliding window with a maximal size of 31 words. Hence, sentences are padded at the beginning and the end with a vector of 15 zeros (a common practice for the padding special character). The window sizes for the convolutions are in the set $\{1, 2, 3\}$ and 300 feature maps are used for each window size. After each convolutional layer with orthogonal weights initialization, a *ReLu* non-linear layer is applied. We employ dropout with a probability of 0.5 after the embedded window of text since they contain most of the parameters and as a consequence, the possibility of being responsible for overfitting. A dropout of 0.3 is also applied after the concatenation of the convolutions. The size of the position embeddings is equal to 50, similarly to [27]. We use the Google News word embeddings pre-trained with *word2vec* (size = 300).

For the Character CNN, we consider a maximum length of 1,024 for a sequence of characters: longer sequences are trimmed and shorter ones are padded with zeros. The window sizes for the convolutions are in the set $\{2, \ldots, 10\}$, with 300 feature maps. The convolutional layer non-linearity and initialization are the same as for the Word CNN. The size of the character embeddings is 300. These embeddings are initialized based on a normal distribution and trained on the event detection task. A dropout of 0.5 is applied after the embedded characters. When jointly trained, in the early fusion model, the features obtained after convolutions from both models are concatenated and, similarly to the Word CNN, a dropout of 0.3 is applied afterward, before the softmax layer. We encode all the characters except space.

We train both networks (Word and Character CNNs) with Adam optimizer. During the training, we optimize the embedding tables (i.e., word, position, and character embeddings) to achieve the optimal states. Finally, for training, we use a batch size of 256 for the Word CNN and 128 for the Character CNN. When they are trained jointly in the early fusion model, we use a batch size of 128. All these hyperparameters were optimized by a grid search on the development set.

5.3   Results

We compare our model with several neural-based models proposed for the same task that do not use external resources, namely: a set of CNN-based models including a CNN model without any additional features [27], the dynamic multi-pooling CNN model of [2], the non-consecutive CNN of [26], and the Graph CNN proposed by [29]; a set of RNN-based models, represented by the bidirectional joint RNN model of [25], the DLRNN model of [8] and the DEEB-RNN model of [38] that both rely on a document representation, and the work of [19], based on a Gated Cross-Lingual Attention mechanism. Our reference models also include the hybrid model proposed by [9], the model exploiting arguments through an attention mechanism of [20], and the GAIL-ELMo model of [36], based on GANs. We do not consider models that are using other external resources such as [1], [18], or [35], since we only rely on the input text in our model. We also compare this model with four baselines based on the BERT language model, applied in a similar way to [3] for the NER task, with the recommended hyperparameters: a learning rate of 2e−5 and the split of sentences into chunks of 128 tokens.

The best performance (75.8 F1 on the test set) is achieved by combining word and position embeddings with the character-level features using a late fusion strategy. This performance relates to improvements that have been reported on other tasks when concatenating word embeddings with the output from a character-level CNN for Part-of-Speech tagging [6] and NER [4]. From Table 3, we can also outline that adding character embeddings in a late fusion strategy outperforms all the word-based models, including complex architectures such as the graph CNN and the models based on the BERT language model. Among

Table 3. Evaluation of our models and comparison with state-of-the-art systems for event detection on the blind test data. [†]beyond sentence level, [+]with gold arguments.

Approaches	Precision	Recall	F1
Word CNN [27] (without entities)	71.9	63.8	67.6
Dynamic multi-pooling CNN [2]	75.6	63.6	69.1
Joint RNN [25]	66.0	73.0	69.3
RNN with document context[†] [8]	77.2	64.9	70.5
Non-consecutive CNN [26]	na	na	71.3
Attention-based[+] [20]	78.0	66.3	71.7
GAIL-ELMo [36]	74.8	69.4	72.0
Gated cross-lingual attention [19]	78.9	66.9	72.4
Graph CNN [29]	77.9	68.8	73.1
Hybrid NN [9]	84.6	64.9	73.4
DEEB-RNN3[†] [38]	72.3	75.8	74.0
BERT-base-uncased + LSTM [33]	na	na	68.9
BERT-base-uncased [33]	na	na	69.7
BERT-base-uncased [7]	67.2	73.2	70.0
BERT-QA [7]	71.1	73.7	72.4
DMBERT [34]	77.6	71.8	74.6
DMBERT+Boot [34]	77.9	72.5	75.1
BERT-base-uncased	71.7	68.5	70.0
BERT-base-cased	71.3	72.0	71.7
BERT-large-uncased	72.1	72.9	72.5
BERT-large-cased	69.3	**77.2**	73.1
Word CNN (replicated)	71.4	65.9	68.5
Character CNN	71.7	41.2	52.3
Word + Character CNN - early fusion	**88.6**	61.9	72.9
Word + Character CNN - late fusion	87.2	67.1	**75.8**

BERT models, it is worth noticing that the cased models perform better than the uncased ones, which confirms that the character morphology is important for the task, maybe because capitalization is connected to the recognition of named entities, which are usually considered important to detect event mentions.

However, we can see that the character embeddings are not sufficient on their own: using only the Character CNN leads to the smallest recall among all the considered approaches. However, its precision is high (71.7), which makes this model fairly reliable about the triggers it retrieves. Given this observation, we can compare the two integration strategies, early and late fusions. In the case of early fusion, where the two models are trained jointly, we notice that the precision is the highest among all the compared models. We assume that in the

Table 4. Examples of new triggers found with the Word+Character CNN (late fusion).

Event type	New triggers correctly found	Trigger words in training data
End-position	*Steps*	*Step*
Extradite	*Extradited*	*Extradition*
Attack	*Wiped*	*Wipe*
Start-org	*Creating*	*Create*
Attack	*Smash*	*Smashed*
End-position	*Retirement*	*Retire*

joint approach, the power of representation of morphological properties provided by the characters is overtaking the influence of the word and positions embedding, and the combination reproduces the imbalance between precision and recall observed for the Character CNN, the recall being the lowest among all the models except the *Character CNN*. In the case of the late fusion, since we have more control over the combination and we can give priority to the Character CNN to establish the labels on the trigger candidates retrieved by the Word CNN, the method takes advantage of the high precision of the Character CNN, allowing an increase of the precision from 71.7 to 87.2, while still having a high recall, also increasing the recall of the *Word CNN* model from 65.9 to 67.1. The late fusion integration is therefore able to take into account the complementarities of the two models.

Finally, for more qualitative analysis, we examine the new triggers correctly detected by the Word + Character CNN (late fusion), in comparison with the Word CNN. We observe that among the 37 new correctly found triggers, some are indeed derivational or inflectional variants of known words in the training data, such as illustrated in Table 4. This seems to confirm that the character-based model can capture some semantic information associated with morphological characteristics of the words and manage to detect new correct event mentions that correspond to inflections of known event triggers (i.e., existing in the training data). Also, the fact that the convolution windows in the Character CNN range from 2 to 10 means that character n-grams in the same range are included in the model and contribute to the model's ability to handle different word variations.

6   Conclusion and Perspectives

We have proposed in this article a study of the integration of character embeddings in an event detection neural-based model using a simple CNN model as core architecture and testing early and late fusion strategies to integrate the character-based features. The best results are achieved by combining the word-based features with the character-based features in a late fusion strategy that gives priority to the Character CNN for deciding the event type. This method outperforms more complex approaches such as Graph CNN or adversarial networks and BERT-based models. Our results demonstrate that a convolutional approach for learning character-level features can be successfully applied to event detection and that these features allow overcoming some issues concerning unseen or misspelled words in the test data.

We do not integrate the character information at the embedding level as it is usually done in models considering smaller units such as ELMO with characters, FastText with character n-grams, or BERT with subwords. In a certain way, they implement another kind of early fusion than ours. However, our late fusion approach is complementary and as a perspective, we consider implementing this late fusion framework using more complex models as core models. Another way to deal with the problem of unseen words would be to exploit data augmentation strategies that would focus on increasing the variability about derivational and inflectional variants of event mentions in the training data.

References

1. Bronstein, O., Dagan, I., Li, Q., Ji, H., Frank, A.: Seed-based event trigger labeling: how far can event descriptions get us? In: ACL-IJCNLP, pp. 372–376 (2015)
2. Chen, Y., Xu, L., Liu, K., Zeng, D., Zhao, J.: Event extraction via dynamic multi-pooling convolutional neural networks. In: ACL-IJCNLP 2015, pp. 167–176 (2015)
3. Devlin, J., Chang, M.W., Lee, K., Toutanova, K.: BERT: pre-training of deep bidirectional transformers for language understanding. In: NAACL-HLT 2019, pp. 4171–4186 (2019)
4. Dos Santos, C., Guimarães, V.: Boosting named entity recognition with neural character embeddings. In: Fifth Named Entity Workshop, pp. 25–33 (2015)
5. Dos Santos, C., Zadrozny, B.: Learning character-level representations for part-of-speech tagging. In: 31st International Conference on Machine Learning (ICML-14), pp. 1818–1826 (2014)
6. Dos Santos, C.N., Gatti, M.: Deep convolutional neural networks for sentiment analysis of short texts. In: COLING, pp. 69–78 (2014)
7. Du, X., Cardie, C.: Event extraction by answering (almost) natural questions. In: 2020 Conference on Empirical Methods in Natural Language Processing (EMNLP), pp. 671–683 (2020)
8. Duan, S., He, R., Zhao, W.: Exploiting document level information to improve event detection via recurrent neural networks. In: Eighth International Joint Conference on Natural Language Processing (IJCNLP 2017), pp. 352–361 (2017)
9. Feng, X., Huang, L., Tang, D., Ji, H., Qin, B., Liu, T.: A language-independent neural network for event detection. In: 54th Annual Meeting of the Association for Computational Linguistics, pp. 66–71 (2016)
10. Hong, Y., Zhou, W., Zhang, J., Zhou, G., Zhu, Q.: Self-regulation: employing a generative adversarial network to improve event detection. In: 56th Annual Meeting of the Association for Computational Linguistics, pp. 515–526 (2018)
11. Ji, H., Grishman, R., et al.: Refining event extraction through cross-document inference. In: ACL, pp. 254–262 (2008)
12. Joulin, A., Grave, E., Bojanowski, P., Mikolov, T.: Bag of tricks for efficient text classification. In: 15th Conference of the European Chapter of the Association for Computational Linguistics (EACL 2017), pp. 427–431 (2017)

13. Jozefowicz, R., Vinyals, O., Schuster, M., Shazeer, N., Wu, Y.: Exploring the limits of language modeling. arXiv preprint arXiv:1602.02410 (2016)
14. Kanaris, I., Kanaris, K., Houvardas, I., Stamatatos, E.: Words versus character n-grams for anti-spam filtering. Int. J. Artif. Intell. Tools **16**(06), 1047–1067 (2007)
15. Kim, Y., Jernite, Y., Sontag, D., Rush, A.M.: Character-aware neural language models. In: Thirtieth AAAI Conference on Artificial Intelligence, pp. 2741–2749 (2016)
16. Lample, G., Ballesteros, M., Subramanian, S., Kawakami, K., Dyer, C.: Neural architectures for named entity recognition. In: NAACL-HLT 2016, pp. 260–270 (2016)
17. Li, Q., Ji, H., Huang, L.: Joint event extraction via structured prediction with global features. In: ACL, pp. 73–82 (2013)
18. Li, W., Cheng, D., He, L., Wang, Y., Jin, X.: Joint event extraction based on hierarchical event schemas from FrameNet. IEEE Access **7**, 25001–25015 (2019)
19. Liu, J., Chen, Y., Liu, K., Zhao, J.: Event detection via gated multilingual attention mechanism. In: Thirty-second AAAI Conference on Artificial Intelligence (AAAI-18) (2018)
20. Liu, S., Chen, Y., Liu, K., Zhao, J.: Exploiting argument information to improve event detection via supervised attention mechanisms. In: 55th Annual Meeting of the Association for Computational Linguistics (ACL 2017), pp. 1789–1798 (2017)
21. Luong, M.T., Manning, C.D.: Achieving open vocabulary neural machine translation with hybrid word-character models. In: 54th Annual Meeting of the Association for Computational Linguistics (ACL 2016), pp. 1054–1063 (2016)
22. Ma, X., Hovy, E.: End-to-end sequence labeling via bi-directional LSTM-CNNs-CRF. In: 54th Annual Meeting of the Association for Computational Linguistics, pp. 1064–1074 (2016)
23. Mikolov, T., Chen, K., Corrado, G., Dean, J.: Efficient estimation of word representations in vector space. In: International Conference on Learning Representations (ICLR 2013), Workshop Track (2013)
24. Muller, B., Sagot, B., Seddah, D.: Enhancing BERT for lexical normalization. In: 5th Workshop on Noisy User-generated Text (W-NUT 2019), pp. 297–306 (2019)
25. Nguyen, T.H., Cho, K., Grishman, R.: Joint event extraction via recurrent neural networks. In: NAACL-HLT, pp. 300–309 (2016)
26. Nguyen, T.H., Fu, L., Cho, K., Grishman, R.: A two-stage approach for extending event detection to new types via neural networks. In: ACL 2016, p. 158 (2016)
27. Nguyen, T.H., Grishman, R.: Event detection and domain adaptation with convolutional neural networks. In: ACL-IJCNLP 2015, pp. 365–371 (2015)
28. Nguyen, T.H., Grishman, R.: Modeling skip-grams for event detection with convolutional neural networks. In: EMNLP (2016)
29. Nguyen, T.H., Grishman, R.: Graph convolutional networks with argument-aware pooling for event detection. In: Thirty-Second AAAI Conference on Artificial Intelligence (AAAI 2018) (2018)
30. Pennington, J., Socher, R., Manning, C.D.: GloVe: global vectors for word representation. In: EMNLP, pp. 1532–1543 (2014)
31. Peters, M., et al.: Deep contextualized word representations. In: NAACL-HLT 2018, pp. 2227–2237 (2018)
32. Sun, L., et al.: Adv-BERT: BERT is not robust on misspellings! Generating nature adversarial samples on BERT. arXiv preprint arXiv:2003.04985 (2020)
33. Wadden, D., Wennberg, U., Luan, Y., Hajishirzi, H.: Entity, relation, and event extraction with contextualized span representations. In: EMNLP-IJCNLP 2019, pp. 5784–5789 (2019)

34. Wang, X., Han, X., Liu, Z., Sun, M., Li, P.: Adversarial training for weakly supervised event detection. In: NAACL-HLT 2019, pp. 998–1008 (2019)
35. Yang, S., Feng, D., Qiao, L., Kan, Z., Li, D.: Exploring pre-trained language models for event extraction and generation. In: 57th Annual Meeting of the Association for Computational Linguistics, pp. 5284–5294 (2019)
36. Zhang, T., Ji, H., Sil, A.: Joint entity and event extraction with generative adversarial imitation learning. Data Intell. 1(2), 99–120 (2019)
37. Zhang, X., Zhao, J., LeCun, Y.: Character-level convolutional networks for text classification. In: Advances in Neural Information Processing Systems, pp. 649–657 (2015)
38. Zhao, Y., Jin, X., Wang, Y., Cheng, X.: Document embedding enhanced event detection with hierarchical and supervised attention. In: 56th Annual Meeting of the Association for Computational Linguistics (ACL 2018), pp. 414–419 (2018)

Classification

Sequence-Based Word Embeddings
for Effective Text Classification

Bruno Guilherme Gomes$^{(\boxtimes)}$, Fabricio Murai, Olga Goussevskaia,
and Ana Paula Couto da Silva

Universidade Federal de Minas Gerais, Belo Horizonte, Brazil
{brunoguilherme,murai,olga,ana.coutosilva}@dcc.ufmg.br

Abstract. In this work we present DiVe (Distance-based Vector Embedding), a new word embedding technique based on the Logistic Markov Embedding (LME). First, we generalize LME to consider different distance metrics and address existing scalability issues using negative sampling, thus making DiVe scalable for large datasets. In order to evaluate the quality of word embeddings produced by DiVe, we used them to train standard machine learning classifiers, with the goal of performing different Natural Language Processing (NLP) tasks. Our experiments demonstrated that DiVe is able to outperform existing (more complex) machine learning approaches, while preserving simplicity and scalability.

Keywords: Word embeedings · Logistic Markov embedding · NLP

1 Introduction

Word embedding techniques compute representations of words as vectors in a continuous space in order to capture some notion of similarity between them. More precisely, words from a corpus are mapped onto a low-dimensional Euclidean space, while preserving certain similarity properties of the input data. Learning good word representations has led to breakthroughs in several Natural Language Processing (NLP) tasks, such as document classification [11], sentiment analysis [7], hate speech detection [16], among others.

Embeddings techniques such as Word2Vec [10, 11] and Glove [14] gained popularity for their performance in NLP tasks and for being easy to train. More recently, the generating effective embeddings using deep neural networks became possible through the larger availability of data and of GPU-based computational resources. Notable examples of these techniques are BERT [3] and ELMo [15].

Embedding techniques tend to represent related words, such as "check" and "bank", as points close to each other in space, as they are trained to reconstruct the context in which a word appeared. This characteristic is known as semantic similarity. Although embeddings trained from large corpora containing up to tens of billions of words are available for download on the Internet, difficulty to find pre-trained embeddings for less used languages and problems associated with

E. Métais et al. (Eds.): NLDB 2021, LNCS 12801, pp. 135–146, 2021.
https://doi.org/10.1007/978-3-030-80599-9_12

polysemy (i.e., multiple meanings) can make it beneficial to train embeddings from data specifically related to the task at hand.

In this work, we present DiVe (Distance-based Vector Embedding), a new word embedding technique based on the Logistic Markov Embeddings, a Markovian model [7, chapter 3] originally designed to represent sequences of songs in a playlist [12]. A drawback of the original model is that it is not possible to shift from a relatively restricted universe of songs to the much larger universe of words due to scalability issues related to the computation of the so-called *partition function*. In essence, a partition function is a normalization constant used to ensure that the sum of the probabilities associated with each event of the sample space is one, given a set of observations. Therefore, each partition function is a sum over all the words in the vocabulary. As a first contribution, we use the negative sampling [10] method to approximate the partition function, making DiVe scalable for large datasets.

Second, we generalize LME to consider other distance metrics. Specifically, instead of using either the negative Euclidean distance or the cosine similarity, we investigate the performance of a convex interpolation between the two metrics. Third, we investigate benefits of using a single vs. a dual point model for DiVe. In language models, a "center" word is said to be surrounded by a context, even when the context appears strictly before or after the center word. In the dual point model, each word has two representations, one for when the word is in the center and another for when the word is part of the context, whereas in the single point model, the representation is the same in both cases.

Then, we compare DiVe to 5 word embedding baselines. All techniques are trained on one of 9 different datasets that together represent 6 different classification tasks (hate speech, user review, text polarity, question type, and subjective vs. objective text). From the embeddings locally generated by each technique, we trained standard machine learning classifiers to predict labels of the sentences that compose each of these textual datasets. Also, using the same datasets, we compare DiVe to two state-of-the-art classification techniques based on deep learning (DL) that make use of pre-trained embeddings. We show that DiVe (i) outperforms the five baselines on several datasets and (ii) yields comparable performance to the two DL methods at a much smaller computational cost.

The rest of this paper is organized as follows. Section 2 discusses existing work on word embeddings. In Sect. 3, we define the models and algorithms behind DiVe. Section 4 presents our experimental results. Section 5 concludes the paper. We provide an appendix in an external repository[1] with further details on the analytical model.

2   Related Work

The task of learning word embeddings has received a significant amount of interest in the last years. We discuss three fronts of research related to our work:

---

[1] https://github.com/DiVeWord/DiVeWordEmbedding.

Shallow Window-Based Methods: This body of works studies vector representations of words. The basis of these techniques lies in the local learning of the representations of words within the same context window. The authors of [13] introduced a model that learns word vector representations using a simple neural network architecture for language modeling. *Word2vec* [10] is a more recent technique, based on a two-layered artificial neural network, trained to reconstruct linguistic contexts of words. Following a similar approach, *FastText* [6] presents an extension of Word2Vec by taking into account *information from subwords* to compose the representation of a word. Bayesian Skip-Gram [1] is another word embedding algorithm, based on a Bayesian neural network.

Statistical Estimation of Word Representation: Statistical models have been widely used to tackle NLP tasks, such as part-of-speech [2] and sense disambiguation [16]. In terms of word representation, primarily, many papers sought to capture the similarity between words by the probability that they occur in a sequence [7]. Later, Bayesian models for the semantic representation of words have also been proposed [5,8]. Recently, following a similar perspective, we can highlight *GloVe* [14], which presents an efficient statistical model for grouping words together with their synonyms and allegories.

Pre-trained Deep Learning: On this research front there are architectural neural networks based on seq2seq, LSTM and encode-decode, which can be used in various tasks, such as machine translation, word embedding, sentiment analysis, and question answering. CoVe [9] is a model based on seq2seq (sequence-to-sequence) machine translation, whose learned representation considers the entire input sentence. ELMo [15] is a neural network based on a bi-directional language model (biLM), in which each word presents a contextualized representation. Word vectors are functions learned from the internal states of biLM, which is pre-trained on a large text corpus. Another technique based on biLM is BERT [3], which was also shown to perform well in the task of determining if one sentence follows another.

Finally, we point out the work of Globerson et al. [5] and LME [12], which are based on an approach similar to ours. One of the general aspects that distinguishes our work is how we estimate the partition function and compute distance in space. In this way, DiVe can be seen as an approximation approach to LME.

3   The DiVe Model

Our goal is to estimate a generative model for continuous word representation from sentences of words. Let $\mathcal{D} = \{s^{(1)}, \ldots, s^{(n)}\}$ be a set of sentences and $\mathcal{V}$ be the vocabulary (set of unique words) that composes sentences $s \in \mathcal{D}$. We define $s = (w_1, w_2, \ldots, w_m)$ as a sentence containing m words, where each word $w \in \mathcal{V}$. Hence, we want to obtain a language model from $\mathcal{D}$ that defines a probability distribution over sentences (i.e., maps a sentence s to a probability mass $\Pr(s)$).

A natural approach for modeling language is to decompose sentences into word-to-word transitions, where each word represents a state of a Markov Chain.

The probability of a sentence, comprised by a sequence of adjacent words, is defined as the product of transition probabilities between consecutive words. Using a first order Markov Chain, we can write the probability of sentence s as

$$\Pr(s) = \prod_{i=1}^{k} \Pr(w_i | w_{i-1}). \tag{1}$$

As usual, the conditional probability $\Pr(w_i | w_{i-1})$ is defined to be proportional to a function of the embeddings of the words that characterize the current state w_{i-1}, or context, and the next state w_i. In most embedding techniques (for example, [10,11]), each word $w \in V$ has two vector representations, depending on whether it is used to encode the current or the next state. We refer to this as the dual point model. In this work, we also investigate a simpler variant of this model, called the single point model, where each word is represented by the same vector regardless of whether it is the current or the next state.

3.1   DiVe Single Point Model

In the single point model, we represent each word $w \in V$ as a vector $\mathcal{X}(w) \in \mathbb{R}^d$ for some dimension d. We denote by $f : \mathbb{R}^d \times \mathbb{R}^d \to \mathbb{R}$ some similarity measure between two word vector representations. To obtain a valid conditional distribution, we define $\Pr(w_i | w_{i-1})$ as the normalized value of some non-linear transformation σ applied to the similarity between w_i and w_{i-1}:

$$\Pr(w_i | w_{i-1}) = \frac{\sigma(f(\mathcal{X}(w_i), \mathcal{X}(w_{i-1})))}{Z(w_{i-1})}, \quad \text{for } w_i \in V, \tag{2}$$

where $Z(w_{i-1}) = \sum_{v \in V} \sigma(f(\mathcal{X}(w_v), \mathcal{X}(w_{i-1})))$ is the partition function.

In this work we investigate three choices of functions for the non-linearity σ: sigmoid, tanh and exp. In addition, instead of measuring similarity by the angle between word vectors as usual [6,10,11], we investigate a more flexible way of measuring similarity based on a linear interpolation between the inner product and the negative square Euclidean distance.

First, note that we can express the dot product of vectors v and u in terms of their Euclidean distance $\|v - u\|^2 = (v - u) \cdot (v - u)$ and their norms:

$$\|v - u\|^2 = v \cdot v + u \cdot u - 2(v \cdot u) \Rightarrow v \cdot u = \frac{1}{2}(\|v\|^2 + \|u\|^2 - \|v - u\|^2). \tag{3}$$

On one hand, if we compute the similarity between two embeddings using the RHS of Eq. (3), we are using the dot product as the similarity measure. On the other hand, if ignore the norms of the embeddings, we recover the negative Euclidean distance (times the constant $1/2$). Rather than choosing between the dot product or the negative Euclidean distance as the similarity measure f, we propose the use of a convex combination of both:

$$f(\mathcal{X}(w_i), \mathcal{X}(w_{i-1})) = -\frac{1}{2}\|\mathcal{X}(w_i) - \mathcal{X}(w_{i-1})\|^2 + \frac{\alpha}{2}\|\mathcal{X}(w_i)\|^2 + \frac{\alpha}{2}\|\mathcal{X}(w_{i-1})\|^2, \tag{4}$$

where $0 \leq \alpha \leq 1$. When α is 0, the similarity measure f is the negative Euclidean distance, and when α is 1, f is (twice) the inner product between word vectors.

We generalize Eq. (1) to consider the case where the context, in this case c_i, of word w_i is formed by the j previous words, i.e., $w_{i-1}, \ldots, w_{i-j}$. This results in a j-th order Markov Chain and thus, the probability of a sentence becomes

$$\Pr(s) = \prod_{i=j}^{k} \Pr(w_i|c_i) = \prod_{i=j}^{k} \frac{\sigma(f(\mathcal{X}(w_i), \mathcal{X}(c_i)))}{Z(c_i)}, \tag{5}$$

where we set $\mathcal{X}(c_i) = \sum_{j \in c_i} \mathcal{X}(j)/|c_i|$ to be the element-wise average of the embeddings of words in c_i. We have also conducted experiments setting $\mathcal{X}(c_i)$ to the element-wise maximum, but we obtained slightly inferior results.

We are now ready to define the cost function to be optimized as the negative likelihood of $\mathcal{D}$ given the embeddings $\mathcal{X}(w)$ for all $w \in \mathcal{V}$:

$$\mathrm{NLL}(\mathcal{D}) = -\sum_{s \in \mathcal{D}} \sum_{i=j}^{k} \log \Pr(w_i|c_i) = -\sum_{s \in \mathcal{D}} \sum_{i=j}^{k} [\log \sigma(f(\mathcal{X}(w_i), \mathcal{X}(c_i))) - \log Z(c_i)] \,.$$

3.2   DiVe Dual Point Model

In the previous section, we described the single point model, that represents each word $w \in \mathcal{V}$ as a d-dimensional vector $\mathcal{X}(w)$. This model has two key limitations. First, natural choices for a similarity function f between two vectors are symmetric and, therefore, even if there are several transitions from w_i to w_j in the corpus $\mathcal{D}$, and no transitions in the opposite direction, the transition probabilities estimated by the model will be the same in both directions. Second, the representation of words can undergo drastic modifications at each stage of learning, making it more difficult to find good representation of words in space.

To overcome these issues, we also consider a dual point model, where each word w_i is represented as a vector pair $(\mathcal{I}(w_i), \mathcal{O}(w_i))$. We call $\mathcal{I}(w_i)$ the "entry vector" of word w_i, and $\mathcal{O}(w_i)$ the "exit vector". The cost function becomes

$$\mathrm{NLL}(\mathcal{D}) = -\sum_{s \in D} \sum_{i=j}^{k} \log \Pr(w_i|c_i) = -\sum_{s \in D} \sum_{i=j}^{k} [\log \sigma(f(\mathcal{O}(w_i), \mathcal{I}(c_i))) - \log Z(c_i)] \,.$$

3.3   Estimating the Partition Function

One of the limitations of LME is the cost of computing the partition function $Z(c)$ exactly for a context c [4]. Since during parameter optimization this computation must be performed several times for each training iteration and at least once for each different context, the resulting complexity is $O(|\mathcal{D}||\mathcal{V}|)$. To address this issue, we resort to negative sampling [11] to approximate the partition function and estimate model parameters more efficiently. The resulting complexity is $O(|\mathcal{D}|k)$, where k is a constant equal to the number of negative samples drawn for each word in $\mathcal{D}$. To compensate for highly unbalanced word frequencies, we adopt the heuristic of sampling a word w in proportion to $\pi_w^{3/4}$ where π_w is the word frequency of w in the corpus.

Using the negative sampling method, the term corresponding to the log of the partition function in Eqs. (6) and (6) is replaced by a sum over the negative instances $\mathcal{V}'$ that were sampled according to the heuristic described above. In the single point model, the new cost function is given by

$$\text{NLL}_{ns}(\mathcal{D}) = -\sum_{s \in D} \sum_{i=j}^{k} \left[\log \sigma(f(\mathcal{X}(w_i), \mathcal{X}(c_i))) - \sum_{v \in \mathcal{V}'} \log \sigma(-f(\mathcal{X}(w_v), \mathcal{X}(c_i))) \right].$$

The corresponding equation for the dual point model is analogous.

Finally, we found that the stochastic gradient algorithm finds a good solution for approximating cost functions for the single and the dual point models. In order to enable the replication of the results in this paper, all the code used in this work, including the baselines is available in a public repository[2].

4   Experiments and Results

4.1   Experimental Setup

We now conduct an experimental study of DiVe, comparing its performance to state-of-the-art word embedding techniques on text classification tasks.

We use the performance of models trained for text classification as a proxy to evaluate the quality of the embeddings obtained by DiVe and by the word embedding baselines: GloVe, Word2vec, fastText, Bayesian Skip Gram and deep learning baselines: ELMo and BERT. We use 9 publicly available datasets:

- **Customer reviews (CR)**: A dataset for binary sentiment classification based on user reviews of 5 products.
- **Hate Speech Twitter Annotations (HSTW)**: A collection of tweets labeled according to 3 categories: sexism, racism, neutral.
- **Polarity of Opinion (PO)**: This data was extracted from Rotten Tomatoes webpages, with reviews marked as "fresh" (positive) and "rotten" (negative).
- **Question Type Classification (QTS)**: This dataset contains questions asked by users, labeled in 6 different categories.
- **Subjectivity and objectivity of sentences (SUBJ)**: A set of sentences containing at least 10 words and labeled as either "subjective" or "objective".
- **IMDB reviews (IM) and (SIM)**: Datasets with large (IM) and small (SIM) number of movie and TV show reviews.
- **Yelp reviews (YR)**: Dataset with sentences from user reviews, about restaurants and bars, labeled with positive or negative sentiment.
- **Amazon reviews (AR)**: A set of sentences labeled with positive or negative sentiment, extracted from Amazon product review.

Table 1 lists the vocabulary size of each dataset, many of them used as benchmarks in prior works [8,16]. We refer to the datasets with $\leq 40K$ words as "small", and as "large" otherwise.

---

[2] https://github.com/DiVeWord/DiVeWordEmbedding.

Table 1. Datasets used in classification tasks ($|V|$ is the vocabulary size).

Acron	Description	$|V|$	# words	Acron	Description	$|V|$	# words
AR	User product review	1 741	5 275	QTS	Question Answering	16 504	30 134
CR	User review polarity	5 176	33 665	SIM	Movie and TV Review	2 933	7 471
HSTW	Hate speech detect	23 739	155 804	SUBJ	Subjectivity and objectivity	20 745	121 366
IM	Movie and TV Review	74 337	3 124 867	YR	Food review polarity	1 919	5 563
PO	Sentence polarity	18 179	114 485				

For the word embedding classification task, we consider 8 "shallow" classifiers implemented on scikit-learn[3]: probabilistic models Logistic Regression (LR), Quadratic Discriminant Analysis (QDA), Linear Discriminant Analysis (LDA) and Naive Bayes (NB); structural models Support Vector Classification (Linear-SVM) and K-Nearest Neighbors (KNN); the ensemble model Random Forest (RF); and a Neural Network (NN). Moreover, we used two deep learning techniques as baselines: BERT and ELMo (see our repository for setup details).

The experiment setup is as follows. For each combination of dataset, word embedding technique and classifier, we use 5-fold cross-validation by learning the word embeddings on 4 folds in an unsupervised fashion, then training a classifier using these embeddings and the labels associated with each sentence, and finally testing on the left-out fold. We then take the performance to be the average weighted F_1 score over the 5 folds. For the deep learning baselines, we performed the 5-fold cross-validation and average weighted F_1 score for evaluation.

Ideally, we would also use cross-validation to jointly optimize the hyperparameters. However, due to the computational demands of running experiments with several large datasets, number of the combinations of embedding techniques and classifiers and the cost of tuning the deep networks, we used fixed values for the hyperparameters. For a fair comparison, we fixed the number of dimensions and context size respectively to 400 and 5 to train DiVe, Word2Vec, Glove, Bayesian Skip Gram and FastText. For BERT and ELMo we did not change the default network settings to represent text, with 1024 dimensions.

It is clear that the quality of the learning representations plays a major role in the classification performance. Since our interest here is to evaluate the embeddings produced by each technique, we argue that not tuning the hyperparameters of the classifiers is not a major problem. In fact, this allows us to better evaluate the robustness of the resulting embeddings.

4.2 Comparison of DiVe's Variants

We compare Dive's single and dual point models while keeping the dimension of the embeddings fixed. Note that, in the **dual point** model, each word is represented by twice as many numbers as in the **single point** model. Therefore, we expect the former model to yield better performance in more complex tasks,

---

[3] http://scikit-learn.org/stable/index.html.

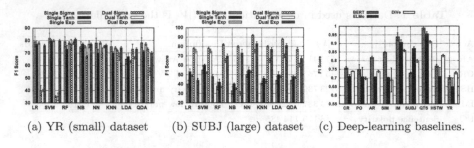

(a) YR (small) dataset   (b) SUBJ (large) dataset   (c) Deep-learning baselines.

Fig. 1. Comparing Dive variants using F1 accuracy.

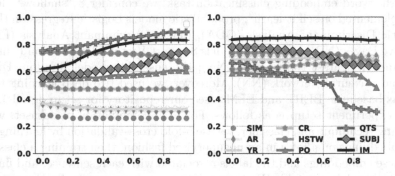

Fig. 2. Impact of DiVe's hyperparameter α (left: dual-point; right: single-point).

but also to require more training data. We also consider the impact of the choice of the activation function – **sigmoid()**, **tanh()** and **exp()**.

We compared the performance of the six models on each dataset. The results were very consistent among "small" datasets and among "large" ones. Hence, we present the results for two representative cases, YR (small) and SUBJ (large).

Figures 1a and 1b show the results for the YR and the SUBJ datasets, respectively. The groups in the x-axis indicate the classifier. And, within each group, a bar corresponds to one of DiVe's variants. The height of each bar is the average F_1 score and the whiskers represent 95%-confidence intervals. In general, we observe that the single point model significantly outperforms the dual counterpart on the small dataset, and among the single point variants, the sigmoid function yields the best results. Conversely, the dual point model significantly outperforms the single counterpart on the large dataset, but among the dual point variants, the sigmoid function is still the best choice. For this reason, in the next experiments we fix the activation function to be the sigmoid(.).

4.3   Analysis of Parameter α in Similarity Function f

The similarity function f, defined in (4), is an interpolation between the negative Euclidean distance ($\alpha = 0$) and the inner product ($\alpha = 1$). In this section, we use the experimental setup described in Sect. 4.1 to investigate the impact of α

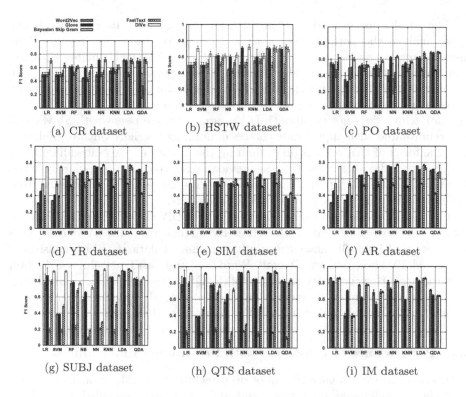

Fig. 3. Embeddings' performance on text classification.

on the tasks' performance. More precisely, we vary α from 0 to 1 in increments of 0.05 and compute the resulting F_1 score. As indicated before, we fix the activation function f to be the sigmoid(). Figures 2(left) and 2(right) compare results of DiVe Single and Dual models. In both cases we observe a large variation in terms of F_1 depending on α. For example, for the QTS dataset, the F_1 score has almost 30% variation for the Single Point model, and 10.5% variation for the Dual Point model, and for PO dataset 12% for Single Point and almost 10% for Dual Point. This shows that α can significantly influence an estimator's accuracy, for example, in some datasets the best embedding are obtained when $\alpha = 1$ can also lead to very poor results (see IM single point model). On the other hand, $\alpha = 0$ is not ideal either (see subj with dual point model). Then, we believe of setting $\alpha = 0.5$ yields a good trade-off between performance and simplicity, and it avoids additional hyper-parameters.

4.4 Performance of Classifiers with Trained Embeddings

Now, we compare the quality of the embeddings obtained with DiVe to Word2Vec, Glove, Bayesian SkipGram and FastText techniques. The embeddings were trained on the specific dataset whose sentences we want to classify.

The results for each dataset are shown in Figs. 3a–i. In Figs. 3a, b and c, we analyze the performance of text classification from user reviews, hate speech detection and sentence polarity, respectively. DiVe yields higher F_1 scores than the baselines for nearly all classifiers. DiVe's performance is also less variable across classifiers than the other embedding techniques. In particular, other embeddings often result in a poor performance when combined with SVM (e.g., Fig. 3c), which does not occur with DiVe. Some of these issues with SVM could be circumvented with appropriate choices of nonlinear kernels, but we emphasize that the focus of this work is on evaluating the quality of the embeddings.

The small datasets consist of user reviews extracted from popular websites. The results obtained for them are shown in Figs. 3d, e and f. We observe that DiVe presented higher F_1 with almost all classifiers.

In Figs. 3g and h, we observe once again that DiVe's performance varies less across classifiers than that of the other techniques and that SVM can yields poor results. Figure 3i shows the results for the IM dataset, which consists of movie and TV show reviews. Overall, Word2Vec and FastText achieved the best performances. However, with exception of the QDA classifier, DiVe's embeddings resulted in very similar F_1 scores. On this dataset, all embedding techniques, except for GloVe, suffered with the SVM issues described above.

4.5   Performance of Classifiers with Pre-trained Embeddings

In this section, we evaluate results of deep techniques ELMo[4] and BERT[5]. We used these baselines as pre-trained embeddings, as recommended in the literature [3,15]. They were trained on a large dataset and used for classification tasks. Yet, prediction using either technique is very computationally expensive. In some cases, several hours of GPU/TPU processing were needed.

We compare the performance of the deep learning techniques with a simple Logistic Regression classifier trained from DiVe's embeddings when $\alpha = 0.5$. Figure 1c summarizes the results obtained using both techniques on all 9 datasets. DiVe outperforms ELMo in 4 classification tasks (CR, AR, HSTW and YR) and BERT in 3 classification tasks (SUBJ, HSTW and YR).

We emphasize that both BERT and ELMo have approximately 100 million parameters, thus requiring much longer training times than DiVe. For each technique, the average time of 5 training sessions carried out in each dataset, on a computer with an Intel Xeon CPU@2.40 GHz, 128G of RAM.

In order to put both time requirements and performance into perspective, in Fig. 4, we present scatterplots of these dimensions for each dataset. We state that one method "dominates" the other on a dataset when it appears above (better performance) and to the left (smaller training time) of the latter. We observe that while DiVe often dominates other shallow methods, no other method – either shallow or deep – dominates DiVe on any of the datasets. Furthermore, the F1 score achieved by DiVe is almost always close to that achieved by BERT

---

[4] https://allennlp.org/elmo.
[5] https://github.com/google-research/bert.

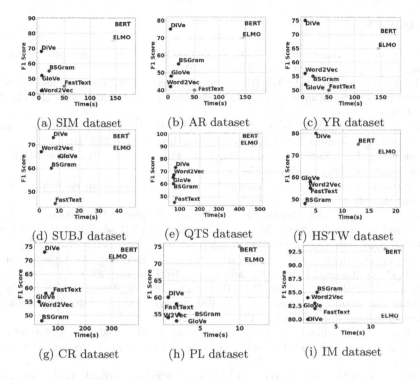

Fig. 4. Scatterplots of training time vs. F1 for all techniques on each dataset.

and ElMo (except in the PL dataset) and, in some cases, even superior to that (see YR and HTSW datasets).

Finally, we conclude that, even though DiVe is a relatively simpler technique and easier to train than the state-of-the-art deep learning solutions, it was able to outperform these more complex techniques.

5   Conclusion

In this work we presented DiVe, a novel word embedding technique based on a variation of the Markovian statistical model. In order to address the scalability problems that arise due to the cost of computing the partition function, we proposed a sampling approach to approximate the latter. Moreover we evaluated a new way of measuring similarity between word vectors, based on a linear interpolation between the inner product and the square Euclidean distance function. Through extensive experiments we demonstrated the efficiency of DiVe on 9 datasets that represent 6 different text classification tasks: hate speech, user review, text polarity, question type, and subjective and objective text. Finally, using the obtained embeddings, we trained shallow and deep machine learning classifiers to predict labels of the sentences that compose each of these textual datasets. DiVe outperformed existing approaches in several tasks.

References

1. Brazinskas, A., Havrylov, S., Titov, I.: Embedding words as distributions with a bayesian skip-gram model. In: COLING (2018)
2. Cheng, J., Druzdzel, M.J.: AIS-BN: an adaptive importance sampling algorithm for evidential reasoning in large bayesian networks. J. Artif. Intell. Res. **13**, 155–188 (2000)
3. Devlin, J., Chang, M., Lee, K., Toutanova, K.: BERT: pre-training of deep bidirectional transformers for language understanding. In: NAACL-HLT (2019)
4. Figueiredo, F., Ribeiro, B., Almeida, J.M., Faloutsos, C.: Tribeflow: Mining & predicting user trajectories (2015)
5. Globerson, A., Chechik, G., Pereira, F., Tishby, N.: Euclidean embedding of co-occurrence data. J. Mach. Learn. Res. **8** (2007)
6. Joulin, A., Grave, E., Bojanowski, P., Douze, M., Jégou, H., Mikolov, T.: Fasttext.zip: compressing text classification models. CoRR abs/1612.03651 (2016). http://arxiv.org/abs/1612.03651
7. Jurafsky, D., Martin, J.H.: Speech and Language Processing: An Introduction to NLP, Computational Linguistics, and Speech Recognition (2009)
8. Maas, A.L., Daly, R.E., Pham, P.T., Huang, D., Ng, A.Y., Potts, C.: Learning word vectors for sentiment analysis. In: The 49th Annual Meeting of the Association for Computational Linguistics (2011)
9. McCann, B., Bradbury, J., Xiong, C., Socher, R.: Learned in translation: contextualized word vectors. In: Advances in Neural Information Processing Systems (2017)
10. Mikolov, T., Chen, K., Corrado, G., Dean, J.: Efficient estimation of word representations in vector space. CoRR abs/1301.3781 (2013)
11. Mikolov, T., Sutskever, I., Chen, K., Corrado, G.S., Dean, J.: Distributed representations of words and phrases and their compositionality. In: Advances in Neural Information Processing Systems (2013)
12. Moore, J.L., Joachims, T., Turnbull, D.: Taste space versus the world: an embedding analysis of listening habits and geography. In: ISMIR (2014)
13. Okita, T.: Neural probabilistic language model for system combination. In: COLING (2012)
14. Pennington, J., Socher, R., Manning, C.D.: Glove: global vectors for word representation. In: EMNLP (2014)
15. Peters, M.E., et al.: Deep contextualized word representations. In: NAACL-HLT (2018)
16. Xia, Y., Cambria, E., Hussain, A., Zhao, H.: Word polarity disambiguation using bayesian model and opinion-level features. Cognit. Comput. **7**(3), 369–380 (2014). https://doi.org/10.1007/s12559-014-9298-4

BERT-Capsule Model for Cyberbullying Detection in Code-Mixed Indian Languages

Krishanu Maity[(✉)] and Sriparna Saha

Department of Computer Science and Engineering,
Indian Institute of Technology, Patna, Patna, India
{krishanu_2021cs19,sriparna}@iitp.ac.in

Abstract. In this work, we have created a benchmark corpus for cyber-bullying detection against children and women in Hindi-English code-mixed language. Both these languages are the medium of communication for a large majority of India, and mixing of languages is widespread in day-to-day communication. We have developed a model based on BERT, CNN along with GRU and capsule networks. Different conventional machine learning models (SVM, LR, NB, RF) and deep neural network based models (CNN, LSTM) are also evaluated on the developed dataset as baselines. Our model (BERT+CNN+GRU+Capsule) outperforms the baselines with overall accuracy, precision, recall and F1-measure values of 79.28%, 78.67%, 81.99% and 80.30%, respectively.

Keywords: Cyberbullying · Code-Mixed (Hindi+English) · MuRIL BERT · Capsule networks

1 Introduction

Cyberbullying is defined through malicious tweets, texts or other social media posts via various digital technologies as the serious, intentional and repeated actions of a person's cruelty towards others [13]. Cyberbullying outcomes can differ from transient fear to suicide. So, automatically detecting cyberbullying at its initial stage is a crucial step to prevent its outcomes. State of the art research primarily concentrates on cyberbullying detection for the English language. Indigenous languages have not been given much attention due to the lack of proper datasets. Code-mixing (CM) is the process of fluid alternation between two or more languages in a conversation [9]. It is a natural process of embedding linguistic units such as sentences, words or morphs of one language into the speech of another [8].

Data released by the National Crime Records Bureau showed that the cases of cyberbullying against women or children have increased by 36% from 2017 to 2018 in India[1]. In India the majority of text conversations in social media

---

[1] https://ncrb.gov.in/en/crime-india-2018-0.

© Springer Nature Switzerland AG 2021
E. Métais et al. (Eds.): NLDB 2021, LNCS 12801, pp. 147–155, 2021.
https://doi.org/10.1007/978-3-030-80599-9_13

platform are in the form of Hindi, English and Hinglish. Hinglish is nothing but the representation of Hindi words in Roman script. We have created a Hindi-English code-mixed annotated (Bully/Non-bully) dataset for cyberbullying detection specially related to children and women.

We have developed a model based on BERT [5], CNN, GRU and Capsule network. During our study, we have used MuRIL BERT[2] (Multilingual Representations for Indian Languages), pre-trained on 17 Indian languages and their transliterated counterparts. In recent years, capsule network [11] has gained much attention not only in the computer vision domain but also in NLP domain due to its ability to learn hierarchical relationships between consecutive layers by using an iterative dynamic routing strategy. The main contributions of this work are as follows:

1. We create a new Hindi-English code-mixed annotated (Bully/Non-bully) dataset for cyberbullying detection specially related to children and women.
2. We have developed a model based on BERT, CNN, GRU and capsule network for detecting cyberbully from code-mixed tweets.
3. We have considered traditional machine learning models (Support Vector Machine (SVM), Logistic Regression (LR), Naive Bayes (NB), Random Forest (RF)) and deep neural network based models (CNN, LSTM) as baselines and our model outperforms all the baselines with a significant margin.

2   Related Works

With the advancement of NLP, a large number of research has been conducted on cyberbullying detection on English language as compared to other languages. Dinakar et al. [6] introduced a machine learning based cyberbullying detection model trained on YouTube comments corpus (4500 instances) based on sexuality, racism and intelligence contents. Reynolds et al. [10] used the data obtained from the Formspring.me website, a formatted question-and-answer website for cyberbullying detection. In 2017, Badjatiya et al. [1] experimented with a dataset of 16K annotated tweets with three labels *sexist, racist, and Nan*. In 2020, Balakrishnan et al. [2] proposed a model for cyberbullying detection based on Twitter users' psychological features and machine learning techniques. Bohra et al. [3] created a Hindi-English code-mixed dataset consisting of 4575 tweets annotated with hate speech and normal speech. Gupta et al. [7] proposed a deep gated recurrent unit (GRU) architecture for entity extraction in code-mixed Indian languages.

From literature review, we have observed that there is no existing corpus for detecting cyberbullying against children and women in Hindi-English code-mixed language.

---

[2] https://tfhub.dev/google/MuRIL/1.

3 Code-Mixed Cyberbully-Annotated Corpora Development

3.1 Data Collection

With the help of Twitter Search API[3], we have collected tweets from Twitter. We have scraped approximately 90K raw tweets between July 2020 to November 2020 based on specific hashtags and keywords related to women's attacks like MeToo, r*ndi, JusticeForSushantSinghRajput, nepotism, IndiaAgainstAbuse, AliaBhatt, bitch etc.

3.2 Data Annotation

After preprocessing of raw tweets, we perform manual annotation of the dataset. Two human annotators having linguistic background and proficiency in both Hindi and English, carried out the data annotation task. For annotation, we follow the guidelines used in Hee et al. [14]. Some examples of the annotated tweets are shown in Table 1. To check the quality of annotation carried out by two annotators, we have calculated the inter-annotator agreement (IAA) using Cohen's Kappa coefficient. Kappa score is 0.85, which proves that data is of acceptable quality. After data preprocessing, we have kept 5062 number of tweets in our corpus. Out of 5062 tweets in our corpus, 2456 were labeled as nonbully and the remaining 2606 tweets were labeled as bully.

Table 1. Samples from annotated dataset

Tweets	Class
T1: Kuch bengali se baat kiya kaar phir Main bhe guwahati gaya tha ak baar beautiful place ha **Translation**: I went to Guwahati after discussing with few Bengali people, it's a beautiful place	Non-bully
T2: Aurat mard brbr hai yh modern concept nikl do khud k dmg sy **Translation**: Woman men are all equal, let this modern concept leave from mind itself	Bully
T3: han g bhai address likh lo, jider tumari maa aur behn soyee huee hai the **Translation**: Yes brother please write the address, wherever your mother and sister were sleeping	Bully
T4: tum itne simple ho isliye sob tumko chuthiya banate he **Translation**: You are so simple, that's why everyone makes you fool	Non-bully

---

[3] https://developer.twitter.com/en/docs/twitter-api/v1/tweets/search/api-reference/get-search-tweets.

4   Methodology for Cyberbullying Detection

Our model (BERT+CNN+GRU+Capsule), drawn in Fig. 1, is a variant of the BERT-Caps [12]. We have also examined some baseline models based on the traditional machine learning algorithms (SVM, LR, NB, RF) and compared them with our model.

4.1   BERT

Bidirectional Encoder Representations from Transformers (BERT) [5] is a Transformer-based [15] language model developed by the Google AI research team. Let the input sentence $X = \{x_1, x_2, ...x_n\}$ be the sequence of n input tokens where n represents the maximum sentence length. We feed the input sentence X to BERT model. It returns two types of outputs, i.e., the pooled output of shape $[batch\ size, 768]$, which represents the entire input sequences and a sequence output of shape $[batch\ size, max\ seq\ length, 768]$ with representations from each input token. Let $W_B \in \mathbb{R}^{n \times D}$ be the embedding matrix obtained from the BERT model for input X where $D = 768$ is the embedding dimension of each token.

4.2   N-Gram Convolutional Layer

The output from the BERT model $W_B^{n \times D}$ is then passed through convolution layers to extract the N-gram feature map. Let $F_a \in \mathbb{R}^{K_1 \times D}$ be the learnable filter where K_1 is the N-gram size. Filter F_a performs an element-wise dot product over each possible word-window, $w_{i:i+k_1-1}$ to get feature map, $\mathbf{c^a} \in \mathbb{R}^{n-K_1+1}$. A feature map c_i^a is generated after convolution by $c_i^a = f(w_{i:i+k_1-1} * F_a + b)$, where f is a non linear activation function with bias b. After applying t number of different filters of the same N-gram size, one can generate t feature maps, which can be rearranged as $\mathbf{C} = [\mathbf{c_1}, \mathbf{c_2}, \mathbf{c_3}, ...\mathbf{c_t}] \in \mathbb{R}^{n-K_1+1 \times t}$.

4.3   Bi-Directional GRU Layer

To learn semantic dependency-based features, we passed t-channel feature vector $\mathbf{C}$ through a bi-directional Gated Recurrent Units (GRUs) [4]. Bi-directional GRU sequentially encodes these feature maps into hidden states to capture long-term dependencies in the tweet as, $\overrightarrow{h}_t = \overrightarrow{GRU}(c_t, h_{t-1})$, $\overleftarrow{h}_t = \overleftarrow{GRU}(c_t, h_{t+1})$, where each convoluted feature map c_t is mapped to a forward hidden state $\overrightarrow{h}_t$ and backward hidden state $\overleftarrow{h}_t$ by invoking $\overrightarrow{GRU}$ and $\overleftarrow{GRU}$, respectively. Finally $\overrightarrow{h}_t$ and $\overleftarrow{h}_t$ are concatenated to get a single hidden state representation h_t, $\left[h_t = \overrightarrow{h}_t, \overleftarrow{h}_t\right]$. The final hidden state matrix is obtained as,

$\mathbf{H} = [\mathbf{h_1}, \mathbf{h_2}, \mathbf{h_3}, ...\mathbf{h_t}] \in \mathbb{R}^{t \times 2d}$, where d is the dimension of hidden state.

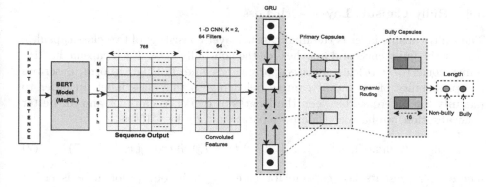

Fig. 1. BERT+CNN+GRU+Capsule architecture.

4.4 Primary Capsule Layer

Primary capsules hold a group of neurons to represent each element in the feature maps as opposed to a scalar, in order to preserve the instantiated parameters such as the local order of words and semantic representations of words. Let $p_i \in \mathbb{R}^d$ denote the instantiated parameters of a capsule, where d is the dimension of the capsule. By sliding each kernel K_i, over the GRU generated hidden state matrix H, we have a sequence of capsules, p_i. A channel P_i in the primary capsule layer is the list of capsules p_i, described as $P_i = g(H * K_i + b)$ where g is a squashing function with bias b. For all R such channels, the generated capsule feature map can be compiled as $P = [P_1, P_2, P_3, \ldots\ldots P_R]$.

4.5 Dynamic Routing Between Capsules

The fundamental idea of dynamic routing is to build a non-linear map in an iterative way, assuring that the lower label capsule has a strong connection to an appropriate capsule in the next layer. This algorithm increases or decreases the connection strength for each potential higher label capsule and by this way, it not only detects whether a feature is present in any position of the text or not, but also keeps the spatial information about the feature. Let u_i be a capsule in layer l. A capsule v_j in layer $l + 1$ is calculated as:

$$v_j = g(\sum_i S_{ij}\hat{u}_{j|i}) \ \ and \ \ \hat{u}_{j|i} = W_{ij}u_i \tag{1}$$

where a predicted vector $\hat{u}_{j|i} \in \mathbb{R}^d$ is calculated from the capsule u_i, W_{ij} is a weight matrix, g is a non-linear squashing function which restricted the length of the capsule in the range of $[0, 1]$ and S_{ij} is a coupling coefficient iteratively updated by the dynamic routing algorithm [11].

4.6 Bully Capsule Layer with Loss

The bully capsule layer is the final capsule layer consisting of two class capsules, one for the bully class and another for the non-bully class. Each capsule has 16-dimensional (d = 16) instantiated parameters, and its length (norm) describes the probability of the input sample belonging to this class label. In order to magnify the difference between the lengths of two class capsules and for better generalization, we have considered separate margin loss [16] as,

$$L_e = T_e \, max(0, m^+ - \parallel v_e \parallel)^2 \, + \lambda \, (1 - T_e) \, max(0, \parallel v_e \parallel - m^-)^2 \qquad (2)$$

where v_e represents the capsule for category e. In our problem, e is either bully or non-bully. Top and bottom margins are represented by $m^+ = 0.9$ and $m^- = 0.1$, respectively. λ is used for down-weighting of the classes which are not present.

5 Experimental Results and Analysis

Out of 5562 instances in our proposed dataset, we have randomly selected 75% of data for training, 15% for validation, and the remaining 15% for testing. We have used Scikit-Learn 0.22.2 to implement machine learning algorithms. Keras 2.3.1 with TensorFlow as a backend is used to implement deep learning-based models. We have conducted all the experiments ten times and reported the average results.

5.1 Comparison with the Baselines

We have introduced the following baselines for comparison with our model.

1. **BERT Embedding+SVM (Baseline-1):** The pooled output of MuRIL BERT with dimension 768 is fed to SVM classifier for predictions. Hyperparameters of SVM: regularization parameter C = 0.8; kernel = linear; class weight = balanced; tolerance = 1e−3.
2. **BERT Embedding+LR (Baseline-2):** The pooled output of MuRIL BERT with dimension 768 is given to LR model as an input. Hyperparameters of LR: penalty = l1; class weight = balanced; solver = liblinear.
3. **BERT Embedding+NB (Baseline-3):** The pooled output of MuRIL BERT with dimension 768 is fed to NB classifier for predictions.
4. **BERT Embedding+RF(Baseline-4):** The pooled output of MuRIL BERT with dimension 768 is given to LR model as an input. Hyperparameters of LR: criterion = "gini", max features = "auto".
5. **BERT+LSTM (Baseline-5):** A sequence of words with 768 embedding vectors generated from BERT model is sent to the LSTM layer with 64 hidden states. Outputs of the LSTM layer are then passed through a softmax layer for prediction. Hyperparameters used are: batch size = 32; optimizer = Adam; loss = categorical cross-entropy; dropout probability = 0.5

6. **BERT+CNN (Baseline-6):** The sequence output from the BERT model is passed through 1-D convolution layers. We have considered 64 filters with filter sizes 1 and 2. After performing the average pooling operation, we have concatenated the feature maps and passed them through fully connected layers with 60 neurons followed by a soft-max layer.

7. **BERT+CNN+Capsule (Baseline-7):** In this baseline, BERT's output is passed through a 1D CNN layer with filter sizes 1, 2 and the number of filters for each size = 64.

8. **BERT+LSTM+Capsule (Baseline-8):** Sequence output of BERT model is sent to the Bidirectional LSTM layer with 64 hidden states. Hidden state matrix generated from LSTM is then passed through the capsule network for prediction.

9. **BERT+GRU+Capsule (Baseline-9):** This is identical to Baseline-8, the only exception is hare LSTM is replaced by GRU.

Table 2. Evaluation results of cyberbully detection attained by the baselines and the proposed approach

Model	Accuracy	Precision	Recall	F1 Score
BERT Embedding + SVM (Baseline-1)	73.93	71.79	78.61	75.04
BERT Embedding + LR (Baseline-2)	72.26	70.74	75.6	73.11
BERT Embedding + NB (Baseline-3)	69.29	68.02	72.47	70.18
BERT Embedding + RF (Baseline-4)	71.86	71.23	73.06	72.14
BERT + LSTM (Baseline-5)	76.18	74.20	79.89	76.94
BERT + CNN (Baseline-6)	77.28	77.87	77.12	77.45
BERT + CNN + Capsule (Baseline-7)	77.70	75.75	77.43	76.58
BERT + LSTM + Capsule (Baseline-8)	78.18	78.24	80.75	78.48
BERT + GRU + Capsule (Baseline-9)	78.33	76.19	78.22	77.19
BERT+CNN+GRU+Capsule	**79.28**	**78.67**	**81.99**	**80.30**

Table 2 presents the results attained by all the baselines and the proposed model in terms of accuracy, precision, recall, and F1-score. Methods from both machine learning (baseline - 1, 2, 3, 4) and deep learning (baseline - 5, 6, 7, 8, 9) have been taken into account in our baselines. It can be concluded from the table that our proposed model produced better results than all other baselines by a significant margin. Compared to the best baseline, i.e., baseline-9, our model showed almost 1% improvement in accuracy. We can conclude that BERT Embedding+SVM (Baseline-1) achieves higher accuracy (73.93%) than other machine learning-based baselines. We have also examined that baseline-7 and baseline-8 outperform baseline-6 and baseline-5 with accuracy values of 0.42% and 2%, respectively. This improvement in accuracy suggests that the inclusion of a capsule network greatly enhances the performance. If we look closely at baseline-7 and 8, we can see that the only discrepancy between these two baselines is separate recurrent network usages, i.e., LSTM vs. GRU. From the result table, we can analyze that baseline-8 marginally outperforms baseline-7. All the reported results are statistically significant as we have performed statistical t-test at 5% significance level.

6   Conclusion and Future Work

In this paper, we have developed a benchmark corpus for cyberbullying identification against children and women in code-mixed Indian languages. From Twitter, we have crawled Hindi-English code-mixed tweets and, after pre-processing, we have manually annotated 5062 number of tweets. Hindi and English are selected because these languages are the most preferred mode of communication in India. We have developed a model based on four deep learning models: BERT, CNN, GRU, and Capsule networks. We have examined that the inclusion of capsule networks with other deep learning models (CNN, LSTM or GRU) significantly enhances the classifier's performance. Experimental results showed that our model BERT+CNN+GRU+Capsule produced better results than all other baselines by a significant margin. In future, we would like to develop a multitasking framework for cyberbullying detection, where sentiment and emotion detections can act as auxiliary tasks.

References

1. Badjatiya, P., Gupta, S., Gupta, M., Varma, V.: Deep learning for hate speech detection in tweets. In: Proceedings of the 26th International Conference on World Wide Web Companion, pp. 759–760 (2017)
2. Balakrishnan, V., Khan, S., Arabnia, H.R.: Improving cyberbullying detection using twitter users' psychological features and machine learning. Comput. Secur. **90**, 101710 (2020)
3. Bohra, A., Vijay, D., Singh, V., Akhtar, S.S., Shrivastava, M.: A dataset of Hindi-English code-mixed social media text for hate speech detection. In: Proceedings of the Second Workshop on Computational Modeling of People's Opinions, Personality, and Emotions in Social Media, pp. 36–41 (2018)
4. Cho, K., Van Merriënboer, B., Bahdanau, D., Bengio, Y.: On the properties of neural machine translation: Encoder-decoder approaches. arXiv preprint arXiv:1409.1259 (2014)
5. Devlin, J., Chang, M.W., Lee, K., Toutanova, K.: Bert: pre-training of deep bidirectional transformers for language understanding. arXiv preprint arXiv:1810.04805 (2018)
6. Dinakar, K., Reichart, R., Lieberman, H.: Modeling the detection of textual cyberbullying. In: Proceedings of the International Conference on Weblog and Social Media 2011. Citeseer (2011)
7. Gupta, D., Ekbal, A., Bhattacharyya, P.: A deep neural network based approach for entity extraction in code-mixed Indian social media text. In: Proceedings of the Eleventh International Conference on Language Resources and Evaluation (LREC 2018) (2018)
8. Muysken, P., Muysken, P.C., et al.: Bilingual Speech: A Typology of Code-mixing. Cambridge University Press (2000)

9. Myers-Scotton, C.: Duelling Languages: Grammatical Structure in Codeswitching. Oxford University Press (1997)
10. Reynolds, K., Kontostathis, A., Edwards, L.: Using machine learning to detect cyberbullying. In: 2011 10th International Conference on Machine learning and applications and workshops, vol. 2, pp. 241–244. IEEE (2011)
11. Sabour, S., Frosst, N., Hinton, G.E.: Dynamic routing between capsules. arXiv preprint arXiv:1710.09829 (2017)
12. Saha, T., Jayashree, S.R., Saha, S., Bhattacharyya, P.: Bert-caps: a transformer-based capsule network for tweet act classification. IEEE Trans. Comput. Soc. Syst. **7**(5), 1168–1179 (2020)
13. Smith, P.K., Mahdavi, J., Carvalho, M., Fisher, S., Russell, S., Tippett, N.: Cyber-bullying: its nature and impact in secondary school pupils. J. child Psychol. Psychiatr. **49**(4), 376–385 (2008)
14. Van Hee, C., Verhoeven, B., Lefever, E., De Pauw, G., Daelemans, W., Hoste, V.: Guidelines for the fine-grained analysis of cyberbullying. Technical Report, version 1.0. Technical Report LT3 15–01, LT3, Language and Translation ... (2015)
15. Vaswani, A., et al.: Attention is all you need. In: Advances in Neural Information Processing Systems, pp. 5998–6008 (2017)
16. Xiao, L., Zhang, H., Chen, W., Wang, Y., Jin, Y.: Mcapsnet: capsule network for text with multi-task learning. In: Proceedings of the 2018 Conference on Empirical Methods in Natural Language Processing, pp. 4565–4574 (2018)

Multiword Expression Features
for Automatic Hate Speech Detection

Nicolas Zampieri[✉], Irina Illina, and Dominique Fohr

University of Lorraine, CNRS, INRIA, Loria, 54000 Nancy, France
{nicolas.zampieri,irina.illina,dominique.fohr}@loria.fr

Abstract. The task of automatically detecting hate speech in social media is gaining more and more attention. Given the enormous volume of content posted daily, human monitoring of hate speech is unfeasible. In this work, we propose new word-level features for automatic hate speech detection (HSD): multiword expressions (MWEs). MWEs are lexical units greater than a word that have idiomatic and compositional meanings. We propose to integrate MWE features in a deep neural network-based HSD framework. Our baseline HSD system relies on Universal Sentence Encoder (USE). To incorporate MWE features, we create a three-branch deep neural network: one branch for USE, one for MWE categories, and one for MWE embeddings. We conduct experiments on two hate speech tweet corpora with different MWE categories and with two types of MWE embeddings, word2vec and BERT. Our experiments demonstrate that the proposed HSD system with MWE features significantly outperforms the baseline system in terms of macro-F1.

Keywords: Social media · Hate speech detection · Deep learning

1 Introduction

Hate speech detection (HSD) is a difficult task both for humans and machines because hateful content is more than just keyword detection. Hatred may be implied, the sentence may be grammatically incorrect and the abbreviations and slangs may be numerous [12]. Recently, the use of machine learning methods for HSD has gained attention, as evidenced by these systems: [8,13]. [9] performed a comparative study between machine learning models and concluded that the deep learning models are more accurate. Current HSD systems are based on natural language processing (NLP) advances and rely on deep neural networks (DNN) [11].

Finding the features that best represent the underlying hate speech phenomenon is challenging. Early works on automatic HSD used different word representations, such as a bag of words, surface forms, and character n-grams with machine learning classifiers [17]. The combination of features, such as n-grams, linguistic and syntactic turns out to be interesting as shown by [12].

© Springer Nature Switzerland AG 2021
E. Métais et al. (Eds.): NLDB 2021, LNCS 12801, pp. 156–164, 2021.
https://doi.org/10.1007/978-3-030-80599-9_14

In this paper, we focus our research on the automatic HSD in tweets using DNN. Our baseline system relies on Universal Sentence Embeddings (USE). We propose to enrich the baseline system using word-level features, called *multiword expressions* (MWEs) [14]. MWEs are a class of linguistic forms spanning conventional word boundaries that are both idiosyncratic and pervasive across different languages [3]. We believe that MWE modelling could help to reduce the ambiguity of tweets and lead to better detection of HS [16]. To the best of our knowledge, MWE features have never been used in the framework of DNN-based automatic HSD. Our contribution is as follows. First, we extract different MWE categories and study their distribution in our tweet corpora. Secondly, we design a three-branch deep neural network to integrate MWE features. Finally, we experimentally demonstrate the ability of the proposed MWE-based HSD system to better detect hate speech: a statistically significant improvement is obtained compared to the baseline system. We experimented on two tweet corpora to show that our approach is domain-independent.

2   Proposed Methodology

In this section, we describe the proposed HSD system based on MWE features. This system is composed of a three-branch DNN network and combines global feature computed at the sentence level (USE embeddings) and word-level features: MWE categories and word embeddings representing the words belonging to MWEs.

Universal sentence encoder provides sentence level embeddings. The USE model is trained on a variety of data sources and demonstrated strong transfer performance on a number of NLP tasks [2]. The HSD system based on USE obtained the best results at the SemEval2019 campaign (shared task 5) [8]. This power of USE motivated us to use it to design our system.

MWE Features. A multiword expression is a group of words that are treated as a unit [14]. For example, the two MWEs *stand for* and *get out* have a meaning as a group, but have another meaning if the words are taken separately. MWEs include idioms, light verb constructions, verb-particle constructions, and many compounds. We think that adding information about MWE categories and semantic information from MWEs might help for the HSD task.

In our work, we focus on social media data. These textual data are very particular, may be grammatically incorrect and may contain abbreviations or spelling mistakes. For this type of data, there are no state-of-the-art approaches for MWE identification. A specific MWE identification system is required to parse MWEs in social media corpora. As the adaptation of an MWE identification system for a tweet corpus is a complex task and as it is not the goal of our paper, we decided to adopt a lexicon-based approach to annotate our corpora in terms of MWEs. We extract MWEs from the STREUSLE web corpus (English online reviews corpus), annotated in MWEs [15]. From this corpus, we create an MWE lexicon composed of 1855 MWEs which are classified into 20 lexical categories. Table 1 presents these categories with examples. Each tweet of our

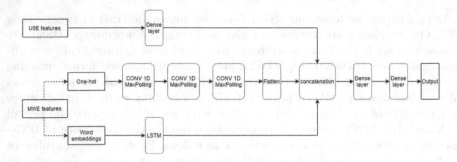

Fig. 1. Proposed hate speech detection system using USE and MWE features.

tweet corpora is lemmatized and parsed with the MWE lexicon. Our parser tags MWEs and takes into account the possible discontinuity of MWEs: we allow that one word, not belonging to the MWE, can be present between the words of the MWE. If, in a sentence, a word belongs to two MWEs, we tag this word with the longest MWE. We do not take into account spelling or grammatical mistakes. We add a special category for words not belonging to any MWE.

HSD System Proposal. In this part, we describe our hate speech detection system using USE embeddings and MWE features. As USE is a feature at the sentence level and MWE features are at the word level, the architecture of our system is composed of a neural network with three branches: two branches are dedicated to the MWE features, the last one deals with USE features. Figure 1 shows the architecture of our system.

In the first branch, we associate to each word of the tweet the number of the MWE category (one-hot encoding). This branch is composed of 3 consecutive blocks of CNN (Conv1D) and MaxPooling layers. Previous experiments with different DNN structures and the fast learning of CNN allow us to focus on this architecture. The second branch takes into account the semantic context of words composing MWE. If a given tweet has one or several MWEs, we associate a word embedding to each word composing these MWEs. We believe that the semantic meaning of MWEs is important to better understand and model them. This branch uses one LSTM layer. We propose to use two types of word embeddings: static where a given word has a single embedding in any context, or dynamic, where a given word can have different embeddings according to his long-term context. We experiment with word2vec and BERT embeddings [4,10]. BERT uses tokens instead of words. Therefore, we use the embedding of each token composing the words of the MWEs. We think that using two branches to model MWEs allows us to take into account complementary information and provides an efficient way of combining different features for a more robust HSD system.

The last branch, USE embedding, supplies relevant semantic information at the sentence level. The three branches are concatenated and went through two dense layers to obtain the output. The output layer has as many neurons as the number of classes.

3   Experimental Setup

3.1   Corpora

The different time frames of collection, the various sampling strategies, and the targets of abuse induce a significant shift in the data distribution and can give a performance variation on different datasets. We use two tweets corpora to show that our approach is domain-independent: the English corpus of SemEval2019 task 5 subTask A (called *HatEval* in the following) [1] and Founta corpora [5]. We study the influence of MWE features on the HatEval corpus, and we use the Founta corpus to confirm our results. Note that these corpora contain different numbers of classes and different percentages of hateful speech. We evaluate our models using the official evaluation script of SemEval shared task 5[1] in terms of macro-F1. It is the average of the F1 scores of all classes.

Table 1. MWE categories with examples from STREUSLE corpus [15] and the number of occurrences of MWEs. The train set of HatEval. The column *Hateful (Non-hateful)* represents MWE occurences that appear only in hateful (non-hateful) tweets. The column *Both* represents MWE occurrences that appear in hateful and non-hateful tweets.

	MWE categories	Examples	Hateful	Non-hateful	Both
MWE5	Adjective	*dead on*	9	8	255
	Adverb	*once again*	1	5	194
	Discourse	*thank you*	12	15	401
	Nominal	*tax payer*	25	36	189
	Adposition phrase (idiomatic)	*on the phone*	9	36	134
VMWE5	Inherently adpositional verb	*stand for*	11	21	447
	Full light verb construction	*have option*	9	10	36
	Verbal idioms	*Give a crap*	14	24	384
	Full verb-particle construction	*take off*	11	20	387
	Semi verb-particle construction	*walk out*	6	18	153
	Auxiliary	*be suppose to*	4	0	475
	Coordinating conjunction	*and yet*	1	0	8
	Determiner	*a lot*	1	2	242
	Infinitive marker	*to eat*	0	0	12
	Adposition	*apart from*	3	13	573
	Non-possessive pronoun	*my self*	0	3	11
	Subordinating conjunction	*even if*	0	0	28
	Cause light verb construction	*give liberty*	1	0	0
	Symbol	*A+*	0	0	0
	Interjection	*lo and behold*	0	0	0

---

[1] https://github.com/msang/hateval/tree/master/SemEval2019-Task5/evaluation.

HatEval Corpus. In the HatEval corpus, the annotation of a tweet is a binary value indicating if HS is occurring against women or immigrants. The corpus contains 13k tweets. We use standard corpus partition in training, development, and test set with 9k, 1k, and 3k tweets respectively. Each set contains around 42% of hateful tweets. The vocabulary size of the corpus is 66k words.

We apply the following pre-processing for each tweet: we remove mentions, hashtags, and URLs. We keep the case unchanged. We use this pre-processing because the systems using this pre-processing obtained the best results at the SemEval2019 task 5.

For train and development sets, we keep only tweets that contain at least two words. Thus, we obtain 8967 tweets for the training set and 998 tweets for the development set. We split the training part into two subsets, the first one (8003 tweets) to train the models, and the second one (965 tweets) for model validation. In the test set, we keep all tweets after pre-processing, even empty tweets. We tag empty tweets as non-hateful.

Founta corpus contains 100k tweets annotated with normal, abusive, hateful, and spam labels. Our experiments focus on HSD, so we decided to remove spams and we keep around 86k tweets. The vocabulary size of the corpus is 132k words. We apply the same pre-processing as for the HatEval corpus. We divide the Founta corpus into 3 sets: train, development, and test with 60%, 20%, and 20% respectively. As for the HatEval corpus, we use a small part of training as the validation part. Each set contains about 62%, 31%, and 6% of normal, abusive, and hateful tweets.

3.2 System Parameters

Our baseline system utilizes only USE features and corresponds to Fig. 1 without MWE branches. The system proposed in this article uses USE and the MWE features as presented in Fig. 1[2]. For the USE embedding, we use the pre-trained model provided by google[3] (space dimension is 512) without fine-tuning.

We tag the MWE of each tweet using the lexicon, presented in the Sect. 2. If an MWE is found, we put the corresponding MWE category for all words of the MWE. To perform fine-grained analysis, we decided to select MWE categories that have more than 50 occurrences (arbitrary value) and occurrences appear less than 97% in hate and non-hate tweets at the same time. We obtain 10 MWE categories: called MWE5 and VMWE5 which are respectively the first and second part of Table 1. VMWE5 is composed of Verbal MWE categories and MWE5 with the rest of the categories. The training part of the HatEval corpus contains 1551 occurrences of VMWE5 and 1329 occurrences of MWE5. During our experiments, we experiment with all MWE categories presented in Table 1 (containing 19 categories: 18 categories, and a special category for words not belonging to any MWE) and with the combination of VMWE5 and MWE5 (10 MWE categories and a special category).

---

[2] https://github.com/zamp13/MWE-HSD.
[3] https://tfhub.dev/google/universal-sentence-encoder-large/3.

Concerning the MWE one-hot branch of the proposed system, we set the number of filters to 32, 16, and 8 for the 3 Conv1D layers. The kernel size of each CNN is set to 3. For the MWE word embedding branch, we set the LSTM layer to 192 neurons. For BERT embedding, we use pre-trained uncased BERT model from [4] (embedding dimension is 768). The BERT embeddings are extracted from the last layer of this model. For word2vec embedding, we use the pre-trained embedding of [7]. This model is trained on a large tweet corpus (embedding dimension is 400). In our systems, each dense layer contains 256 neurons.

For each system configuration, we train 9 models with different random initialization. We select the model that obtains the best result on the development set to make predictions on the test set.

4   MWE Statistics

We first analyze the distribution of the MWEs in our corpora. We observe that about 25% of the HatEval training tweets contain at least one MWE and so the presence of MWE can influence the HSD performance.

As a further investigation, we analyze MWEs appearing per MWE category and for hate/non-hate classes. In the training set of the HatEval corpus our parser, described in Sect. 2, annotated 4257 MWEs. Table 1 shows MWEs that appear only in hateful or non-hateful tweets or both in HatEval training part. We observe that some MWE categories, as *symbol* and *interjection*, do not appear in HatEval training set. We decided to not use these two categories in our experiments. Most of the categories appear in hateful and non-hateful tweets. For the majority of MWE categories, there are MWEs that occur only in hateful speech and MWEs that occur only in non-hateful tweets.

Finally, we analyze the statistics of each MWE category for hate and non-hate classes. As in HatEval the classes are almost balanced, there is no bias due to imbalanced classes. We observe that there are no MWE categories used only in the hateful speech or only in the non-hateful speech excepted for the *cause light verb construction* category, but this category is underrepresented. We note that there is a difference between the use of MWEs in the hateful and the non-hateful tweets: MWEs are used more often in non-hateful speech. These observations reinforce our idea that MWE features can be useful for hate speech detection.

5   Experimental Results

The goal of our experiments is to study the impact of MWEs on automatic hate speech detection for two different corpora: HatEval and Founta. We carried out experiments with the different groups of MWE categories: MWEall, including all MWE categories, and the combination of VMWE5 and MWE5.

Table 2 displays the macro-F1 on HatEval and Founta test sets. Our baseline system without MWE features, called *USE* in Table 2, achieves a 65.7% macro-F1 score on HatEval test set. Using MWE features with word2vec or BERT embeddings, the system proposed in this paper performs better than the baseline. For instance, on HatEval, MWEall with BERT embedding configuration achieves the **best result** with 66.8% of macro-F1. Regarding Founta corpus, we observe a similar result improvement: the baseline system achieves 72.2% and systems with MWE features obtain scores ranging from 72.4% to 73.0% of macro-F1. It is important to note that according to a matched pair test in terms of accuracy with 5% risk [6], the systems using MWE features and word2vec or BERT embeddings *significantly* outperform the baseline system on the two corpora. Finally, the proposed system with MWEall and BERT embedding for HatEval outperforms the state-of-the-art system FERMI submitted at HatEval competition (SemEval task 5): 66.8% for our system versus 65% for FERMI of macro-F1 [8].

To analyze further MWE features, we experiment with different groups of MWE categories: VMWE5, MWE5, and MWEall. Preliminary experiments with the two-branch system with USE and word embeddings branches only gave a marginal improvement compared to the baseline system. Using the three-branch neural network with only VMWE5 or MWE5 instead of MWEall seems to be interesting only for word2vec embedding. With BERT embedding it is better to use MWEall categories. Finally, the use of all MWEs could be helpful rather than the use of a subgroup of MWE categories. Comparing word2vec and BERT embeddings, dynamic word embedding performs slightly better than the static one, however, the difference is not significant.

We compare the confusion matrices of two systems: the baseline system and the proposed one with MWEall and BERT embeddings. On the HatEval, the proposed system classifies better non-hateful tweets than the baseline system. In contrast, on Founta our system is more accurate to classify hateful tweets. We think that the balance between the classes plays an important role: in the case of HatEval corpus, the classes are balanced, in the case of Founta, the classes are unbalanced.

To perform a deeper analysis, we focus our observations on only the tweets from the test sets containing at least one MWE: 758 tweets from the HatEval test set and 3508 tweets from the Founta test set. Indeed, according to Sect. 4, there is about 25% of tweets containing MWEs. The second part of Table 2 shows that the results are consistent with those observed previously in this section, and the obtained improvement is more important.

Table 2. The first part represents F1 and macro-F1 scores (%) on *HatEval* and *Founta* test sets. The second part represents F1 and macro-F1 scores (%) on tweets containing at least one MWE in *HatEval* and *Founta* test sets.

Features	HatEval			Founta			
	F1		Macro-F1	F1			Macro-F1
	Hateful	Non-hate		Norm	Abus	Hate	
All test set							
USE	64.9	66.4	65.7	94.2	87.8	34.6	72.2
USE, MWEall, word2vec	64.5	68.2	66.3	93.8	86.9	36.5	72.4
USE, VMWE5, MWE5, word2vec	66.1	67.0	66.5	93.9	87.1	37.2	72.7
USE, MWEall, BERT	64.2	69.4	**66.8**	94.0	87.1	37.5	72.9
USE, VMWE5, MWE5, BERT	64.8	68.2	66.5	93.8	86.9	38.2	**73.0**
Tweets containing at least one MWE							
USE	67.8	62.3	65.0	91.1	94.1	41.6	75.6
USE, MWEall, word2vec	71.7	61.4	66.6	91.4	86.9	44.6	**76.5**
USE, MWEall, BERT	73.9	61.3	**67.6**	90.9	94	43.3	76.1

6   Conclusion

In this work, we explored a new way to design a HSD system for short texts, like tweets. We proposed to add new features to our DNN-based detection system: mutliword expression features. We integrated MWE features in a USE-based neural network thanks to a neural network of three branches. The results were validated on two tweet corpora: HatEval and Founta. The models we proposed yielded significant improvements in macro-F1 over the baseline system (USE system). Furthermore, on HatEval corpus, the proposed system with MWEall categories and BERT embedding significantly outperformed the state-of-the-art system FERMI ranked first at the SemEval2019 shared task 5. These results showed that MWE features allow to enrich our baseline system. The proposed approach can be adapted to other NLP tasks, like sentiment analysis or automatic translation.

References

1. Basile, V., et al.: SemEval-2019 task 5: multilingual detection of hate speech against immigrants and women in twitter. In: Proceedings of the 13th International Workshop on Semantic Evaluation, pp. 54–63. ACL (2019)
2. Cer, D., et al.: Universal sentence encoder for English. In: Proceedings of the 2018 Conference on Empirical Methods in Natural Language Processing: System Demonstrations, pp. 169–174. ACL (2018)
3. Constant, M., et al.: Multiword expression processing: a survey. Computational Linguistics, pp. 837–892 (2017)
4. Devlin, J., Chang, M.W., Lee, K., Toutanova, K.: BERT: pre-training of deep bidirectional transformers for language understanding. In: NAACL-HLT (2019)

5. Founta, A., et al.: Large scale crowdsourcing and characterization of twitter abusive behavior (2018)
6. Gillick, L., Cox, S.J.: Some statistical issues in the comparison of speech recognition algorithms. In: International Conference on Acoustics, Speech, and Signal Processing, vol. 1, pp. 532–535 (1989)
7. Godin, F.: Improving and interpreting neural networks for word-level prediction tasks in natural language processing. Ph.D. thesis, Ghent University, Belgium (2019)
8. Indurthi, V., Syed, B., Shrivastava, M., Chakravartula, N., Gupta, M., Varma, V.: FERMI at SemEval-2019 task 5: using sentence embeddings to identify hate speech against immigrants and women in Twitter. In: Proceedings of the 13th International Workshop on Semantic Evaluation, pp. 70–74. ACL (2019)
9. Lee, Y., Yoon, S., Jung, K.: Comparative studies of detecting abusive language on Twitter. In: Proceedings of the 2nd Workshop on Abusive Language Online (ALW2), p. 101–106 (2018)
10. Mikolov, T., Chen, K., Corrado, G., Dean, J.: Efficient estimation of word representations in vector space. In: ICLR Workshop Papers (2013)
11. Mozafari, M., Farahbakhsh, R., Crespi, N.: A BERT-based transfer learning approach for hate speech detection in online social media. In: Cherifi, H., Gaito, S., Mendes, J.F., Moro, E., Rocha, L.M. (eds.) COMPLEX NETWORKS 2019. SCI, vol. 881, pp. 928–940. Springer, Cham (2020). https://doi.org/10.1007/978-3-030-36687-2_77
12. Nobata, C., Tetreault, J., Thomas, A., Mehdad, Y., Chang, Y.: Abusive language detection in online user content. In: Proceedings of the 25th International Conference on the World Wide Web, pp. 145–153. International World Wide Web Conferences Steering Committee (2016)
13. Pamungkas, E.W., Cignarella, A.T., Basile, V., Patti, V.: Automatic identification of misogyny in English and Italian tweets at Evalita 2018 with a multilingual hate lexicon. In: EVALITA@CLiC-it (2018)
14. Sag, I.A., Baldwin, T., Bond, F., Copestake, A., Flickinger, D.: Multiword expressions: a pain in the neck for NLP. In: Gelbukh, A. (ed.) CICLing 2002. LNCS, vol. 2276, pp. 1–15. Springer, Heidelberg (2002). https://doi.org/10.1007/3-540-45715-1_1
15. Schneider, N., Smith, N.A.: A corpus and model integrating multiword expressions and supersenses. In: Proceedings of the 2015 Conference of the North American Chapter of the Association for Computational Linguistics: Human Language Technologies, pp. 1537–1547. ACL (2015)
16. Stanković, R., Mitrović, J., Jokić, D., Krstev, C.: Multi-word expressions for abusive speech detection in Serbian. In: Proceedings of the Joint Workshop on Multiword Expressions and Electronic Lexicons, pp. 74–84. ACL (2020)
17. Waseem, Z., Hovy, D.: Hateful symbols or hateful people? Predictive features for hate speech detection on Twitter. In: Proceedings of the NAACL Student Research Workshop, pp. 88–93. ACL (2016)

Semantic Text Segment Classification of Structured Technical Content

Julian Höllig[1]([✉]), Philipp Dufter[2], Michaela Geierhos[1], Wolfgang Ziegler[3], and Hinrich Schütze[2]

[1] Research Institute CODE, Bundeswehr University Munich, Neubiberg, Germany
{julian.hoellig,michaela.geierhos}@unibw.de
[2] Center for Language and Information Processing, LMU Munich, Munich, Germany
philipp@cis.lmu.de
[3] Information Management and Media, Karlsruhe University of Applied Sciences, Karlsruhe, Germany
wolfgang.ziegler@hs-karlsruhe.de

Abstract. Semantic tagging in technical documentation is an important but error-prone process, with the objective to produce highly structured content for automated processing and standardized information delivery. Benefits thereof are consistent and didactically optimized documents, supported by professional and automatic styling for multiple target media. Using machine learning to automate the validation of the tagging process is a novel approach, for which a new, high-quality dataset is provided in ready-to-use training, validation and test sets. In a series of experiments, we classified ten different semantic text segment types using both traditional and deep learning models. The experiments show partial success, with a high accuracy but relatively low macro-average performance. This can be attributed to a mix of a strong class imbalance, and high semantic and linguistic similarity among certain text types. By creating a set of context features, the model performances increased significantly. Although the data was collected to serve a specific use case, further valuable research can be performed in the areas of document engineering, class imbalance reduction, and semantic text classification.

Keywords: Semantic text classification · Context features · Technical documentation

1 Introduction

The area of technical documentation is highly relevant in the industry, due to the legal need for technical information, such as user manuals, alongside commercial products [4]. There are established standards to ensure efficiency and quality in the document creation process. One important standard is the uniform assignment of XML tags to text segments. Some of these segments, such as notes, commands or warnings, contain semantics, while others, like continuous text or generic lists, are non-semantic. Semantic text segments are indicated by

© Springer Nature Switzerland AG 2021
E. Métais et al. (Eds.): NLDB 2021, LNCS 12801, pp. 165–177, 2021.
https://doi.org/10.1007/978-3-030-80599-9_15

the XML name tag. Depending on the communicative goal of the segment type, the respective tags are connected to rules, which ensure consistent structure and layout in the documentation [4]. Besides supporting readability of the contents, this provides the prerequisites for intelligent information processing.

When multiple authors, eventually located at multiple sites, write documents for the same customer, or even work together on a single document, divergent tag assignments are likely to happen. Some text segments are semantically very close, which might as well lead to different tagging. Our research focuses on the automated validation of the tagging process through the support of machine learning, and thus on the improvement and standardization of highly structured content, including all the associated benefits.

There is much prior research on text classification, especially on applications for social media [2,5,15], and online product customer reviews [6,7,11]. As opposed to this type of content, technical content is highly structured, emotionally unbiased, and usually follows writing guidelines. Consequently, important features for the classification remain in the communicative style, and in contextual patterns. Therefore, we extracted a set of context features specific to structured content, and tailored to the standards of the data source.

Our study makes several contributions: first, we developed a new concept for validating automated tagging, which enables intelligent and automated information processing; second, we created a comprehensive, high-quality dataset, as a basis for further research on the use case, or similar scenarios; third, we designed context features to increase model performance and to save resources.

2   Related Work

Text classification is a common area in computational linguistics and has been applied to diverse domains such as health [2], law [8], finance [20], and social media [5,15]. The text input size reaches from document, chapter, and paragraph, to even sentence level. The few works done at the interface between machine learning and technical documentation deal with classification of relatively large, self-contained units of content, for example on chapter [12,13] or document level [9]. Writing in all conscience, this work is the first dealing with paragraph-level text classification in the technical documentation domain.

Oevermann and Ziegler [13] conducted a comprehensive study, where they applied traditional machine learning models to categorize product component-related text blocks in technical documents. They also used real-world data, which was manually tagged by professionals such as technical writers, or content experts. While they used data from the engineering sector, the data for our work is software-related. They applied the vector space model as baseline, and tf-idf weighting for single words and word groups, where word n-grams of two and three achieved the best performances. The classification was done by finding the highest cosine similarity between a document vector and the class vectors.

Since the invention of BERT in 2018 [1], the model has been frequently applied to various text classification tasks and compared to traditional machine learning methods, which mostly used tf-idf feature extraction. The superiority of BERT was found in most cases. González-Carvajal and Garrido-Merchán [6] conducted several binary- and multi-classification tasks on movie and hotel reviews, where BERT achieved better performance than traditional models, including a support vector machine (SVM). In our work, we use similar-sized text data, and we apply the same models, but to a fairly unexplored domain. Lund [9] contradicts the results found by González-Carvajal and Garrido-Merchán [6] by achieving parity of a tf-idf model and BERT. Lund applied BERT to technical documents following the product life cycle such as installation, operation, maintenance, troubleshooting, and disposal. In our work, we also match BERT's performance with a tf-idf model, but only by using additional context features.

Di Iorio et al. [3] followed the goal of standardized document structures to achieve consistent layouts. But instead of applying machine learning, they approached the task with an algorithm based on a pattern rules concept. This concept was developed by abstracting XML patterns from large amounts of documents. Hereby, the authors identified common structural and content-related characteristics in and between typical XML elements that applied to all documents. Unlike in our work, the semantic value of the contents was irrelevant. Examples for patterns were blocks, containers, or fields. While the authors considered structural patterns as the basis for styling decisions, we are considering the semantics as more important. For instance, our documents contain tables with entries of definitions, and while we would capture the actual definitions, the authors from this work would capture the whole table. Due to the strong abstraction, this approach is not as fine-granular as ours, but generalization can be achieved more easily.

3   Data

3.1   Data Collection

Data Source. The dataset for this work was scraped from the SAP Help Portal[1], an open-source online documentation platform containing a high number of user manuals for different SAP products. The documentation is created and maintained in a content management system, and structured according to DITA.[2] DITA is a popular XML standard, which is frequently used in the technical documentation industry.

Segment Type Definition. We defined ten text segment types by manually examining documents in the SAP Help Portal across different products. Hereby, we used the underlying CSS classes in the HTML code to identify the different segment types. The following semantic text segment types are contained in the dataset: Command, Definition, Example, Note, Recommendation, Reminder, Restriction, Tip, Warning, Shortdescription.

---

[1] https://help.sap.com/viewer/index.
[2] https://www.oasis-open.org/committees/tc_home.php?wg_abbrev=dita.

Data Quality. Although in no official cooperation, SAP confirmed the defined segment types and gave insight in their content creation process. The contents in the SAP Help Portal are written by experts in the field of technical writing, who are supported by an editorial guide stating tagging standards and writing style recommendations. After data collection, we reviewed 100 random samples of each segment type to validate the data quality. Review criteria were the semantic correctness and data cleanness. Apart from a few outliers (text missing, incorrect tag assignments), the quality was good.

Scraping Process. For the data collection process, we built a web scraper using the Python framework *selenium*. Selenium offers user-like interaction with web content by taking control of the browser [17]. The segment types were identified in the underlying CSS class of the web page, and retrieved via XPath expressions. Example 1 illustrates this process.

Example 1. The example shows how to scrape segment type *Warning*.

```
driver.find_elements_by_xpath("//section/descendant::aside[@class =
    'note note caution']")
```

The object `driver` is a WebDriver element at the currently active browser window. The function `find_elements_by_xpath` finds the required elements via a XPath expression. In this example, the driver traverses the DOM tree and looks for all HTML elements descending from *section*, being tagged with *aside* and containing the class attribute *note note caution*.

3.2   Dataset Description

Data Statistics. The dataset was randomly split into training (70%), test (20%), and validation (10%) sets. Figure 1(left) shows the count of collected text segments per class (=type), and for each set, with a total count of **86,450**, after postprocessing. There are huge divergences between the class counts, which are visualized in Figure 1(right). This strong class imbalance caused issues in the following classification experiments.

Data Access. We provide the datasets on Github[3] in three json files (train.json, val.json, test.json) in postprocessed form. On request, we also provide all data before preprocessing in one big XML file, which contains all semantic and non-semantic segments collected. Along with the XML file, we provide a Python module *dataset.py* for data cleaning, sampling, context feature extraction, and transformation into *pandas* data frames. The data cleaning involves removal of inline HTML tags, HTML-generated newlines/tabs/spaces, 'None' values, and duplicates. The sampling involves oversampling on minority and undersampling on majority classes to enable the model to learn small classes [18].

---

[3] https://github.com/juhoUnibw/semSegClass.

Total count: 86,450	Class counts		
	Train	Val	Test
Commands	20,081	2,231	9,180
Notes	11,604	1,289	5,336
Shortdescriptions	11,498	1,278	5,409
Examples	4,746	528	625
Warnings	3,123	347	459
Definitions	2,756	306	368
Tips	2,151	239	291
Recommendations	981	109	156
Reminders	558	62	69
Restrictions	529	59	82

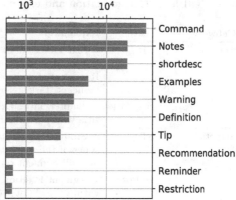

Fig. 1. Dataset: overview of the class counts (left), and the relative class distributions (right)

4 Methods

4.1 Feature Extraction

In addition to the presented text segments, we developed nine context features, which were added to all models in different combinations. They can be categorized in topical, structural, environmental, and grammatical features. Table 1 states the categories, underlying features, and short explanations. All features were normalized for values between zero and one.

4.2 Models

Model Selection. We evaluated the use case through a series of experiments with a deep learning and a traditional model. Hereby, we applied the transformer model BERT (Bidirectional Encoder Representations from Transformers) from the *transformers* library by Huggingface[4], and a linear SVM (Support Vector Machine) from the *sklearn* library. As deep learning model, we chose BERT due to its state-of-the-art performances in many text classification tasks [16]. BERT also uses the so-called *self-attention* mechanism to capture long-range

---

[4] https://huggingface.co/transformers/model_doc/bert.html#tfbertforsequenceclassi fication.

Table 1. Categorization and explanation of extracted context features

Category	Feature	Explanation
Topical	TF	=Text Function. States the text function of the chapter where the segment was found (e.g. instructing, descriptive). Represented as binary feature over all text function categories (concept, task, reference, topic)
	chapTitle	=Chapter Title. Extracts a tf-idf representation of the chapter title where the segment was found
Structural	ST	= Sibling Types. States the predecessor and successor segment types of the current segment. Represented as binary feature over all segment types
	segPos	=Segment Position. States the position of the current segment within the chapter (0 = start, 1 = end)
	chapPos	=Chapter Position. States the position of the current segment within the document (0 = start, 1 = end)
Environmental	nSeg	=Number of Segments. States the number of segments within the chapter where the segment was found
	CS	=Content Share. Measures how much of the chapter content is owned by the segment (chars segment/chars chapter)
	semDistr	=Semantic Distribution. Measures the semantic quantity and diversity within the chapter where the segment was found (number of semantic segments + number of unique semantic segments)
Grammatical	POS	=Part-of-speech. Extracts a tf-idf representation of the segment text after it was transformed into part-of-speech tags

dependencies in text sequences [19], which we assumed to be helpful in finding complex writing style patterns. As traditional model we chose linear SVM because it shows the best results for text classification tasks between several traditional models [10]. Alternative deep learning and traditional models such as ALBERT, Gradient Boosting, Decision Tree, and Bagging performed worse than the presented models in first experiments, which is why they were deprecated.

Model Design. In order to use the pre-trained BERT model, a classification layer was built on top of the traditional BERT architecture. During training, the additional context features were appended to the output of the pooler layer of BERT, which is the sequence representation fed into the classification layer. The weight embedding matrix was adapted to the increased number of features. To access the pooler layer, the source code of the feed forward function was extracted from the transformers library and modified accordingly. For the SVM, the context features were appended as numpy array to the text representation.

Hyperparameter Selection. Table 2 shows the hyperparameter configurations we used with the models. BERT's configuration is recommended by Akshay Prakash [14]. For the linear SVM model, we validated different combinations to find the optimal one.

Table 2. Hyperparameters used with BERT (left) and SVM (right)

BERT		Linear SVM	
Batch size	16	Stemming	Yes
Learning rate	2e-5	Stop words	English
Epochs	4	Tolerance	0.0001
Max. sequence length	128	Max. features	20,000
Loss function	Cross-entropy	Loss function	Hinge-loss

4.3 Sampling

An oversampling of factor 2 was applied to the minority classes (*Definition, Example, Warning, Tip, Recommendation, Reminder, Restriction*), and an undersampling factor of 0.6 to the majority classes (*Command, Note, Short-description*). The samplings were only applied to the training and validation sets, the test set retains the real-world data distribution.

5 Experiments

5.1 Setup

In a series of experiments with eleven setups, we evaluated the classification of the ten presented semantic segment types through application of the presented models. In setup 1, bare text input was used for modeling. In setups 2–10 the impacts of the presented context features were evaluated. In setup 11, the best combinations of features for BERT and SVM were modeled. For the SVM, we additionally modeled bare context features, to compare their performance against bare text features.

The experiments were evaluated both quantitatively and qualitatively. In the quantitative evaluation, we present the overall performances of each experiment, measured in accuracy (1) and macro-average (5). The accuracy specifies the fraction of correctly classified text segments, the macro-average reflects the average success of all classes. We also specify the individual performances of each class in F1-score (4) for the best models. The exact definitions of the measures are as follows (where TP stands for True Positives, FP for False Positives, FN for False Negatives, N for the count of test samples, and C for the number of classes):

$$Accuracy = \frac{TP + TN}{N} \tag{1}$$

$$Precision = \frac{TP}{TP + FP} \tag{2}$$

$$Recall = \frac{TP}{TP + FN} \tag{3}$$

$$F1 = 2 * \frac{Precision * Recall}{Precision + Recall} \tag{4}$$

$$Macro\text{-}average = \frac{\sum_{i=1}^{C} F1}{C} \tag{5}$$

In the qualitative evaluation, we reveal some challenging aspects of the classification task by analyzing the test samples that were hard to distinguish for the model.

5.2   Quantitative Evaluation

Table 3 shows the overall model performances of all experiments. The highest macro-average (60%) was achieved by BERT combined with the context features 'Text Function' and 'Siblings Type'. The highest accuracy (88%) was achieved by SVM combined with all context features. Overall, these models match in performance, which indicates that superior text embedding in deep learning models can be equalized by using context features in traditional models.

Table 3. Overall model performance comparison across all setups (values in %)

	Accuracy		Macro AVG	
	BERT	SVM	BERT	SVM
Setup 1: Text only	83	77	54	45
Setup 2: +TF	84	79	59	48
Setup 3: +chapTitle	84	78	54	50
Setup 4: +ST	83	87	54	54
Setup 5: +segPos	83	79	56	46
Setup 6: +chapPos	83	77	56	46
Setup 7: +nSeg	83	77	54	46
Setup 8: +CS	84	77	55	45
Setup 9: +semDistr	83	77	54	47
Setup 10: +POS	84	78	54	46
Setup 11: +Best combination	84	**88**	**60**	56
Only context features	–	85	–	46

Table 4 shows the individual class performances of the two best models. The five best classes, which are the same for both models, are marked in bold. The other classes show low performances, due to high semantic (and linguistic) similarities, and the negative class imbalance influence. These challenges are further examined in the following subsection.

Table 4. Class performances (values in %) of the two best models: BERT and SVM from setup 11 (PREC=Precision, REC=Recall, F1=F1-score)

	BERT+TF+ST			SVM+ALL			
	PREC	REC	F1	PREC	REC	F1	Count
Command	95	94	95	94	96	95	9,180
Definition	87	69	77	96	98	97	368
Example	75	78	77	60	72	66	625
Note	77	76	77	81	75	78	5,336
Recommendation	46	45	45	35	33	34	156
Reminder	71	30	42	14	10	12	69
Restriction	41	28	33	42	22	29	82
Tip	30	39	34	30	26	28	291
Warning	34	44	38	23	25	24	459
Shortdescription	85	85	85	97	98	97	5,409
Accuracy	84			88			21,975
Macro AVG	60			56			

5.3   Qualitative Evaluation

Here, we examined individual test predictions to understand the challenges of the classification task. The basis of our analysis was 100 random test samples of each segment type, of which 50 samples were correctly, and 50 samples were incorrectly predicted. In the following, we focus on examples of conflicting segment type pairs, where the one type was mispredicted as the other type.

Command vs. Note. Example 2 shows the command-like syntax of type *Command*, but the wording legitimates the type *Note*, due to the phrase 'Make sure'. Without further context, it is difficult to say whether the true class *Command* or predicted class *Note* should actually be correct.

Example 2. "Make sure the SNC PSE is still the selected PSE."

Shortdescription vs. Definition. *Shortdescription* and *Definition* both contain descriptions of some kind, which is why they have similar linguistic patterns. The sample in Example 3 of type *Definition* could also be used in a *Shortdescription* segment, for example, to introduce a chapter.

Example 3. "The options that describe the operation of an object, which are viewable in the workspace when you open the object".

Note vs. Warning. Generally, the *Note* type is semantically very similar to the other note-like classes *Tip, Recommendation, Restriction, Reminder*, or *Warning*. The terms 'should' and 'have to' in Example 4 justify the falsely predicted *Warning* as well as the correct type *Note*.

Example 4. "The following checks and steps should be performed on all hosts of the affected sap HANA system. They have to be executed as the root user in the Linux shell".

Restriction vs. Note. Example 5 shows a sample of the type *Restriction*, which was predicted as *Note*. In this case, one could argue that the term 'available only' indicates a restriction. However, the models were trained with far more *Note* than *Restriction* samples (11,604 : 529), so that a single linguistic difference like this seems to be not strong enough to influence the model.

Example 5. "The feature is available only on browsers (desktop/laptop)".

Tip vs. Note. A similar effect can be observed for *Tips*, which often contain the indicator 'you can' in order to animate the reader to act. Example 6 shows such a case, where the sample was predicted as *Note*. In this example, we can observe a mix of segment types, which would be legitimate, but hard to learn for the model. The last sentence for itself could easily belong to type *Command*.

Example 6. "If you have one instructor and you want to authorize that one instructor to teach many learning items, you can do that in the instructor's record. Go to people instructors authorized to teach".

Reminder vs. Command. A challenge of the type *Reminder* is that it can easily be formed by just repeating any statement made at some point in the documentation, while changing the semantics of that statement. The statement in Example 7 shows all linguistic patterns of a *Command* (verb at the start of sentence, imperative form), but the author might have tagged it legitimately as *Reminder* to prevent the reader from missing an action.

Example 7. "Copy and save the client secret as you won't be able to retrieve it later".

Discussion. Most of the challenging test samples belong to one of the sub-note types because they are hard to distinguish from the general *Note* type, both linguistically and semantically. This can be supported by the fact that 57.6% of the mispredicted note sub-type samples were falsely predicted as *Note*. The strong class imbalance in the data adds additional complexity to the classification of these types. In an experiment with equal class distribution, the performances of the sub-note type classes increased, but still remained below the other classes. Consequently, both the class imbalance and close class similarity affect their classification results negatively. Selective random oversampling, multiple SMOTE variants, class re-weighting, and a feature selection method did not improve the performance. The most effective solution to both problems is to merge all note type classes, taking into account the restraints it puts on the use case. For this scenario, the performance for the best model achieved 93%/89%.

6   Conclusion and Future Work

In this paper, we introduced the novel approach of validating document structures by means of machine learning in order to enable intelligent information processing and ensure consistent document layouts. We showed model evaluations with promising results, and revealed the remaining challenges. Moreover, we provided a comprehensive dataset for further research in different areas such as document engineering, text classification, and the handling of imbalanced data, along with baseline results. During the experiments, we discovered the strong impact of context features on the performance of traditional models. Our SVM model, which was originally thought of as baseline model, matched BERT's performance through using context features. We could derive that structured semantic content yields useful underlying contextual patterns besides linguistic features. Thus, choosing traditional models with context features over deep learning models in such a scenario can achieve the same results with significantly less resources.

Our experimental results showed that the best deep learning model (BERT) and the best traditional model (SVM) achieve equal performances. They solve the classification task partially well, with a combined macro-average performance of 83.4% for the classes *Command, Definition, Example, Note,* and *Shortdescription,* and of 31.8% for the classes *Recommendation, Reminder, Restriction, Tip,* and *Warning.* Hereby, the SVM model achieves better performance on the first group of the classes (86.6%), while BERT achieves better performance on the second group of the classes (38.2%).

A big constraint of this work is the semantic and linguistic similarity between the note type elements. Combined with the class imbalance, the generic majority class *Note* is mostly predicted in unclear cases. Merging the sub-note types into a common note-type class, shows the potential of the application, and produces a ready-to-use model for the use case, although restricted to fewer classes.

A substantial advancement of our system would be the automated tagging of unstructured documents, for example, in the context of migration of large document collections to content management systems. Such an application would significantly lower the initial workload of content structure standardization in the industry and therefore, accelerate the process of intelligent content processing and delivery.

Acknowledgments. This work was supported by the Bavarian Research Institute for Digital Transformation and the European Research Council (#740516).

References

1. Devlin, J., Chang, M.W., Lee, K., Toutanova, K.: BERT: pre-training of deep bidirectional transformers for language understanding. In: Proceedings of the 2019 Conference of the North American Chapter of the Association for Computational Linguistics: Human Language Technologies, Volume 1 (Long and Short Papers), pp. 4171–4186. ACL, Minneapolis, June 2019. https://doi.org/10.18653/v1/N19-1423
2. Dhiman, A., Toshniwal, D.: An enhanced text classification to explore health based indian government policy tweets. CoRR abs/2007.06511 (2020)
3. Di Iorio, A., Peroni, S., Poggi, F., Vitali, F.: A first approach to the automatic recognition of structural patterns in XML documents. In: Concolato, C., Schmitz, P. (eds.) ACM Symposium on Document Engineering, DocEng 2012, Paris, France, 4–7 September 2012, pp. 85–94. ACM (2012). https://doi.org/10.1145/2361354.2361374
4. Drewer, P., Ziegler, W.: Technische Dokumentation: Übersetzungsgerechte Texterstellung und Content-Management, pp. 25–27. Vogel Business Media (2011)
5. Fei, G., Liu, B.: Social media text classification under negative covariate shift. In: Màrquez, L., Callison-Burch, C., Su, J., Pighin, D., Marton, Y. (eds.) Proceedings of the 2015 Conference on Empirical Methods in Natural Language Processing, EMNLP 2015, Lisbon, Portugal, 17–21 September 2015, pp. 2347–2356. ACL (2015). https://doi.org/10.18653/v1/d15-1282
6. González-Carvajal, S., Garrido-Merchán, E.C.: Comparing BERT against traditional machine learning text classification. CoRR abs/2005.13012 (2020)
7. Gräbner, D., Zanker, M., Fliedl, G., Fuchs, M.: Classification of Customer Reviews based on Sentiment Analysis. In: Fuchs, M., Ricci, F., Cantoni, L. (eds.) ENTER 2012, pp. 460–470. Springer, Vienna (2012). https://doi.org/10.1007/978-3-7091-1142-0_40
8. Lee, J.S., Hsiang, J.: Patent classification by fine-tuning BERT language model. World Patent Inf. **61**, 101965 (2020). https://doi.org/10.1016/j.wpi.2020.101965
9. Lund, M.: Duplicate detection and text classification on simplified technical English. Dissertation, Linköping University (2019). http://urn.kb.se/resolve?urn=urn:nbn:se:liu:diva-158714
10. Manning, C.D., Raghavan, P., Schütze, H.: Introduction to Information Retrieval. Cambridge University Press, Cambridge (2008)
11. Nicholls, C., Song, F.: Improving sentiment analysis with part-of-speech weighting. In: 2009 International Conference on Machine Learning and Cybernetics, vol. 3, pp. 1592–1597 (2009). https://doi.org/10.1109/ICMLC.2009.5212278
12. Oevermann, J.: Reconstructing semantic structures in technical documentation with vector space classification. In: Martin, M., Cuquet, M., Folmer, E. (eds.) SEMANTiCS 2016, Leipzig, Germany, 12–15 September 2016. CEUR Workshop Proceedings, vol. 1695. CEUR-WS.org (2016)
13. Oevermann, J., Ziegler, W.: Automated classification of content components in technical communication. Comput. Intell. **34**(1), 30–48 (2018)
14. Prakash, A.: Fine-tuning BERT model using PyTorch, December 2019. https://medium.com/@prakashakshay90/f34148d58a37

15. Pratama, B.Y., Sarno, R.: Personality classification based on Twitter text using naive bayes, KNN and SVM. In: 2015 International Conference on Data and Software Engineering (ICoDSE), pp. 170–174 (2015). https://doi.org/10.1109/ICODSE.2015.7436992

16. Raj, B.S.: Understanding BERT: is it a game changer in NLP? (2019). https://towardsdatascience.com/7cca943cf3ad

17. Stewart, S., Burns, D. (eds.): W3C Recommendation, chap. WebDriver. W3C, August 2020. https://www.w3.org/TR/webdriver/

18. Tan, P.N., Steinbach, M., Kumar, V.: Introduction to Data Mining, p. 306. Addison-Wesley Longman Publishing Co., Inc., USA (2005)

19. Vig, J.: Deconstructing BERT, Part 2: visualizing the inner workings of attention, January 2019. https://towardsdatascience.com/60a16d86b5c1

20. Wang, W., Liu, M., Zhang, Y., Xiang, J., Mao, R.: Financial numeral classification model based on BERT. In: Kato, M.P., Liu, Y., Kando, N., Clarke, C.L.A. (eds.) NTCIR 2019. LNCS, vol. 11966, pp. 193–204. Springer, Cham (2019). https://doi.org/10.1007/978-3-030-36805-0_15

On the Generalization of Figurative Language Detection: The Case of Irony and Sarcasm

Lorenzo Famiglini[1], Elisabetta Fersini[1]([✉]), and Paolo Rosso[2]

[1] University of Milano-Bicocca, Milano, Italy
elisabetta.fersini@unimib.it
[2] PRHLT Research Center, Universitat Politcnica de Valncia, Valencia, Spain

Abstract. The automatic detection of figurative language, such as irony and sarcasm, is one of the most challenging tasks of Natural Language Processing (NLP). In this paper, we investigate the generalization capabilities of figurative language detection models, focusing on the case of irony and sarcasm. Firstly, we compare the most promising approaches of the state of the art. Then, we propose three different methods for reducing the generalization errors on both in- and out-domain scenarios.

Keywords: Irony and sarcasm detection · Generalization capabilities

1 Introduction

During the last decade, several models have been introduced in the research panorama to recognize few rhetorical figures, and in particular to identify those elements that discriminate, in a significant way, what is sarcastic or ironic from what is not. In particular, sarcasm and irony detection has been defined as a classification problem, where the ground truth is a dichotomy variable 0 and 1, where 0 means that text is not a rhetorical figure, otherwise is an ironic or sarcastic statement. The irony and sarcasm detection problem has been widely addressed in the literature, where a plethora of computational approaches have been proposed ranging from the earlier techniques based on linguistic patterns [3,12], to the more recent ones based on neural architecture [6,14] or combination of both [4]. Although all of these approaches in the state of the art represent a fundamental step towards the modeling of irony and sarcasm, less effort has been dedicated to measure and improve the generalization capabilities of the models when considering both in- and out-domain vocabularies. In order to address this problem, we investigate three main research questions:

(R1) What are the most representative linguistic features for identifying sarcasm and irony patterns?
(R2) How can we exploit transformer-based and emotional-based embeddings to train accurate irony and sarcasm detection models? In particular, are pure neural models better than approaches based also on linguistic features?
(R3) What are the generalisation capabilities of the developed models?

© Springer Nature Switzerland AG 2021
E. Métais et al. (Eds.): NLDB 2021, LNCS 12801, pp. 178–186, 2021.
https://doi.org/10.1007/978-3-030-80599-9_16

Contribution. We addressed the above-mentioned research questions, by comparing the most promising approaches of the state of the art, and by proposing several approaches, based on embeddings and ensembles methods, for reducing the generalization errors on both in- and out-domain scenarios. In particular, the main contributions of the paper are:

1. A comparative analysis of the state of the art models for irony/sarcasm detection to determine their generalising capabilities, highlighting the most representative features for discriminating irony and sarcasm from others;
2. A novel methodology, based on the combination of multiple output encoder layers of the BERTweet model [7], for creating a more contextualized sentence embeddings, called BERTweet-Features based;
3. A novel model based on the emotional features of DeepMoji [2], built on the concept of self-attention layer, called DeepMoji Features-based;
4. A novel model, called Ensemble of Ensembles, where machine learning classifiers trained on several aspects of the text identify various patterns of irony and sarcasm.

All the developed models are available at https://github.com/MIND-Lab/GIS.

2   State-of-the-art Models for Irony and Sarcasm Detection

The first objective of this paper is to carry out a comparative analysis of different models, which are the most promising approaches in the state of the art for irony and sarcasm detection. To this purpose, we considered the following models:

- **Machine Learning classifiers**, i.e. XGBoost, AdaBoost, HistGradient-Boosting, Logistic Regression and Random Forest trained on embeddings (extracted from BertTweet and reduced on the basis of Principal Component Analysis maintaining 95% of the variance) together with a set of hand-crafted features. In particular, we considered Part-OF-Speech (POS) tags, pragmatic particles (PP), including emoji, punctuation, initialisms and onomatopoeic figures, and finally the polarity of the text (POL). All the possible combinations of these features have been evaluated.
- **Bayesian Model Averaging (BMA)**, initially presented in [9], which combines the models introduced above to finally derive an ensemble of traditional classifiers.
- **DeepMoji**, presented in [2], focused emotional information encoded by a recent transformer-based architecture named RCNN-Roberta.
- **RCNN-Roberta**, presented in [8], consists of a RoBERTa pre-trained transformer followed by a bidirectional LSTM layer (BiLSTM).

3   Proposed Models

3.1   BERTweet Features-Based Model (BERTweet-FB)

We firstly introduce in Fig. 1 the proposed BERTweet Features-based model, which is based on the outputs encoding layers of the original BERTweet model. The BERTweet Features-based model[1] stems from the following question: how can we exploit, in a flexible way, the sentence embeddings of the outputs of each encoding layer of a Transformer? To address this question, an architecture has been developed that focuses on the last four output layers of BERTweet. Instead of concatenating the various layers, they are joined by inserting the concept of flexibility, i.e. contextualised weights for the task to be analysed. In this case, the input tensor has a size of N × 4 × 1 × 768, where N denotes the number of training examples and the second dimension is associated with each input layer of the model. The next layers are based on the reduction of the number of channels to obtain a single one in order to merge the different information obtained from the different features' levels. They are developed on the basis of 1D Convolutions, self-attention layers and residual connections.

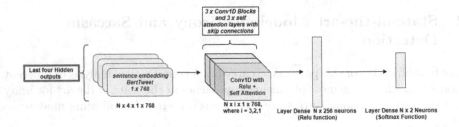

Fig. 1. BERTweet Features-based model.

3.2   DeepMoji Features-Based Model (DeepMoji-FB)

The DeepMoji Features-based model[2], presented in Fig. 2, takes as input a tensor of a dimension N × 1 × 2304. Each instance is the emotional embedding generated by the original DeepMoji model.

---

[1] Sarcasm task: batch size 64, learning rate 0.0001, optimizer AdamW and 80 epochs.
    Irony task: batch size 32, learning rate 0.00002, optimizer AdamW, and 100 epochs.
[2] Sarcasm task: batch size 32, learning rate 0.00001, optimizer Adam and 25 epochs.
    Irony task: batch size 32, learning rate 0.0002, optimizer Adam, and 35 epochs.

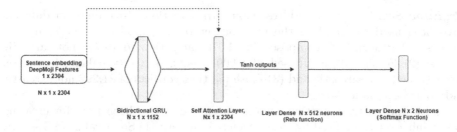

Fig. 2. DeepMoji features-based

In this case, different information is used for developing a new model for irony and sarcasm detection. The architecture of the DeepMoji Features-based model is slightly different from the BERTweet Features-based model. Indeed, the sentence embedding is fed into a Bidirectional GRU [1] layer. The most important aspect is that a new sentence embedding is generated using a skip connection between the input embedding and the output embedding generated by the BiGRU.

3.3 Ensemble of Ensembles (EoE)

The last model that we proposed is based on the combination of Bayesian Model Averaging, DeepMoji-FB and BERTweet-FB, by means of Soft/Hard classification. We will call this model 'Ensemble of ensembles' (EoE). The proposed EoE relies on a simple concept: exploiting several models, trained on different aspects of the text, to create a composition of models that better identifies the meaningful pattern of irony and sarcasm. Therefore, we created an ensemble that includes BMA, BERTweet-FB and DeepMoji-FB. Two different *classification strategies* have been evaluated: hard classification and soft classification. In particular, hard classification determines the final label of each testing instance by using the most frequent predicted label (i.e. majority voting), while soft classification selects the final label according to the sum of the marginal probability distributions given by each model.

4 Experimental Settings

In order to understand if the compared models are characterized by good generalization capabilities, we created the training and the test set (for both irony and sarcasm detection tasks), to make possible two different experimental scenarios: (1) train and test models using posts drawn from the same dataset to investigate the in-domain performance and (2) train and test models using posts coming from two different datasets, to estimate out-domain capabilities.

Training Set. In order to address **sarcasm** detection, three different datasets have been used for creating the training set to be supplied to the compared models: (1) Ptacek [10], composed of 14.070 sarcastic and 16.718 not sarcastic tweets; (2) Fersini [3], composed of 8.000 tweets, perfectly balanced between sarcastic and not sarcastic and (3) Gosh [5], that consists of 21.292 not sarcastic and 18.488 sarcastic tweets.

Regarding **irony** detection, two main datasets have been used for creating the training set to be used by the considered models: (1) SemEval-2018 Task 3A [13], specifically task 3A, composed of 1898 ironic and 1904 not ironic tweets; (2) Reyes [11], which consists of 10,000 ironic tweets, and 30,000 non-ironic posts about Politics, Humour, and Education. For irony detection, in order to compare the results of the proposed models with the state of the art, we considered the constrained and unconstrained settings defined at SemEval-2018 Task 3A. For the unconstrained scenario, we created a training set composed of the training released for SemEval-2018 Task 3A and the training of the Reyes dataset. The unconstrained settings will allow us to understand if, by introducing more variance in the training data (SemEval 2018 + Reyes), the models will maintain/improve their prediction capabilities on the test set (SemEval 2018). For the constrained settings, only the training set of the SemEval-2018 challenge has been used to train the models, and to be then validated on the SemEval 2018 test set.

Test Set. As far as **sarcasm** is concerned, two different test sets were selected: (1) Ghosh [5], which consists of 1975 samples, i.e. 975 labelled as non-sarcastic and 1000 labelled as sarcastic. This test set is used for in-domain validations; (2) Riloff [12], composed of 1956 tweets, i.e. 1648 non-sarcastic and 308 sarcastic posts. This test set is used for out-domain validations. Concerning **irony** detection, due to the limited number of available datasets, only the test set of [13] Task 3 A was chosen, with a total of 784 tweets, of which 473 as non-ironic and 311 as ironic. This test set is used for both constrained and unconstrained experimental settings. In the experiments, *Accuracy, Precision, Sensitivity*" and $F_1 - Measure$ are reported as the main measures of comparison among the models.

5   Results and Discussion

The first experiment regards the identification of the most representative features for identifying sarcasm and irony patterns (R1). To this purpose, Machine Learning classifiers introduced in Sect. 2, have been trained considering both embeddings (extracted from BERTweets and reduced by means of PCA) and hand-crafted features. The hyper-parameters of each model have been optimized using a k-folds cross-validation based on random search and considering accuracy as the target metric to optimize.

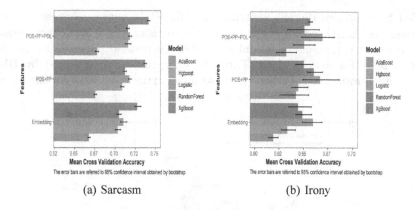

(a) Sarcasm   (b) Irony

Fig. 3. Comparison of feature contribution. The accuracy achieved by traditional machine learning models are reported, together with their confidence interval at 95%.

In Fig. 3, we report the most significant combinations of features considered by the traditional Machine Learning models. It is interesting to note that, for sarcasm detection, adding hand-creafted features related to pragmatic particles, part of speech and polarity, to the embeddings leads the models to achieve a significant improvement of F1 score with a 95% confidence level. However, this improvement emerges only in the case of sarcasm, while for the irony detection task, adding these features to the baseline of the embeddings, does not seem to discriminate better the information related to irony.

Regarding the remaining two research questions (R2 and R3), we compared the results of all the considered models, focusing on both in- and out-domain distributions. Figure 4, reports the results achieved in terms of F1-Measure.

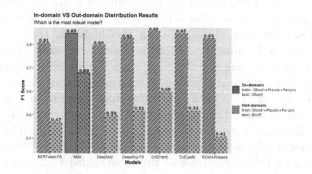

Fig. 4. Generalization abilities for sarcasm detection.

It is important to underline that the state of the art models, i.e. BMA, DeepMoji and RCNN-Roberta, achieve very good performance in the case of an in-domain distribution of the test set. However when processing a test set sampled from an out-domain distribution, the F1-Measure decreases by 40%. This suggests that the state of the art models focus on particular characteristics of the training set, and are not able to identify generic patterns for sarcasm that still hold for unseen data. The only exception is represented by BMA, which is much more robust when the unseen test data come from an out-domain distribution.

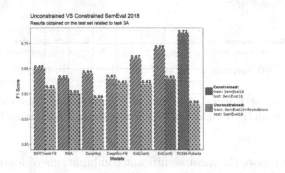

Fig. 5. Generalization abilities for irony detection.

Concerning the irony detection task, since only in-domain data are available for testing (SemEval 2018 test set), we compared the models in terms of constrained and unconstrained settings. When addressing an unconstrained task, where the training set is composed of tweets from different datasets (i.e. SemEval and Reyes-Rosso), the performance of the various models deteriorates significantly with respect to the constrained task, where the training and the testing data come from the same (SemEval 2018) distribution. By comparing the results reported in Fig. 5, it emerges that all the models are not able to capture the features that can discriminate what is irony from what is not, denoting therefore reduced generalization capabilities. We can also highlight that even if RCNN-Roberta is the best performing model for the constrained task, when introducing more variance in the training set, the model is no longer able to generalize well. On the contrary, the proposed EoE model emerges as more robust than others.

Regarding irony detection both in a constrained and unconstrained settings, we report in Tables 1 and 2 the comparison of our best performing model (EoE with soft classification) with the systems ranked in the official SemEval 2018 competition. We can highlight that the proposed model, in the constrained case (Table 1), is ranked third (the rank of the constrained task was based on F1-Measure).

Table 1. Ranking SemEval Task 3A, constrained

	Accuracy	Precision	Sensitivity	F₁-Measure
UCDCC	**0.797**	**0.788**	0.669	**0.724**
THUNGN	0.735	0.630	0.801	0.705
Ensemble of ensembles (soft)	0.693	0.681	0.692	0.690
NTUA-SLP	0.732	0.654	0.691	0.672
WLV	0.643	0.532	**0.836**	0.650

For the unconstrained task (Table 2), the results obtained by our EoE model are much better, highlighting that the proposed model outperforms the other teams that participated in the challenge (also in this case the rank of the unconstrained task was based on F1-Measure).

Table 2. Ranking SemEval Task 3A, unconstrained

	Accuracy	Precision	Sensitivity	F₁-Measure
Ensemble of ensembles (soft)	0.612	**0.661**	0.653	**0.631**
NonDicevo-SulSerio	**0.679**	0.583	0.666	0.622
INAOE-UPV	0.651	0.546	0.714	0.618
RM@IT	0.649	0.544	0.714	0.618
Valen'TO	0.598	0.496	**0.781**	0.607

The results reported above highlight that the proposed EoE model is quite robust even when considering more variance in the training data. In fact, EoE is not only ranked third in the constrained settings (with 0.69 of F₁-Measure), but it is placed first in the unconstrained scenario with a reduced drop of performance with respect to the constrained one.

6   Conclusions

The proposed models and the comparative analysis presented in this paper about irony and sarcasm has provided several insights. For the case of sarcasm, the models that achieved the best generalization are based on linguistic features, showing their robustness in the case of out-domain scenarios. Regarding irony, as the sample size increases, the performance of the models are reduced significantly. This shows that the structures of these models are not able to identify general information related to irony, but only focus on specific in-domain aspects.

References

1. Cho, K., et al.: Learning phrase representations using RNN encoder-decoder for statistical machine translation. In: Proceedings of the 2014 Conference on Empirical Methods in Natural Language Processing (EMNLP), pp. 1724–1734 (2014)
2. Felbo, B., Mislove, A., Søgaard, A., Rahwan, I., Lehmann, S.: Using millions of emoji occurrences to learn any-domain representations for detecting sentiment, emotion and sarcasm. In: Proceedings of Empirical Methods in Natural Language Processing, pp. 1615–1625 (2017)
3. Fersini, E., Pozzi, F.A., Messina, E.: Detecting irony and sarcasm in microblogs: the role of expressive signals and ensemble classifiers. In: 2015 IEEE International Conference on Data Science and Advanced Analytics (DSAA), pp. 1–8. IEEE (2015)
4. Ghanem, B., Karoui, J., Benamara, F., Rosso, P., Moriceau, V.: Irony detection in a multilingual context. In: Jose, J.M., et al. (eds.) ECIR 2020. LNCS, vol. 12036, pp. 141–149. Springer, Cham (2020). https://doi.org/10.1007/978-3-030-45442-5_18
5. Ghosh, A., Veale, T.: Fracking sarcasm using neural network. In: Proceedings of the 7th Workshop on Computational Approaches to Subjectivity, Sentiment and Social Media Analysis, pp. 161–169 (2016)
6. Kayalvizhi, S., Thenmozhi, D., Kumar, B.S., Aravindan, C.: Ssn_nlp@ idat-fire-2019: irony detection in Arabic tweets using deep learning and features-based approaches. In: FIRE (Working Notes), pp. 439–444 (2019)
7. Nguyen, D.Q., Vu, T., Nguyen, A.T.: BERTweet: A pre-trained language model for English tweets. In: Proceedings of the 2020 Conference on Empirical Methods in Natural Language Processing: System Demonstrations, pp. 9–14 (2020)
8. Potamias, R.A., Siolas, G., Stafylopatis, A.G.: A transformer-based approach to irony and sarcasm detection. Neural Comput. Appl. **32**(23), 17309–17320 (2020)
9. Pozzi, F.A., Fersini, E., Messina, E.: Bayesian model averaging and model selection for polarity classification. In: Métais, E., Meziane, F., Saraee, M., Sugumaran, V., Vadera, S. (eds.) NLDB 2013. LNCS, vol. 7934, pp. 189–200. Springer, Heidelberg (2013). https://doi.org/10.1007/978-3-642-38824-8_16
10. Ptáček, T., Habernal, I., Hong, J.: Sarcasm detection on Czech and English Twitter. In: Proceedings of the 25th International Conference on Computational Linguistics, pp. 213–223 (2014)
11. Reyes, A., Rosso, P., Veale, T.: A multidimensional approach for detecting irony in Twitter. Lang. Resour. Eval. **47**(1), 239–268 (2013)
12. Riloff, E., Qadir, A., Surve, P., Silva, L.D., Gilbert, N., Huang, R.: Sarcasm as contrast between a positive sentiment and negative situation. In: Proceedings of the 2013 Conference on Empirical Methods in Natural Language Processing, EMNLP, pp. 704–714 (2013)
13. Van Hee, C., Lefever, E., Hoste, V.: SemEval-2018 task 3: irony detection in English tweets. In: Proceedings of The 12th International Workshop on Semantic Evaluation, pp. 39–50 (2018)
14. Zhang, S., Zhang, X., Chan, J., Rosso, P.: Irony detection via sentiment-based transfer learning. Inf. Process. Manage. **56**(5), 1633–1644 (2019)

Extracting Facts from Case Rulings Through Paragraph Segmentation of Judicial Decisions

Andrés Lou[1]([✉]), Olivier Salaün[2], Hannes Westermann[3], and Leila Kosseim[1]

[1] CLaC Lab, Department of Computer Science and Software Engineering,
Concordia University, Montréal, QC H3G 1M8, Canada
{andres.lou,leila.kosseim}@concordia.ca
[2] RALI, Université de Montréal, Montréal, QC H3T 1J4, Canada
olivier.salaun@umontreal.ca
[3] CyberJustice Lab, Université de Montréal, Montréal, QC H3T 1J4, Canada
hannes.westermann@umontreal.ca

Abstract. In order to justify rulings, legal documents need to present facts as well as an analysis built thereon. In this paper, we present two methods to automatically extract case-relevant facts from French-language legal documents pertaining to tenant-landlord disputes. Our models consist of an ensemble that classifies a given sentence as either Fact or non-Fact, regardless of its context, and a recurrent architecture that contextually determines the class of each sentence in a given document. Both models are combined with a heuristic-based segmentation system that identifies the optimal point in the legal text where the presentation of facts ends and the analysis begins. When tested on a dataset of rulings from the *Régie du Logement* of the city of ANONYMOUS, the recurrent architecture achieves a better performance than the sentence ensemble classifier. The fact segmentation task produces a splitting index which can be weighted in order to favour shorter segments with few instances of non-facts or longer segments that favour the recall of facts. Our best configuration successfully segments 40% of the dataset within a single sentence of offset with respect to the gold standard. An analysis of the results leads us to believe that the commonly accepted assumption that, in legal documents, facts should precede the analysis is often not followed.

Keywords: Legal document · Text classification · Text segmentation

1 Introduction

Understanding the rationale behind a particular ruling made by a judge is a complex task that requires formal legal training in the relevant case law. Nevertheless, the rulings are still made using traditional methods of human discourse

Supported by the CLaC Lab and the CyberJustice Lab at the University of Montréal.

E. Métais et al. (Eds.): NLDB 2021, LNCS 12801, pp. 187–198, 2021.
https://doi.org/10.1007/978-3-030-80599-9_17

and reasoning, including default logic [WLRA15], deontic logic [DVRG15], and rhetoric [Wal95]. In particular, knowing all the relevant facts surrounding a case is of the utmost importance to understand the outcome of a ruling, as they are necessary to arrive at the best-possible decision, since facts are what give way to what is usually called the *Best Evidence* [Ste05,Nan87].

Within a ruling, however, determining what constitutes a fact, as opposed to other types of content that motivate and illustrate a judge's decision, is also a matter that requires formal training. Figure 1 shows a sample from our dataset. As the figure shows, a variety of linguistic factors, such as specialised terminology, textual structure, linguistic register, as well as domain knowledge, make the ruling stray from more general-domain texts. As such, fact extraction from legal texts is time consuming, expensive, and requires legal expertise. Additionally, as [WWAB19] have shown, even amongst trained experts, inter-annotator agreement tends to be low. For example, in the task of labelling a corpus of rulings with a pre-established set of labels on the ruling's subject matter, [WWAB19] reported low inter-annotator agreement and suggested that its cause might be the general vagueness and lack of explicit reasoning in the texts.

This paper proposes an automatic method to identify and segment facts in texts of rulings. The extraction of facts from a ruling is performed by classifying each sentence in the text as either belonging to the facts or not; this is followed by the identification of the boundary between the segment that holds the facts and the segment that holds everything else. To this end, we use two different approaches based on Deep Learning (DL) to perform the sentence classification task: a **sentence ensemble classifier**, which individually takes each sentence in the corpus and classifies it either as fact or non-fact, regardless of its context, and a **recurrent architecture**, which encodes each document in the corpus as a binary string, where each sentence is classified as fact or non-fact as a function of its context[1].

2   Related Work

The use of Natural Language Processing (NLP) to mine judicial corpora is not new; however, very little work has used neural methods, as most of the cited literature uses rules or hand-crafted features to perform their stated tasks. [dMWVE06] developed a parser that automatically extracts reference structures in and between legal sources. A few years later, [dMW09] developed a model based on syntactic parsing that automatically classifies norms in legislation by recognising typical sentence structures. [DMM+12] proposed a shared task on dependency parsing of legal texts and domain adaptation where participants compared different parsing strategies utilising the DeSR parser [Att06] and a weighted directed graph-based model [ST10]. [GAC+15] demonstrated the feasibility of extracting argument-related semantic information and used it to improve document retrieval systems. [WHNY17] introduced an annotated dataset based

---

[1] The source code is publicly available at https://gitlab.com/Feasinde/fact-extraction-from-legal-documents.

Original sentence	English translation	Tag
Comme le mandat fourni à l'audience par Madame <NAME> émane de <ORG>. qui n'est pas le véritable locateur, ce mandat n'est pas conforme à l'article 72 de la Loi sur la Régie du logement.	*As the mandate provided at the hearing by Ms. <NAME> emanates from <ORG>, who is not the real landlord, the mandate does not comply with section 72 of the Act respecting the Régie du logement.*	Fact
Rien ne prouve par ailleurs que <NAME> est employée de <ORG> et <ORG> puisque ces compagnies ne lui ont pas donné de mandat.	*There is also no evidence that <NAME> is an employee of <ORG> and <ORG> since these companies did not give him a mandate.*	Fact
Ceci étant dit, revenons à l'argumentation de Monsieur <NAME> et de Madame <NAME> voulant que toutes ces compagnies soient liées entre elles et puissent représenter l'autre sans plus de formalité.	*That said, let us return to the argument of <NAME> and <NAME> that all of these companies are linked and can represent the other without further formality.*	non-Fact
Cet argument ne tient pas; en effet, même si les compagnies sont dirigées par les mêmes personnes et que l'une soit l'actionnaire majoritaire de l'autre, il n'en demeure pas moins qu'il s'agit d'entités juridiques distinctes.	*This argument does not hold; even if the companies are managed by the same people and one is the majority shareholder of the other, the fact remains that they are separate legal entities.*	non-Fact

Fig. 1. Example of sentence classification as either stating a case Fact or non-Fact. (English translations provided by the authors. Proper names redacted.)

on propositional connectives and sentence roles in veterans' claims, wherein each document is annotated using a typology of propositional connectives and the frequency of the sentence types that led to adjudicatory decisions. [SA17] used Conditional Random Fields (CRF) to extract sentential and non-sentential content from decisions of United States courts, and also extract functional and issue-specific segments [SA18], achieving near-human performance. Finally, [DVRG15] performed rule-extraction from legal documents using a combination of a syntax-based system using linguistic features provided by WordNet [Mil95], and a logic-based system that extracts dependencies between the chunks of their dataset. Their work is closely related to our task; however, whereas [DVRG15] base their model on syntactic and logical rulesets, we base our model on Recurrent and Convolutional Neural models, introduce our own segmentation heuristic, and use independent and contextual word embeddings [MCCD13, MMO+19].

Outside the frame of legal texts, semantic sentence classification has recently achieved new benchmarks thanks to the application of neural methods. The use Recurrent Neural Networks (RNN), and their derivative recurrent architectures (in particular encoder-decoder models [CVMG+14]), has steadily produced results that outperform traditional models in many NLP tasks such as Machine Translation, Question Answering and Text Summarisation (eg the Attention mechanism [BCB14] and the Sequence-to-Sequence model [SVL14], two of the most important architectural developments in RNNs). Sentence classification using Deep Learning (DL) was first proposed by [Kim14], who used Convolutional Neural Networks (CNN) and showed that their models outperformed classical models on many standard datasets. Their work was quickly followed by the application of RNN architectures for similar tasks, including those of [LXLZ15, ZZL15], and [ZQZ+16]. A major breakthrough was achieved with the Transformer [VSP+17], whose non-recurrent, Multi-Head Attention architecture allowed for richer language model representations that capture many more linguistic features than the original attention mechanism. Subsequently, Google's BERT [DCLT18] has given way to a whole new family of language models able to produce state-of-the-art contextual embeddings for both individual tokens in a sentence and for the sentence itself. The following paragraphs will explain these architectures in detail.

3   Methodology

3.1   The Dataset

The current work was developed as part of the JusticeBot project. JusticeBot aims to provide a gateway to law and jurisprudence for lay people [WWAB19] through a chatbot where users can seek remedies to terminate their lease because of landlord-tenant disputes. The chatbot was developed using a corpus of 1 million written decisions in French, provided by the *Régie du Logement* of the city of Montréal. One of the numerous tasks related to the development and training of the chatbot is the extraction of case-related facts from a given document in the corpora.

The dataset used for fact extraction consists of a subset of the *Régie du Logement*'s corpus and includes 5,605 annotated rulings; these were selected from the original dataset because they include an explicit separation between two distinct sections: Facts and Analysis, as determined by the original author of the ruling (the judge), and delimited by appropriate headings. These two sections have been used as gold standard annotations to train and test our model. The **Facts** section should consist of all the case-relevant facts on which the ruling is supported, while the **non-Facts** should contain the analysis and discussion of the facts that ultimately lead to the resolution presented in the document. Table 1 shows statistics of the dataset.

As Table 1 shows, Fact and non-Fact segments are similar both in terms of average number of words per sentence and average number of sentences per segment, which is why the process of detecting either cannot rely on simple word or sentence statistics.

Table 1. Statistics of the dataset

Number of documents	5,605
Total number of sentences	454,210
Total number of sentences in Facts segments	239,077
Av number of sentences in Facts segments	36.25
Total number of sentences in non-Facts segments	215,133
Av number of sentences in non-Facts segments	32.62

3.2 Sentence Classification

Given a document, the first step in our approach is to represent its contents as a binary sequence, where sentences that include case-related facts are represented by continuous sub-sequences of 1's and the sentences containing everything else are represented by continuous sub-sequences of 0's. In order to produce this binary encoding of the document, we examine two methods: a **sentence ensemble classifier** method, and a **recurrent architecture** method.

The Sentence Ensemble Classifier. The sentence ensemble classifier method processes a given document as a collection of sentences whose classes are independent of one another. Each sentence in the corpus, regardless of the document in which it is found, is classified as either *fact* or *non-fact* using an ensemble model consisting of the combination of a Gated Recurrent Unit network (GRU) [CVMG+14]) and a Convolutional Neural Network (CNN). The process is illustrated in Fig. 2. In the recurrent part of the classifier (Fig. 2a), a tokenised sentence is passed through an Word2Vec embedding layer [MCCD13] whose outputs are passed into a stack of GRU layers, producing a context vector $h_T^{(R)}$; In the CNN part of the classifier (Fig. 2b), the input is a tensor of size $1 \times T \times k$, where T is the sequence length, and k is the size of the word vector. The output feature maps are passed through a 1-D Max Pool layer that produces an output vector $h_T^{(C)}$. The concatenation of $h_T^{(R)}$ and $h_T^{(C)}$ (Fig. 2c) is passed through an affine layer with a softmax activation function to produce the probability of the sentence being Facts.

The Recurrent Architecture. The recurrent architecture approach tries to determine whether a sentence is classified as *fact* or *non-fact* by using the sentences around it. We experimented with two different models to process a given document: a bidirectional GRU and an Encoder-Decoder model using an Attention mechanism [BCB14]. The process is illustrated in Fig. 3. The input document is split into sentences, and each sentence is vectorised using the Camem-BERT language model [MMO+19]. The bidirectional GRU model (Fig. 3a) produces an output ($\overrightarrow{\mathbf{h}_i}$ and $\overleftarrow{\mathbf{h}_j}$) at each time step as a function of the previous and following steps, and the sentence input; after this, the output is passed through

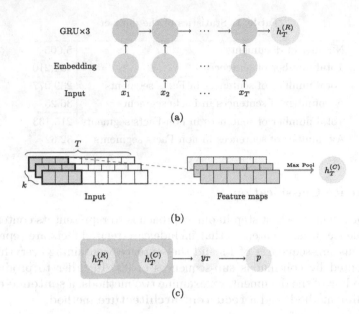

Fig. 2. Architecture of the sentence ensemble classifier.

a affine layer with a binary output. The encoder of the attention mechanism (Fig. 3b) is similar to this architecture, except the output of the bidirectional GRU is gathered as an input matrix **H** and passed to the decoder such that each time step output becomes a weighted function of every other token in the input sequence. The networks are trained so that the outputs correspond to the binary representation of the document.

3.3   Text Segmentation

Once each sentence is classified as fact (1) or non-fact (0), the next step is to optimally divide the sequence into two substrings, each representing the *Facts* and *non-Facts* segments of the ruling. Figure 4 illustrates this. To perform the segmentation, we establish the following propositions:

- Let L represent the number of sentences document.
- Let L_f represent the number of sentences in its Facts segment.
- Let n_f represent the number of Facts sentences found in L_f (such that $n_p \leq L_f$).
- Define $p_f = \frac{n_f}{L_f}$ as the *purity* of L_f.

Maximising L_f is equivalent to maximising the recall of facts in the segmentation, and maximising p_f ensures the segmentation corresponds as closely as possible to the gold standard. Hence, our approach aims at maximising both L_f and p_f. This can be described as the following optimisation problem:

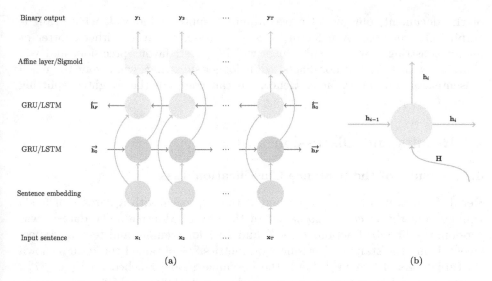

Fig. 3. Architecture of the LSTM/GRU and Attention Encoder-Decoder method

$$\overbrace{}^{L_f}$$

11110111111100000100001

$$\underbrace{}_{L}$$

Fig. 4. Visual representation of the segmentation of the binary string where 1 refers to a sentence classified as *fact* and 0 as *non-fact*. In this example, $L_f = 13$, $L = 24$ and $n_f = 12$.

$$\max J(L_f) = \max \left(\alpha L_f + \beta p_f \right) \tag{1}$$

where J is a loss function of L_f, and $\alpha, \beta \in \mathbb{R}$ are arbitrary weights representing the importance of each term. We can rewrite and differentiate Eq. 1 to find an expression that optimises L_f:

$$L_f = \frac{\beta}{\alpha} p_f \tag{2}$$

Equation 2 indicates that there is a linear relation between the purity (p_f) of a substring and its length (L_f). Therefore, for all possible substrings of length L_{f_i} in the original string, we select the one that maximises p_f. Since any substring comprised exclusively of 1's will have a trivial purity $p_{f_i} = 1$, we select L_{f_i} that maximises p_f such that $p_f \neq 1$.

By considering the problem of finding the optimal segmentation point as extracting the substring with the highest purity from a binary representation of the document, our model can compute a splitting index, l, which can be empirically weighted by a factor, $\gamma \propto \frac{\beta}{\alpha}$ in order to favour either shorter or longer substrings. Shorter substrings will be purer, favour precision, and will contain few instances of non-Facts, while longer substrings will favour the recall of sentences containing facts. Hence, we can compute the weighted splitting index: $L_\gamma = \gamma l$

4   Results and Discussion

4.1   Results of the Sentence Classification Task

Recall from Sect. 3.1 that we evaluated the approach with a dataset of 5,605 rulings from the *Régie de Logement* of the city of Montréal. The dataset was randomly split into fractions of 90% and 10% for training and testing respectively. Using the standard classification metrics, we obtained the results shown in Table 2. As shown in Table 2, the ensemble model reached an F_1 of 77%, where as the GRU and the Attention models reached 99% and 90% respectively. Given the low performance of the ensemble, we experimented with a data augmentation technique. We used part-of-speech lexical substitution for data augmentation (PLSDA) [XCL+20], generating new sentences by randomly replacing POS-annotated tokens in a given sentence with syntactically identical synonyms. We used the spaCy tokeniser and annotator [HM17] and WordNet [Mil95]. We observed no considerable improvement in our performance when doubling the number of training instances.

As Table 2, the recurrent architecture's improved performance suggests that contextually determining whether a sentence is *fact* or *non-fact* is a much better approach than assuming individual sentences are independently distributed from one another.

Table 2. Intrinsic performance of the sentence classification task

Model	Accuracy	Precision	Recall	F_1
Ensemble	0.79	0.78	0.76	0.77
Ensemble (PLSDA)	0.79	0.78	0.76	0.77
GRU	0.98	0.98	0.99	0.99
Attn	0.92	0.91	0.92	0.90

4.2   Results of the Text Segmentation Task

We evaluated the text segmentation on the test set of documents (660) using the headings separating Facts and non-Facts as gold standard and using both the augmented sentence ensemble classifier and the recurrent architecture methods.

Given a value of γ, for each document, we compute its corresponding splitting index l_{pred} and split the text according to the weighted splitting index L_γ. We then compute the percentage of sentences by which the resulting text is off compared to the gold standard of number of sentences in the Facts section. Results are shown in Table 3.

Table 3. Different values of γ and the number of sentence by which the predicted segmentation is off with respect to the gold standard. Bold indicates the number of documents obtained at the expected splitting index for $\gamma = 1$.

	Offset										
γ	< −4	-4	-3	-2	-1	0	1	2	3	4	> 4
0.6	602	25	16	8	3	3	2	0	1	0	0
0.8	570	25	30	7	13	6	1	3	3	0	2
1	529	19	26	21	15	11	13	8	3	5	10
1.2	525	21	24	22	16	9	11	8	6	8	10
1.4	520	17	26	21	16	11	13	6	9	7	14

(a) Sentence Ensemble Classifier

	Offset										
γ	< −4	-4	-3	-2	-1	0	1	2	3	4	> 4
0.6	517	44	60	23	10	3	1	0	0	0	2
0.8	319	74	75	89	77	16	1	3	2	1	3
1	176	8	2	6	30	**408**	16	1	4	0	9
1.2	137	2	7	7	30	445	17	0	4	1	10
1.4	112	2	2	9	33	468	16	3	3	2	10

(b) Recurrent Architecture: GRU

	Offset										
γ	< −4	-4	-3	-2	-1	0	1	2	3	4	> 4
0.6	518	42	57	21	11	5	2	0	0	1	3
0.8	326	70	73	90	74	15	1	5	0	0	6
1	179	8	4	5	46	**385**	15	4	4	0	10
1.2	140	2	12	7	50	406	14	8	9	2	10
1.4	112	2	6	7	66	413	17	10	11	1	15

(c) Recurrent Architecture: Attention

Table 3 shows the number of documents that fall within a distance (in sentences) from the expected index, given a weighting value of γ. For example, as shown in Table 3b, for the GRU recurrent architecture at $\gamma = 1$, 408 of the documents (61%) are segmented exactly where the gold standard indicates, while 454 documents $(30+408+16, 68\%$ of the test dataset) fall within a single sentence of difference with respect to the gold standard; nevertheless, 176 documents (27%) have their index underestimated and fall short of the target, having more than 4 sentences fewer than the gold standard. Increasing the value of γ favours the recall of sentences annotated as Facts, but the percentage of documents whose segmentation falls short by more than 4 sentences does not decrease as quickly as the percentage of overestimated segmentations; for $\gamma = 1.4$, the number of underestimated segmentations by a margin greater than 4 is still 112 (17%).

For the different values of γ, the distribution of offsets presents a large number of underestimated splitting indices, which suggests that the distribution of fact sentences and non-fact sentences does not actually follow our base assumption, namely, that facts should always give way to analyses. The gold standard expects us to find many more facts after the predicted splitting index, weighted or otherwise, which suggests that some cases either contain an imbalance of facts and analyses or contain facts and analysis interspersed with each other on a larger scale than expected.

5   Conclusion

This paper presented a method to automatically extract case-relevant facts in French-language legal documents pertaining to tenant-landlord disputes using text segmentation. We used two approaches based on classifying the sentences of a given document as either *facts* or *non-facts*: considering each sentence as independent from all others, and using the context in which the sentence is found to predict its class. Subsequently, we used a metric based on the purity of the facts substring to find an optimal splitting index and perform the segmentation.

Experiments with French-language rulings of the *Régie du Logement* of the city of Montréal produced a significant number of underestimations (up to 27%); this seems to indicate that the standard assumption that the discourse structure should be such that all facts will precede the analysis is not always followed. Indeed, our text segmentation approach, based on the heuristic of maximising the density of facts on the purported facts segment of the ruling, has shown that the distribution of facts is not usually concentrated in the first segment of the text.

Our work has considered sentences as the unit of classification; a sentence that contains facts is considered fact-bearing even if it might also contain analysis. Future work might explore a more fine-grained intra-sentence analysis in order to find smaller fact-bearing units than sentences. Additionally, future work should also involve the classification and rearrangement of sentences, perhaps by means of standard automatic summarisation techniques [NM12], in order to produce coherent paragraphs that both maximise the purity of a substring and the recall of facts. Finally, rather than segmenting legal documents as a single fact-analysis block, it might be worth considering breaking them down into smaller fact-analysis constituents.

Acknowledgment. The authors would like to thank the anonymous reviewers for their comments on an earlier version of this paper.

References

[Att06]   Attardi, G.: Experiments with a multilanguage non-projective dependency parser. In: Proceedings of the Tenth Conference on Computational Natural Language Learning (CoNLL-X), pp. 166–170, June 2006

[BCB14]   Bahdanau, D., Cho, K., Bengio, Y.: Neural machine translation by jointly learning to align and translate. arXiv preprint arXiv:1409.0473, May 2014

[CVMG+14] Cho, K., et al.: Learning phrase representations using RNN encoder-decoder for statistical machine translation. arXiv preprint arXiv:1406.1078, September 2014

[DCLT18]   Devlin, J., Chang, M.-W., Lee, K., Toutanova, K.: BERT: pre-training of deep bidirectional transformers for language understanding. arXiv preprint arXiv:1810.04805 (2018)

[DMM+12]   Dell'Orletta, F., Marchi, S., Montemagni, S., Plank, B., Venturi, G.: The SPLeT-2012 shared task on dependency parsing of legal texts. In: Proceedings of the 4th Workshop on Semantic Processing of Legal Texts, pp. 42–51, May 2012 [dMW09] de Maat, E., Winkels, R.: A next step towards automated modelling of sources of law. In: Proceedings of the 12th International Conference on Artificial Intelligence and Law, pp. 31–39. ACM, June 2009

[dMWVE06] de Maat, E., Winkels, R., Van Engers, T.: Automated detection of reference structures in law. Front. Artif. Intell. Appl. 41 (2006)

[DVRG15] Dragoni, M., Villata, S., Rizzi, W., Governatori, G.: Combining natural language processing approaches for rule extraction from legal documents. In: Pagallo, U., Palmirani, M., Casanovas, P., Sartor, G., Villata, S. (eds.) AICOL 2015-2017. LNCS (LNAI), vol. 10791, pp. 287–300. Springer, Cham (2018). https://doi.org/10.1007/978-3-030-00178-0_19

[GAC+15] Grabmair, M., et al.: Introducing LUIMA: an experiment in legal conceptual retrieval of vaccine injury decisions using a UIMA-type system and tools. In: Proceedings of the 15th International Conference on Artificial Intelligence and Law, pp. 69–78. ACM, June 2015

[HM17] Honnibal, M., Montani, I.: spaCy 2: natural language understanding with bloom embeddings. Convolutional Neural Netw. Incremental Parsing. 7(1) (2017, to appear)

[Kim14] Kim, Y.: Convolutional neural networks for sentence classification. arXiv preprint arXiv:1408.5882, September 2014

[LXLZ15] Lai, S., Xu, L., Liu, K., Zhao, J.: Recurrent convolutional neural networks for text classification. In: Proceedings of the 29th AAAI Conference on Artificial Intelligence, pp. 2267–2273, February 2015

[MCCD13] Mikolov, T., Chen, K., Corrado, G., Dean, J.: Efficient estimation of word representations in vector space (2013)

[Mil95] Miller, G.A.: WordNet: a lexical database for English. Commun. ACM 38(11), 39–41 (1995)

[MMO+19] Martin, L., et al.: CamemBERT: a Tasty French Language Model. arXiv e-prints, arXiv:1911.03894, November 2019

[Nan87] Nance, D.A.: The best evidence principle. Iowa Law Rev. 73, 227 (1987)

[NM12] Nenkova, A., McKeown, K.: A survey of text summarization techniques. In: Aggarwal, C., Zhai, C. (eds.) Mining Text Data, pp. 43–76. Springer, Heidelberg (2012). https://doi.org/10.1007/978-1-4614-3223-4_3

[SA17] Savelka, J., Ashley, K.D.: Using conditional random fields to detect different functional types of content in decisions of united states courts with example application to sentence boundary detection. In: Workshop on Automated Semantic Analysis of Information in Legal Texts, p. 10 (2017)

[SA18] Savelka, J., Ashley, K.D.: Segmenting US court decisions into functional and issue specific parts. In: JURIX: The 31st International Conference on Legal Knowledge and Information Systems, pp. 111–120 (2018)

[ST10] Sagae, K., Tsujii, J.-I.: Dependency parsing and domain adaptation with data-driven LR models and parser ensembles. In: Bunt, H., Merlo, P., Nivre, J. (eds.) Trends in Parsing Technology, pp. 57–68. Springer, Heidelberg (2010). https://doi.org/10.1007/978-90-481-9352-3_4

[Ste05] Stein, A.: Epistemological corollary. In: Foundations of Evidence Law. Oxford University Press (2005)

[SVL14] Sutskever, I., Vinyals, O., Le, Q.V.: Sequence to sequence learning with neural networks. In: Advances in Neural Information Processing Systems, pp. 3104–3112, December 2014

[VSP+17] Vaswani, A., et al.: Attention is all you need. In: Advances in Neural Information Processing Systems, pp. 5998–6008 (2017)

[Wal95]   Wald, P.M.: The rhetoric of results and the results of rhetoric: judicial writings. Univ. Chicago Law Rev. **62**(4), 1371–1419 (1995)

[WHNY17] Walker, V.R., Han, J.H., Ni, X., Yoseda, K.: Semantic types for computational legal reasoning: propositional connectives and sentence roles in the veterans' claims dataset. In: Proceedings of the 16th Edition of the International Conference on Artificial Intelligence and Law, pp. 217–226 (2017)

[WLRA15] Walker, V.R., Lopez, B.C., Rutchik, M.T., Agris, J.L.: Representing the logic of statutory rules in the United States. In: Araszkiewicz, M., Płeszka, K. (eds.) Logic in the Theory and Practice of Lawmaking. LL, vol. 2, pp. 357–381. Springer, Cham (2015). https://doi.org/10.1007/978-3-319-19575-9_13

[WWAB19] Westermann, H., Walker, V.R., Ashley, K.D., Benyekhlef, K.: Using factors to predict and analyze landlord-tenant decisions to increase access to justice. In: Proceedings of the 17th International Conference on Artificial Intelligence and Law, pp. 133–142. ACM, June 2019

[XCL+20] Xiang, R., Chersoni, E., Long, Y., Lu, Q., Huang, C.-R.: Lexical data augmentation for text classification in deep learning. In: Goutte, C., Zhu, X. (eds.) Canadian AI 2020. LNCS (LNAI), vol. 12109, pp. 521–527. Springer, Cham (2020). https://doi.org/10.1007/978-3-030-47358-7_53

[ZQZ+16] Zhou, P., Qi, Z., Zheng, S., Xu, J., Bao, H., Xu, B.: Text classification improved by integrating bidirectional LSTM with two-dimensional max pooling. arXiv preprint arXiv:1611.06639, April 2016

[ZZL15]   Zhang, X., Zhao, J., LeCun, Y.: Character-level convolutional networks for text classification. In: Advances in Neural Information Processing Systems, pp. 649–657, December 2015

Detection of Misinformation About COVID-19 in Brazilian Portuguese WhatsApp Messages

Antônio Diogo Forte Martins[1]([✉]), Lucas Cabral[1],
Pedro Jorge Chaves Mourão[2], José Maria Monteiro[1], and Javam Machado[1]

[1] Federal University of Ceará, Fortaleza, Ceará, Brazil
{diogo.martins,jose.monteiro,javam.machado}@lsbd.ufc.br,
lucascabral@aridalab.dc.ufc.br
[2] Universidade Estadual do Ceará, Fortaleza, Ceará, Brazil
pedro.mourao@aluno.uece.br

Abstract. During the coronavirus pandemic, the problem of misinformation arose once again, quite intensely, through social networks. In many developing countries such as Brazil, one of the primary sources of misinformation is the messaging application WhatsApp. However, due to WhatsApp's private messaging nature, there still few methods of misinformation detection developed specifically for this platform. Additionally, a MID model built to Twitter or Facebook may have a poor performance when used to classify WhatsApp messages. In this context, the automatic misinformation detection (MID) about COVID-19 in Brazilian Portuguese WhatsApp messages becomes a crucial challenge. In this work, we present the COVID-19.BR, a data set of WhatsApp messages about coronavirus in Brazilian Portuguese, collected from Brazilian public groups and manually labeled. Besides, we evaluated a series of misinformation classifiers combining different techniques. Our best result achieved an F1 score of 0.778, and the analysis of errors indicates that they occur mainly due to the predominance of short texts. When texts with less than 50 words are filtered, the F1 score rises to 0.857.

Keywords: Misinformation detection · Fake news detection · Natural language processing · WhatsApp · COVID-19

1 Introduction

During the coronavirus pandemic, the problem of misinformation arose once again, quite intensely, through social networks. In April 2020, the United Nations (UN) declared that there is a "dangerous misinformation epidemic", responsible for the spread of harmful health advice and false solutions. The misinformation

---

Supported by CAPES and LSBD.

concept can be understood as a process of intentional production of a communicational environment based on false, misleading, or decontextualized information to cause a communicational disorder [11].

Currently, the main tool used to spread misinformation is WhatsApp instant messaging application. Through this application, misinformation can deceive thousands of people in a short time, bringing great harm to public health. A very relevant WhatsApp feature is the public groups which are accessible through invitation links published on popular websites and social networks. Each group can put together a maximum of 256 members and they usually have specific topics for discussion, very similar to social networks. Thus, these public groups have been used to spread misinformation.

In this context, the automatic misinformation detection (MID) about COVID-19 in Brazilian Portuguese WhatsApp messages becomes a crucial challenge. In a wide definition, MID is the task of assessing the appropriateness (truthfulness, credibility, veracity, or authenticity) of claims in a piece of information [11]. However, due to WhatsApp's private messaging nature, there are still few MID methods developed specifically for this platform. Additionally, a MID model built to Twitter or Facebook may have a poor performance when used to classify WhatsApp messages. A model's performance is highly dependent on the linguistic patterns, topics, and vocabulary present in the data used to train it. Nevertheless, the linguistic patterns found in WhatsApp messages are quite different from those found in Facebook and Twitter [12]. Thus, despite the scientific community's efforts, there is still a need for a large-scale corpus containing WhatsApp messages in Portuguese about COVID-19.

In order to fill this gap, we built a large-scale, labeled, anonymized, and public data set formed by WhatsApp messages in Brazilian Portuguese (PT-BR) about coronavirus pandemic, collected from public WhatsApp groups. Then, we conduct a series of classification experiments using different machine learning methods to build an efficient MID for WhatsApp messages. Our best result achieved an F1 score of 0.778 due to the predominance of short texts.

2   Related Work

Several works attempt to detect misinformation in different languages and platforms. Most of them use news in English or Chinese languages. Further, Websites and social media platforms with easy access are amongst the main data sources used to build misinformation data sets.

The study presented in [2] proposes a misleading-information detection model that relies on several contents about COVID-19 collected from the World Health Organization, UNICEF, and the United Nations, as well as epidemiological material obtained from a range of fact-checking websites. The research presented in [1] proposed a set of machine learning techniques to classify information and misinformation. In [6], the authors introduced CoVerifi, a web application that combines both the power of machine learning and the power of human feedback to assess the credibility of news about COVID-19. The study presented in

[4] proposed a multimodal multi-image system that combines information from different modalities in order to detect fake news posted online.

3   Data Set Design

An important aspect to consider while developing a MID method for WhatsApp messages in Brazilian Portuguese is the necessity of a large-scale labeled data set. However, there is no corpus for Brazilian Portuguese with these characteristics as far as we know. Besides, due to its private chat purpose, WhatsApp does not provide a public API to automatically collect data. Thus, build this data set is a technical, also ethical challenge. For this reason, we used a methodology similar to [3, 10] to build a large-scale labelled corpus of WhatsApp messages in Brazilian Portuguese.

In order to create the data set presented in this paper, we collected messages from open WhatsApp groups. These groups were found by searching for "chat.whatsapp.com/" on the Web. Next, we analyzed the theme and purpose of each group found previously. Then, we selected 236 public groups. After this, we joined these groups and started collecting messages. Each collected message is stored in a row of the data set. Finally, we select a message subset called "viral messages". We defined "viral messages" as identical messages with more than five words that appear more than once in the data set. It is important to highlight that sensitive attributes such as user name, cell phone number and group name were anonymized using hash functions. Figure 1 shows an extract from our data set after anonymization and before data labeling. Our data set has 228061 WhatsApp messages from users and groups from all over Brazil.

	id	date	hour	ddi	country	country_iso3	ddd	state	group	midia	url	characters	words	viral	sharings	text
0	146759200457638065	07/04/20	04:07	55	BRASIL	BRA	21	Rio de Janeiro	2020_1	0	0	9	1	0	1	Morreram?
1	146759200457638065	07/04/20	04:07	55	BRASIL	BRA	21	Rio de Janeiro	2020_1	0	0	24	4	0	1	Olá novato, se apresente
2	5788106393468158140	07/04/20	04:07	55	BRASIL	BRA	21	Rio de Janeiro	2020_1	0	0	9	2	0	1	há tempos
3	146759200457638065	07/04/20	04:07	55	BRASIL	BRA	21	Rio de Janeiro	2020_1	0	0	13	2	0	1	Legião Urbana
4	5788106393468158140	07/04/20	04:13	55	BRASIL	BRA	21	Rio de Janeiro	2020_1	0	0	6	1	0	1	Indios

Fig. 1. Extract from the collected data before the labeling process.

In order to build a high-quality corpus, data labelling is another hard challenge since we have to specify if the text is true or false based on trusted sources, such as specialized journalists or fact-checking sites. So, we conducted the data labeling process entirely manually. A human specialist checked each message's content and determined if it contains or not misinformation. Since this process is time-consuming, we chose to label only unique messages containing the following keywords: "covid", "coron", "virus", "china", "chines", "cloroquin", "vacina". The resulting data set now has 2899 unique messages. We labeled all these messages with the general misinformation definition adopted in [11] labeling them as 0 if the message does not contain misinformation and 1 if it contains misinformation. Three annotators, two computer science masters students and one sociologist, conducted the labeling process. We solved labeling disagreements executing a collective review round.

Our labeling process was based on the following steps. If the text contains verifiable untrue claims, we annotate it as misinformation. We made use of trustful Brazilian fact-checking platforms such as *Agência Lupa*[1] and *Boatos.org*[2]. If the text contains imprecise, biased, alarmist, or harmful claims that cannot be proven, we annotate it as misinformation. If the text is short and accompanied by media content (image, video, or audio), we search on the web for the media content and, if we find the corresponding media, we decide the label based on the previous criteria. If the original media cannot be found, we use the second criterion to label it. And If none of the previous criteria is found in the text, we label it as not containing misinformation.

After the labeling process, we removed messages with only *url* as text content. So, the resulting corpus contains 532 unique messages labeled as misinformation (label 1) and 858 unique messages labeled as non-misinformation (label 0). Table 1 presents basic statistics about the data set.

Table 1. Data set basic statistics.

Statistics	Non-misinformation	Misinformation
Count of unique messages	858	532
Mean and std. dev. of number of tokens	92.02 ± 203.24	167.02 ± 248.02
Minimum number of tokens	1	1
Median number of tokens	20	50
Maximum number of tokens	3100	1666
Mean and std. dev. of shares	2.51 ± 4.85	2.47 ± 3.41

4   Experiments

We have explored multiple combinations between feature extraction from text and classification algorithms. We performed our experiments using k-fold cross-validation with $k = 5$ folds. We also performed a Bayesian optimization over hyperparameters to search the optimal configuration for the best classifiers. Besides, we evaluate different techniques for text feature extraction, but we decided to use traditional Bag-Of-Words (BoW) and TF-IDF text representations in our experiments. Since one of our goals is to define a baseline for automatic MID about the COVID19 in WhatsApp messages in Brazilian Portuguese, these techniques features are suitable for this purpose and have been already used in a wide range of text classification problems.

---

[1] http://piaui.folha.uol.com.br/lupa/.
[2] http://www.boatos.org/.

Our text pre-processing method consists in convert to lowercase, separate emojis with white spaces to avoid generating a new token for each emoji sequence, and maintain only the domain name for *urls*. Because of the lexical diversity of the corpus, the resulting vectors have large dimensions and sparsity. Moreover, we added more variety to our experiments by using different n-gram values. So, we combined these different vectorization techniques (TF-IDF or binary BoW), the n-grams range (unigrams, bigrams, and trigrams), and the extra steps of pre-processing (lemmatization and stop words removal), leading to a total of 12 different feature extraction scenarios.

For each scenario, we performed experiments using nine machine learning classification techniques, already used in several text classification tasks [7]: logistic regression (LR), Bernoulli (if the features are BoW) or Complement Naive-Bayes (if features are TF-IDF) (NB) [5,9], support vector machines with a linear kernel (LSVM), SVM trained with stochastic gradient descent (SGD), SVM trained with an RBF kernel [8] (SVM), K-nearest neighbors (KNN), random forest (RF), gradient boosting (GB), and multilayer perceptron neural network (MLP). At first, all techniques were used with default hyperparameters. Next, we performed a Bayesian optimization to find the optimal hyperparameters for the best combinations of features and classifiers.

Just considering all combinations between features, pre-processing, and classification methods and excluding the Bayesian optimization step, we performed a total of 108 experiments, all of them using k-fold cross-validation with $k = 5$.

In order to evaluate the performance of the experiments and considering we are working with a binary classification task, where non-misinformation represents the negative class and misinformation the positive, we use the following metrics: False positive rate (FPR), Precision (PRE), Recall (REC), and F1-score (F1). Because we use k-fold cross-validation, each metric's mean are collected and will also be presented.

5   Results

For the sake of readability, we included only the results of the top 10 best combinations of classifiers and features extraction techniques. The results presented in the following tables are the metrics' mean after 5 rounds of k-fold cross-validation.

Table 2 summarizes the results for the experiments we run with standard hyperparameters. Analyzing the F1 values, we can observe that the difference is not large, less than 1% from the first to last. We achieved the best results when using BoW and NB. The removal of stop words and lemmatization helped improve some of NB results in the trigram and bigram scenarios. When using TF-IDF and LSVM, we achieved the lowest value of FPR among the top 5 results. The best result was obtained using BoW as feature extractor, bigram, removing stop words and performing lemmatization, and with the NB classifier.

Next, we performed a Bayesian optimization over the hyperparameters to search the optimal configuration for the classifiers. For NB, the best value of *alpha* was 0; for LSVM, the best value of C was 348.61; for SGD, the best value of *alpha* was 0.00185. Table 3 summarizes the results of the experiments with the best hyperparameters. Analyzing the results, we can see that now the best combination of classifier and features extraction techniques is SGD using BoW as feature extractor, trigram, removing stop words, and performing lemmatization (with 0.4% of improvement in F1 and 3.3% of improvement in FPR). Besides, the result shows that, even if we searched for hyperparameters for a specific combination, we improved the SGD classifier performance using different feature extraction methods.

Table 2. Top 10 best combinations of classifiers and features extraction techniques. All presented metrics values are the mean after 5 rounds of cross-validation.

Rank	Experiment	Vocabulary	FPR	PRE	REC	F1
1	BOW-BIGRAM-LEMMA-NB	70986	0.179	0.734	0.840	0.774
2	TFIDF-BIGRAM-LSVM	84189	0.149	0.775	0.780	0.773
3	BOW-UNIGRAM-NB	15165	0.183	0.734	0.833	0.771
4	TFIDF-TRIGRAM-SGD	190376	0.160	0.746	0.804	0.770
5	BOW-TRIGRAM-LEMMA-NB	147900	0.182	0.728	0.836	0.770
6	BOW-UNIGRAM-LEMMA-NB	13039	0.183	0.730	0.836	0.769
7	TFIDF-TRIGRAM-LEMMA-SGD	147900	0.162	0.741	0.808	0.769
8	BOW-BIGRAM-NB	84189	0.181	0.733	0.827	0.768
9	BOW-TRIGRAM-NB	190376	0.178	0.736	0.821	0.768
10	TFIDF-TRIGRAM-MLP	190376	0.152	0.779	0.772	0.768

Table 3. Top 10 best combinations of classifiers and features extraction using the Bayesian optimization hyperparameters. All presented metrics values are the mean after 5 rounds of cross-validation.

Rank	Experiment	Vocabulary	FPR	PRE	REC	F1
1	BOW-TRIGRAM-LEMMA-SGD	147900	0.146	0.771	0.791	0.778
2	BOW-BIGRAM-LEMMA-NB	70986	0.179	0.734	0.840	0.774
3	BOW-UNIGRAM-NB	15165	0.183	0.734	0.833	0.771
4	BOW-TRIGRAM-LEMMA-NB	147900	0.182	0.728	0.836	0.770
5	BOW-UNIGRAM-LEMMA-NB	13039	0.183	0.730	0.836	0.769
6	BOW-BIGRAM-NB	84189	0.181	0.733	0.827	0.768
7	BOW-TRIGRAM-NB	190376	0.178	0.736	0.821	0.768
8	TFIDF-BIGRAM-LEMMA-LSVM	70986	0.159	0.755	0.789	0.766
9	TFIDF-BIGRAM-LEMMA-MLP	70986	0.158	0.765	0.772	0.763
10	TFIDF-BIGRAM-MLP	84189	0.157	0.778	0.756	0.760

Table 4. Top 10 best combinations of classifiers and features extraction for long texts. All presented metrics values are the mean after 5 rounds of cross-validation.

Rank	Experiment	Vocabulary	FPR	PRE	REC	F1
1	BOW-UNIGRAM-NB	14186	0.153	0.846	0.885	0.857
2	BOW-BIGRAM-MLP	77174	0.140	0.862	0.862	0.856
3	BOW-BIGRAM-NB	77174	0.163	0.833	0.892	0.855
4	BOW-TRIGRAM-NB	173315	0.163	0.836	0.888	0.854
5	TFIDF-TRIGRAM-MLP	173315	0.156	0.831	0.888	0.853
6	BOW-BIGRAM-LEMMA-NB	64803	0.168	0.826	0.896	0.852
7	BOW-TRIGRAM-LEMMA-NB	134067	0.172	0.822	0.892	0.848
8	TFIDF-BIGRAM-LSVM	77174	0.169	0.820	0.881	0.844
9	TFIDF-UNIGRAM-LEMMA-MLP	12255	0.176	0.790	0.907	0.842
10	BOW-UNIGRAM-LR	14186	0.170	0.832	0.866	0.841

Lastly, we decided to select only the messages containing 50 or more words from our data set, resulting in a subset of 269 messages with misinformation and 292 messages without misinformation. We repeated all the experiments to analyze the influence of the text length in the prediction. Table 4 shows the results for these experiments. We had a significant performance increase in this scenario, achieving an F1 of 0.857 when using BoW, unigram, and NB as the combination of features and classifier. In terms of FPR, we achieved a result of 0.14 using BoW, bigram, and MLP. By analyzing these results, we can observe that the text length affects the classifiers' performance since there are short messages in our data set linked to external media that contain misinformation.

From our results, we can recognize how difficult it is to perform MID in WhatsApp since our best result was an F1 of 0.778. When considering only long texts, our best F1 result is 0.857.

6   Conclusions

In this work, we presented a large-scale, labeled, and public data set of WhatsApp messages in Brazilian Portuguese about coronavirus pandemic. In addition, we performed a wide set of experiments seeking out to build an efficient solution to the MID problem in this specific context. Our best result achieved an F1 score of 0.778 due to the predominance of short texts. However, when texts with less than 50 words are filtered, the F1 score rises to 0.857. In future work, we pretend to investigate how the metadata associated with the message (senders, timestamps, groups where it was shared, etc.) can be combined with textual features to improve our MID solution's performance. All the experiments and the COVID-19.BR data set are available at our public repository[3].

---

[3] https://gitlab.com/jmmonteiro/misinformation_covid19.

References

1. Choudrie, J., Banerjee, S., Kotecha, K., Walambe, R., Karende, H., Ameta, J.: Machine learning techniques and older adults processing of online information and misinformation: a covid 19 study. Comput. Hum. Behav. **119**, 106716 (2021). https://doi.org/10.1016/j.chb.2021.106716
2. Elhadad, M.K., Li, K.F., Gebali, F.: Detecting misleading information on COVID-19. IEEE Access **8**, 165201–165215 (2020). https://doi.org/10.1109/ACCESS.2020.3022867
3. Garimella, K., Tyson, G.: WhatsApp, doc? A first look at WhatsApp public group data. arXiv preprint arXiv:1804.01473 (2018)
4. Giachanou, A., Zhang, G., Rosso, P.: Multimodal multi-image fake news detection. In: 2020 IEEE 7th International Conference on Data Science and Advanced Analytics (DSAA), pp. 647–654 (2020). https://doi.org/10.1109/DSAA49011.2020.00091
5. Kim, S.B., Han, K.S., Rim, H.C., Myaeng, S.H.: Some effective techniques for naive bayes text classification. IEEE Trans. Knowl. Data Eng. **18**(11), 1457–1466 (2006)
6. Kolluri, N.L., Murthy, D.: CoVerifi: a COVID-19 news verification system. Online Soc. Netw. Media **22**, 100123 (2021). https://doi.org/10.1016/j.osnem.2021.100123
7. Pranckevičius, T., Marcinkevičius, V.: Comparison of naive bayes, random forest, decision tree, support vector machines, and logistic regression classifiers for text reviews classification. Baltic J. Mod. Comput. **5**(2), 221 (2017)
8. Prasetijo, A.B., Isnanto, R.R., Eridani, D., Soetrisno, Y.A.A., Arfan, M., Sofwan, A.: Hoax detection system on Indonesian news sites based on text classification using SVM and SGD. In: 2017 4th International Conference on Information Technology, Computer, and Electrical Engineering (ICITACEE), pp. 45–49. IEEE (2017)
9. Rennie, J.D., Shih, L., Teevan, J., Karger, D.R.: Tackling the poor assumptions of naive bayes text classifiers. In: Proceedings of the 20th International Conference on Machine Learning (ICML-03), pp. 616–623 (2003)
10. Resende, G., Messias, J., Silva, M., Almeida, J., Vasconcelos, M., Benevenuto, F.: A system for monitoring public political groups in WhatsApp. In: Proceedings of the 24th Brazilian Symposium on Multimedia and the Web, WebMedia 2018, pp. 387–390. Association for Computing Machinery, New York (2018). https://doi.org/10.1145/3243082.3264662
11. Su, Q., Wan, M., Liu, X., Huang, C.R.: Motivations, methods and metrics of misinformation detection: an NLP perspective. Nat. Lang. Process. Res. **1**, 1–13 (2020). https://doi.org/10.2991/nlpr.d.200522.001
12. Waterloo, S.F., Baumgartner, S.E., Peter, J., Valkenburg, P.M.: Norms of online expressions of emotion: comparing Facebook, Twitter, Instagram, and WhatsApp. New Media Soc. **20**(5), 1813–1831 (2018)

Sentiment Analysis

Multi-Step Transfer Learning for Sentiment Analysis

Anton Golubev[1](✉)[iD] and Natalia Loukachevitch[2][iD]

[1] Bauman Moscow State Technical University, Moscow, Russia
[2] Lomonosov Moscow State University, Moscow, Russia

Abstract. In this study, we test transfer learning approach on Russian sentiment benchmark datasets using additional train sample created with distant supervision technique. We compare several variants of combining additional data with benchmark train samples. The best results were obtained when the three-step approach is used where the model is iteratively trained on general, thematic, and original train samples. For most datasets, the results were improved by more than 3% to the current state-of-the-art methods. The BERT-NLI model treating sentiment classification problem as a natural language inference task reached the human level of sentiment analysis on one of the datasets.

Keywords: Targeted sentiment analysis · Distant supervision · Transfer learning · BERT

1 Introduction

Sentiment analysis or opinion mining is an important natural language processing task used to determine sentiment attitude of the text. Nowadays most state-of-the-art results are obtained using deep learning models, which require training on specialized labeled datasets. To improve the model performance, transfer learning approach can be used. This approach includes a pre-training step of learning general representations from a source task and an adaptation step of applying previously gained knowledge to a target task.

The most known Russian sentiment analysis datasets include ROMIP-2013 and SentiRuEval2015-2016 [4,10,11] consisting of annotated data on banks and telecom operators reviews from Twitter posts and news quotes. Current best results on these datasets were obtained using pre-trained RuBERT [7,19] and conversational BERT model [3,5] fine-tuned as architectures treating a sentiment classification task as a natural language inference (NLI) or question answering (QA) problem [7].

In this study, we introduce a method for automatic generation of annotated sample from a Russian news corpus using distant supervision technique. We compare different variants of combining additional data with original train samples and test the transfer learning approach based on several BERT models.

© Springer Nature Switzerland AG 2021
E. Métais et al. (Eds.): NLDB 2021, LNCS 12801, pp. 209–217, 2021.
https://doi.org/10.1007/978-3-030-80599-9_19

For most datasets, the results were improved by more than 3% to the current state-of-the-art performance. On SentiRuEval-2015 Telecom Operators Dataset, the BERT-NLI model treating a sentiment classification problem as a natural language inference task, reached human level according to one of the metrics.

2 Related Work

Russian sentiment analysis datasets are based on different data sources [19], including reviews [4,18], news stories [4] and posts from social networks [10,14,15]. The best results on most available datasets are obtained using transfer learning approaches based on Russian BERT-based models [2,3,5,13,19]. In [7], the authors tested several variants of RuBERT and different settings of its applications, and found that the best results on sentiment analysis tasks on several datasets were achieved using Conversational RuBERT trained on Russian social networks posts and comments. Among several architectures, the BERT-NLI model treating the sentiment classification problem as a natural language inference task usually has the highest results.

For automatic generation of annotated data for sentiment analysis task, researchers use so-called distant supervision approach, which exploits additional resources: users' tags, manual lexicons [6,15] and users' positive or negative emoticons in case of Twitter sentiment analysis task [12,15,17]. Authors of [16] use the RuSentiFrames lexicon for creating a large automatically annotated dataset for recognition of sentiment relations between mentioned entities.

3 Russian Sentiment Benchmark Datasets

In our study, we consider the following Russian datasets (benchmarks): news quotes from the ROMIP-2013 evaluation [4] and Twitter datasets from SentiRuEval 2015–2016 evaluations [10,11]. The collection of the news quotes contains opinions in direct or indirect speech extracted from news articles [4]. Twitter datasets from SentiRuEval-2015–2016 evaluations were annotated for the task of reputation monitoring [1,10], which means searching sentiment-oriented opinions about banks and telecom companies.

Table 1 presents the main characteristics of datasets including train and test sample sizes and sentiment classes distributions. It can be seen in Table 1 that the neutral class is prevailing in all Twitter datasets, while ROMIP-2013 data is rather balanced. For this reason, along with the standard metrics of F_1 $macro$ and accuracy, $F_1^{+-} macro$ and $F_1^{+-} micro$ ignoring the neutral class were also calculated. Insignificant part of samples contains two or more sentiment analysis objects, so these tweets are duplicated with corresponding attitude labels [11].

Table 1. Benchmark sample sizes and sentiment class distributions (%).

Dataset	Train sample				Test sample			
	Volume	Posit	Negat	Neutral	Volume	Posit	Negat	Neutral
ROMIP-2013[a]	4260	26	44	30	5500	32	41	27
SRE-2015 Banks[b]	6232	7	36	57	4612	8	14	78
SRE-2015 Telecom[b]	5241	19	34	47	4173	10	23	67
SRE-2016 Banks[c]	10725	7	26	67	3418	9	23	68
SRE-2016 Telecom[c]	9209	15	28	57	2460	10	47	43

[a] http://romip.ru/en/collections/sentiment-news-collection-2012.html
[b] https://drive.google.com/drive/folders/1bAxIDjVz_0UQn-iJwhnUwngjivS2kfM3
[c] https://drive.google.com/drive/folders/0BxlA8wH3PTUfV1F1UTBwVTJPd3c

4   Automatic Generation of Annotated Dataset

The main idea of automatic annotation of dataset for targeted sentiment analysis task is based on the use of a sentiment lexicon comprising negative and positive words and phrases with their sentiment scores. We utilize Russian sentiment lexicon RuSentiLex [9], which includes general sentiment words of Russian language, slang words from Twitter and words with positive or negative associations (connotations) from the news corpus.

As a source for automatic dataset generation, we use 4 Gb Russian news corpus, collected from various sources and representing different themes, which is an important fact that the benchmarks under analysis cover several topics. For creation of the general part of annotated dataset, we select monosemous positive and negative nouns from the RuSentiLex lexicon, which can be used as references to people or companies, which are sentiment targets in the benchmarks. We construct positive and negative word lists and suppose that if a word from the list occurs in a sentence, it has a context of the same sentiment. Examples of such words are presented below (all further examples are translated from Russian):

- positive: *"champion, hero, good-looker"*, etc.;
- negative: *"outsider, swindler, liar, defrauder, deserter"*, etc.

Sentences may contain several seed words with different sentiments. In such cases, we duplicate sentences with labels in accordance with their attitudes. The examples of extracted sentences are as follows:

- positive: *"A MASK is one who, on a gratuitous basis, helps the development of science and art, provides them with material assistance from their own funds"*;
- negative: *"Such irresponsibility—non-payments—hits not only the MASK himself, but also throughout the house in which he lives"*.

To generate the thematic part of the automatic sample, we search for sentences that mention relevant named entities depending on a task (banks or operators) using the named entity recognition model (NER) from DeepPavlov [3] co-occurred with sentiment words in the same sentences. To ensure that an attitude word refers to an entity, we restrict the distance between two words to be not more than four words:

- banks (positive): *"MASK increased its net profit in November by 10.7%"*
- mobile operators (negative): *"FAS suspects MASK of imposing paid services."*

We remove examples containing a particle *"not"* near sentiment word because it could change sentiment of text in relation to target. Sentences with attitude word located in quotation marks were also removed because they could distort the meaning of the sentence being a proper name.

Since the benchmarks contain also the neutral class, we extract sentences without sentiments by choosing among examples selected by NER those that do not contain any sentiment words from the lexicon:

- persons: *MASK is already starting training with its new team.*
- banks: *"On March 14, MASK announced that it was starting rebranding."*
- mobile operators: *"MASK has offered its subscribers a new service."*

To create an additional sample from the raw corpus, we divide raw articles into separate sentences using spaCy sentence splitter library [8]. Too short and long sentences, duplicate sentences (with similarity more than 0.8 cosine measure) were removed. We also take into account the distribution of sentiment words in the resulting sample, trying to bring it as close as possible to uniform. Since negative events are more often included in the news articles, there are much more sentences with a negative attitude in the initial raw corpus than with a positive one. We made automatically generated dataset and source code publicly available[1].

5   BERT Architectures

In our study, we consider three variants of fine-tuning BERT models [5] for sentiment analysis task. These architectures can be subdivided into the single-sentence approach using only initial text as an input and the two-sentence approach [7,20], which converts the sentiment analysis task into a sentence-pair classification task by appending an additional sentence to the initial text.

The sentence-single model represents a vanilla BERT with an additional single linear layer on the top. The unique token *[CLS]* is added for the classification task at the beginning of the sentence. The sentence-pair architecture adds an auxiliary sentence to the original input, inserting the *[SEP]* token between two sentences. The difference between two models is in addition of a linear layer with an output dimension equal to the number of sentiment classes (3): for the sentence-pair model it is added over the final hidden state of *[CLS]* token, while for the sentence-single variant it is added on the top of the entire last layer.

For the targeted sentiment analysis task, there are labels for each object of attitude so they can be replaced by a special token *[MASK]*. Since general

---

[1] https://github.com/antongolubev5/Auto-Dataset-For-Transfer-Learning.

sentiment analysis problem has no certain attitude objects, token is assigned to the whole sentence and located at the beginning.

The sentence-pair model has two kind of architecture based on question answering (QA) and natural language inference (NLI) problems. The auxiliary sentences for each model are as follows:

- pair-NLI: *"The sentiment polarity of MASK is"*
- pair-QA: *"What do you think about MASK?"*

In our study, we use pre-trained Conversational RuBERT[2] from DeepPavlov framework [3] trained on Russian social networks posts and comments which showed better results in preliminary study. We kept all hyperparameters used in [7] unchanged.

Table 2. Results based on using the two-step approach.

Dataset	Model	Accuracy	F_1 macro	F_1^{+-} macro	F_1^{+-} micro
ROMIP-2013	BERT-single	79.95	71.16	85.39	85.61
	BERT-pair-QA	80.21	71.29	85.72	85.93
	BERT-pair-NLI	**80.56**	**71.68**	**86.14**	**86.19**
	Current SOTA	80.28	70.62	85.52	85.68
SRE-2015 Banks	BERT-single	86.06	79.11	64.87	66.73
	BERT-pair-QA	86.34	79.58	65.29	67.02
	BERT-pair-NLI	**87.62**	**80.72**	**68.44**	**71.39**
	Current SOTA	86.88	79.51	67.44	70.09
SRE-2015 Telecom	BERT-single	77.11	69.76	61.89	66.95
	BERT-pair-QA	**78.14**	**70.03**	**64.53**	**68.29**
	BERT-pair-NLI	77.96	69.68	64.52	68.21
	Current SOTA	76.63	68.54	63.47	67.51
SRE-2016 Banks	BERT-single	81.94	74.08	67.24	70.68
	BERT-pair-QA	**84.36**	**77.43**	**72.32**	**74.06**
	BERT-pair-NLI	84.19	75.63	68.52	70.89
	Current SOTA	82.28	74.06	69.53	71.76
SRE-2016 Telecom	BERT-single	75.82	69.78	65.04	74.22
	BERT-pair-QA	77.25	69.71	67.35	76.22
	BERT-pair-NLI	**77.59**	69.84	**68.11**	75.93
	Current SOTA	–	**70.68**	66.40	**76.71**

6   Experiments and Results

We consider fine-tuning strategies to represent training in several steps with intermediate freezing of the model weights and include two following variants:

---

[2] http://docs.deeppavlov.ai/en/master/features/models/bert.html.

– two-step approach: independent iterative training on additional dataset at the first step and on the benchmark training set at the second;
– three-step approach: independent iterative training in three steps using the general part from the additional dataset, the thematic examples from the additional dataset and the benchmark training sets.

During this experiment, we also studied the dependence between the results and the size of additional dataset. It was found that the boundary between extension of automatically generated data and increasing the results was set at a sample size of 27000 (9000 per each sentiment class). Using the two-step approach allowed us to overcome the current best results [7,19] for almost all benchmarks (Table 2).

Table 3. Results based on using the three-step approach.

Dataset	Model	Accuracy	F_1 macro	F_1^{+-} macro	F_1^{+-} micro
ROMIP-2013	BERT-single	80.27	71.78	85.82	86.07
	BERT-pair-QA	80.78	72.09	86.14	86.42
	BERT-pair-NLI	**82.33**	**72.69**	**86.77**	**87.04**
	Current SOTA	80.28	70.62	85.52	85.68
SRE-2015 Banks	BERT-single	87.65	80.79	65.74	67.46
	BERT-pair-QA	87.92	81.12	66.47	68.55
	BERT-pair-NLI	**88.14**	**81.63**	**68.76**	**72.28**
	Current SOTA	86.88	79.51	67.44	70.09
SRE-2015 Telecom	BERT-single	77.85	70.42	62.29	67.38
	BERT-pair-QA	**79.21**	70.94	65.68	69.11
	BERT-pair-NLI	79.12	**71.16**	**65.71**	**70.65**
	Current SOTA	76.63	68.54	63.47	67.51
	Manual	–	–	70.30	70.90
SRE-2016 Banks	BERT-single	83.21	75.31	68.45	71.69
	BERT-pair-QA	**85.59**	**78.93**	**74.05**	**75.12**
	BERT-pair-NLI	85.43	76.85	70.23	72.07
	Current SOTA	82.28	74.06	69.53	71.76
SRE-2016 Telecom	BERT-single	76.79	70.64	66.16	75.27
	BERT-pair-QA	78.42	70.54	**68.65**	**77.45**
	BERT-pair-NLI	**78.62**	**71.18**	69.36	76.85
	Current SOTA	–	70.68	66.40	76.71

For a three-step transfer learning approach, we divided the first step of the previous experiment into two. Thus, the models are trained on the general data, then the weights are frozen and the training continues on the thematic examples retrieved with the list of organizations and NER from DeepPavlov. After the second weights freezing, models are trained on the benchmark training sets.

At this stage we also added sentiment examples to the thematic part of the additional sample via selection thematic sentences containing attitude words. The first step sample contains 18000 general examples and the second sample consists of 9000 thematic examples (both samples are equally balanced across sentiment classes).

The use of the three-step approach combined with an extension of thematic part of the additional dataset improved the results by a few more points (Table 3). One participant of SentiRuEval-2015 evaluation sent the results of manual annotation of the test sample [11]. As it can be seen, BERT-pair-NLI model reaches human sentiment analysis level by $F_1^{+-}micro$.

Some examples are still difficult for the improved models. For example, the following negative sarcastic examples were erroneously classified by all models as neutral:

- *"Sberbank of Russia – 170 years on the queue market!"*;
- *"While we are waiting for a Sberbank employee, I could have gone to lunch 3 times"*.

In the following example with different sentiments towards two mobile operators, the models could not detect the positive attitude towards the Beeline operator:

- *"MTS does not work! Forever out of reach. The connection is constantly interrupted. We transfer the whole family to Beeline."*

7   Conclusion

In this study, we presented a method for automatic generation of an annotated sample from a news corpus using the distant supervision technique. We compared different options of combining the additional data with several Russian sentiment analysis benchmarks and improved current state-of-the-art results by more than 3% using BERT models together with the transfer learning approach. The best variant was the three-step approach of iterative training on general, thematic and benchmark train samples with intermediate freezing of the model weights. On one of benchmarks, the BERT-NLI model treating a sentiment classification problem as a natural language inference task, reached human level according to one of the metrics.

Acknowledgements. The reported study was funded by RFBR according to the research project № 20-07-01059.

References

1. Amigó, E., et al.: Overview of RepLab 2013: evaluating online reputation monitoring systems. In: Forner, P., Müller, H., Paredes, R., Rosso, P., Stein, B. (eds.) CLEF 2013. LNCS, vol. 8138, pp. 333–352. Springer, Heidelberg (2013). https://doi.org/10.1007/978-3-642-40802-1_31

2. Baymurzina, D., Kuznetsov, D., Burtsev, M.: Language model embeddings improve sentiment analysis in Russian. In: Komp'juternaja Lingvistika i Intellektual'nye Tehnologii, pp. 53–62 (2019)
3. Burtsev, M.: DeepPavlov: open-source library for dialogue systems. In: Proceedings of ACL 2018, System Demonstrations, pp. 122–127 (2018)
4. Chetviorkin, I., Loukachevitch, N.: Evaluating sentiment analysis systems in Russian. In: Proceedings of the 4th Biennial International Workshop on Balto-Slavic Natural Language Processing, pp. 12–17 (2013)
5. Devlin, J., Chang, M.W., Lee, K., Toutanova, K.: BERT: Pre-training of Deep Bidirectional Transformers for Language Understanding. arXiv preprint arXiv:1810.04805 (2018)
6. Go, A., Bhayani, R., Huang, L.: Twitter sentiment classification using distant supervision. CS224N project report, Stanford 1(12), 2009 (2009)
7. Golubev, A., Loukachevitch, N.: Improving results on Russian sentiment datasets. In: Filchenkov, A., Kauttonen, J., Pivovarova, L. (eds.) AINL 2020. CCIS, vol. 1292, pp. 109–121. Springer, Cham (2020). https://doi.org/10.1007/978-3-030-59082-6_8
8. Honnibal, M., Montani, I., Van Landeghem, S., Boyd, A.: spaCy: Industrial-strength Natural Language Processing in Python. Zenodo (2020). https://doi.org/10.5281/zenodo.1212303
9. Loukachevitch, N., Levchik, A.: Creating a general Russian sentiment lexicon. In: Proceedings of the Tenth International Conference on Language Resources and Evaluation (LREC 2016), pp. 1171–1176 (2016)
10. Loukachevitch, N., Rubtsova, Y.: Entity-oriented sentiment analysis of tweets: results and problems. In: Král, P., Matoušek, V. (eds.) TSD 2015. LNCS (LNAI), vol. 9302, pp. 551–559. Springer, Cham (2015). https://doi.org/10.1007/978-3-319-24033-6_62
11. Loukachevitch, N., Rubtsova, Y.: SentiRuEval-2016: overcoming time gap and data sparsity in tweet sentiment analysis. In: Proceedings of International Conference Dialog-2016 (2016)
12. Mohammad, S., Salameh, M., Kiritchenko, S.: Sentiment lexicons for Arabic social media. In: Proceedings of the Tenth International Conference on Language Resources and Evaluation (LREC 2016), pp. 33–37 (2016)
13. Moshkin, V., Konstantinov, A., Yarushkina, N.: Application of the bert language model for sentiment analysis of social network posts. In: Russian Conference on Artificial Intelligence. pp. 274–283. Springer (2020)
14. Rogers, A., Romanov, A., Rumshisky, A., Volkova, S., Gronas, M., Gribov, A.: Rusentiment: an enriched sentiment analysis dataset for social media in Russian. In: Proceedings of the 27th International Conference on Computational Linguistics, pp. 755–763 (2018)
15. Rubtsova, Y.: Constructing a corpus for sentiment classification training. Softw. Syst. 109, 72–78 (2015)
16. Rusnachenko, N., Loukachevitch, N., Tutubalina, E.: Distant supervision for sentiment attitude extraction. In: Proceedings of the International Conference on Recent Advances in Natural Language Processing (RANLP 2019), pp. 1022–1030 (2019)
17. Sahni, T., Chandak, C., Chedeti, N.R., Singh, M.: Efficient Twitter sentiment classification using subjective distant supervision. In: 2017 9th International Conference on Communication Systems and Networks (COMSNETS), pp. 548–553 (2017)

18. Smetanin, S., Komarov, M.: Sentiment analysis of product reviews in Russian using convolutional neural networks. In: 2019 IEEE 21st Conference on Business Informatics (CBI), vol. 1, pp. 482–486 (2019)
19. Smetanin, S., Komarov, M.: Deep transfer learning baselines for sentiment analysis in Russian. Inf. Process. Manag. **58**(3) (2021)
20. Sun, C., Huang, L., Qiu, X.: Utilizing BERT for aspect-based sentiment analysis via constructing auxiliary sentence. In: Proceedings of the 2019 Conference of the North American Chapter of the Association for Computational Linguistics: Human Language Technologies, vol. 1, pp. 380–385 (2019)

Improving Sentiment Classification in Low-Resource Bengali Language Utilizing Cross-Lingual Self-supervised Learning

Salim Sazzed[✉]

Old Dominion University, Norfolk, VA 23508, USA
ssazz001@odu.edu

Abstract. One of the barriers of sentiment analysis research in low-resource languages such as Bengali is the lack of annotated data. Manual annotation requires resources, which are scarcely available in low-resource languages. We present a cross-lingual hybrid methodology that utilizes machine translation and prior sentiment information to generate accurate pseudo-labels. By leveraging the pseudo-labels, a supervised ML classifier is trained for sentiment classification. We contrast the performance of the proposed self-supervised methodology with the Bengali and English sentiment classification methods (i.e., methods which do not require labeled data). We observe that the self-supervised hybrid methodology improves the macro F1 scores by 15%–25%. The results infer that the proposed framework can improve the performance of sentiment classification in low-resource languages that lack labeled data.

Keywords: Bangla sentiment analysis · Pseudo-label generation · Cross-lingual sentiment analysis

1 Introduction

Sentiment analysis determines the semantic orientation of an opinion expressed in a text. The rapid growth of user-generated online content necessitates analyzing user's opinions and emotions in textual data for various purposes. Researchers applied both the machine learning-based [1,17] and lexicon-based methods [26] to classify sentiments at various levels of granularity such as binary, 3-class, or 5-class. The supervised ML methods usually exhibit much better performance; however, they require a large volume of annotated data.

English and several other languages enjoy ample resources, such as annotated data for sentiment analysis; however, such resources are not available in resource-constrained languages. The self-supervised approaches can be an effective way to deal with the inadequacy of labeled data in low-resource languages. Instead of manual annotation, the self-supervised learning methods automatically generate pseudo-labels by implicitly learning underlying patterns from the data or utilizing a set of rules.

© Springer Nature Switzerland AG 2021
E. Métais et al. (Eds.): NLDB 2021, LNCS 12801, pp. 218–230, 2021.
https://doi.org/10.1007/978-3-030-80599-9_20

Cross-lingual sentiment classification is another way to deal with resource scarcity issues in low-resource languages. Cross-lingual sentiment classification aims to leverage resources like labeled data and opinion lexicons from a resource-rich language (typically English) to classify the sentiment polarity of texts in a low-resource language. Though cross-lingual approaches have been studied in several low-resource languages [5,16], in Bengali only a few works utilized it for sentiment classification [22] or sentiment lexicon creation [8]. In [22], the performances of various supervised ML classifiers have been compared in a Bengali corpus and corresponding machine-translated English version. The authors found Bengali-English machine translation system had reached some level of maturity; thus could be utilized for cross-lingual sentiment analysis.

In this work, we present a cross-lingual self-supervised methodology for classifying sentiments in unlabeled Bengali text. The proposed self-supervised hybrid methodology combines lexicon-based and supervised ML-based methods. Employing machine translation, we first transform Bengali text to English. Then we leverage prior word-level sentiment information (i.e., sentiment lexicon), a set of rules, and consensus-based filtering to generate accurate pseudo-labels for training a supervised ML classifier. We compare the performance of the proposed method with English lexicon-based sentiment analysis tools, VADER [13], TextBlob[1], and SentiStrength [24] and a Bengali lexicon-based method [21]. We observe that the hybrid approach improves the F1 score by 15% and accuracy by 11% compared to the best lexicon-based method.

1.1 Contributions

The major contributions of this work can be summarized as follows:

- We conduct a comparative performance analysis of Bengali and English lexicon-based methods.
- To elevate the performance of sentiment classification in unlabeled data, we present a cross-lingual self-supervised learning approach.
- We demonstrate how to generate highly accurate pseudo-labels to deal with the lack of labeled data in Bengali.
- We show that by utilizing machine translation and combining lexicon-based and ML-based methods, substantially improved performance can be attained.

2   Related Work

Most of the research in sentiment analysis has been conducted in English and a few other major languages such as Chinese, Arabic, and Spanish. In Bengali, limited research has been performed using corpora collected from various sources such as microblogs, Facebook statuses, and other social media sources [9,18]. Researchers utilized various supervised methods, such as SVM with maximum entropy [7], Naive Bayes (NB) [14], Deep Neural Network [11,25] for Bengali

---

[1] https://textblob.readthedocs.io/en/dev/.

sentiment analysis. The word-embedding-based approach has been explored in [2].

Cross-lingual approaches of sentiment classification have been applied to several low-resource languages. The linked WordNets was used in [4] to bridge the language gap between two Indian languages, Hindi and Marathi. The performance and effectiveness of machine translation systems and supervised methods for multilingual sentiment analysis was investigated in [3] using four languages: English, German, Spanish, and French; three machine translation systems: Google, Bing, and Moses; several supervised algorithms and various types of features. [16] proposed a cross-lingual mixture model (CLMM) that exploits unlabeled bilingual parallel corpus. In [5], authors utilized a machine translation system for projecting resources from English to Romanian and Spanish and obtained a comparative performance. In [10], the authors proposed an end-to-end cross-lingual sentiment analysis (CLSA) model by leveraging unlabeled data in multiple languages and domains. The authors of [27] proposed a learning approach that does not require any cross-lingual labeled data. Their algorithm optimizes the transformation functions of monolingual word-embedding space. The authors of [6] introduced an Adversarial Deep Averaging Network (ADAN) that uses a shared feature extractor to learn hidden representations that are invariant across languages. Their experiments on Chinese and Arabic sentiment classification demonstrated the efficacy of ADAN.

The cross-lingual approach of sentiment analysis in Bengali is still largely unexplored; only a few works investigated it for tasks such as translating English polarity lexicon to Bengali [8], comparing the performance of ML algorithms in Bengali and machine-translated corpus [22]. The authors of [22] utilized two small datasets to compare the performance of supervised ML algorithms in Bengali and machine-translated English corpora. They found supervised ML algorithms showed better performance in the model trained on the translated corpus.

A plethora of studies explored hybrid approaches of sentiment classification; however, most of them utilized labeled or partially labeled datasets. A hybrid method was proposed in [28] for sentiment analysis in Twitter data that does not require any labeled data. The proposed method adopted a lexicon-based approach to label the training examples. The authors of [12] proposed a framework where an initial classifier is learned by incorporating a sentiment lexicon and using generalized expectation criteria. SESS (SElf-Supervised and Syntax-Based method) [29] works in three phases; initially, some documents are classified iteratively based on a sentiment dictionary. Afterward, a machine learning model is trained using the classified documents, and finally, the learned model is applied to the whole data set.

To the best of our knowledge, this work is the first attempt to incorporate the cross-lingual setting with self-supervised learning in Bengali. Compared to the existing self-supervised approaches, the proposed methodology differs in the way we perform pseudo-label generation and selection, training-testing set split, and model training.

3   Dataset and Machine Translation

3.1   Dataset

We use a large annotated review corpus[2] deposited by the author of [20]. The reviews in the corpus represent viewer's opinions toward a number of Bengali dramas. The data collection and annotation procedures were described in [20].

Bengali Reviews	Machine Translation	Polarity
সত্যিই দুর্দান্ত দুর্দান্ত উভয় অংশ ... আমি একজন ভারতীয় ... তব আমি বাংলাদেশী নাটক পছন্দ করি। বিশেষভাবে জিয়াউল ফারুক অপূর্ব স্যার এর ... আমি আপনাকে স্যার ভালবাসি	Both really awesome great parts ... I'm an Indian ... But I like Bangladeshi drama. In particular Ziaul Farooq is a wonderful sir ... I love you sir.	Positive
এই নাটকটি আমার জীবনের সাথে জড়িয়ে আছে। সত্যি কিছু কিছু স্মৃতি থেকে যায়, যা কখনো ভুলা যায় না। এই নাটকটি আমার ভীষণ ভালো লেগেছে! সবাই ভালো থাকবেন।	This drama is embedded in our lives. There are some memories that will never be forgotten. This drama is very like ours! Everyone will be fine.	Positive
আমি কোনো এতো হাসি নাই, যতটা না এই নাটক দেখে হাসছি! পুরো টা সময় শামীম ভাইয়ার অভিনয় উপভোগ করলাম	I didn't laugh at all, until I was laughing at this drama. I enjoyed Shamim Bhai's acting the whole time	Positive
বাংলাদেশের নাটক কেমন যেন দিন দিন দেখার অযোগ্য হয়ে পড়ছে। এসব নাটক পরিবারকে নিয়ে দেখার মত নয়। কি সব ধরনের নাম আর কি সব ভাষা ব্যবহার করে নিজেই বুঝিনা। এইসব নাটক থেকে আমরা কি শিক্ষা নিতে পারি। আমাদের সমাজ পরিবার দেশ সব কিছু ধ্বংস করে দিচ্ছে সব ধরনের কিছু নাটক। এসব পরিচালনা দের জুতাপেটা করে বাংলাদেশ থেকে বিতাড়িত করা হোক।	How is the drama of Bangladesh becoming inaccessible day by day. These plays are not worth watching with the family. Do not use all kinds of names and all the languages myself. What can we learn from these plays? Our society family country is destroying everything drama of all kinds. Be expelled from Bangladesh by wearing these shoes.	Negative
নাটকের কোন কনসেপ্ট নেই, কোনো কাহিনী নেই... অদ্ভুত উটের পিঠে চলছে স্বদেশ।	No Concept of Drama, No Story ... Strange camels running the country.	Negative
বাল আমার।।। ফালতু স্টোরি, ফালতু মেকিং। এসব আজাইরা নাটকের জন্যই দিন দিন বাংলা নাটকের মান কমে যাচ্ছে।	My boy. False Story, False Making. The quality of Bangla drama is declining day by day for these Azira plays.	Negative

Fig. 1. Example of Bengali and translated English reviews with annotations

This review corpus consists of 11807 annotated reviews, where each review contains between 2 to 300 Bengali words. This class-imbalanced dataset comprised of 3307 *negative* and 8500 *positive* reviews. From the annotator ratings, the author observed an inter-rater agreement of around 0.83 based on Cohen's κ. The reviews are highly polar since reviews that are marked as non-subjective by either of the annotators were excluded. Figure 1 shows some examples of Bengali reviews and corresponding English machine translation with annotations.

3.2   Quality of Machine Translation and Sentiment Preservation

To leverage cross-lingual resources, it is required to link the source and target languages. The machine translation (MT) service is one of the most prevalent ways to connect languages. The quality of a machine translation system largely depends on the amount of training data used for model training. Without using an advanced machine translation service built on a huge training dataset, good translation accuracy is not attainable. The authors of [22] utilized Google Translate[3] to translate Bengali reviews to English for cross-lingual sentiment analysis. They manually assessed the quality of the machine translation and observed

---

[2]   https://github.com/sazzadcsedu/BN-Dataset.git.

[3]   https://translate.google.com.

that the quality of the translation varies among reviews. Among 1016 translated Bengali comments, on a Likert scale of 1–5, they assigned 170 comments to a rating of 1, 279 comments as 2, 229 comments as 3, 140 comments as 4, and 198 comments as 5, with an average translation rating of 2.92, which they described as fair. Therefore, in this work, we use Google Translate to translate the Bengali reviews into English.

To investigate the sentiment preservation after machine translation, the author of [20] computed the agreement of the predictions of two highly accurate ML classifiers, logistic regression (LR) and support vector machine (SVM) in Bengali and machine-translated English corpus in a drama review dataset. The author utilized Cohen's kappa and Gwet's AC1 to assess inter-rater agreements. Both SVM and LR show kappa scores above 0.80 and AC1 scores above 0.85 (where a score of 1 refers to perfect agreement). The results indicate sentiment consistency exists between original Bengali and machine-translated English reviews.

The above-mentioned studies suggest that the quality of Bengali-to-English machine translation is fair, and the sentiment is preserved in most cases. Therefore, no manual error correction is employed in the machine-translated reviews. Besides, one of the main objectives of this work is to eliminate manual intervention.

4   Cross-Lingual Sentiment Analysis in Bengali

We apply various approaches of sentiment classification in the Bengali corpus, as shown in Fig. 2.

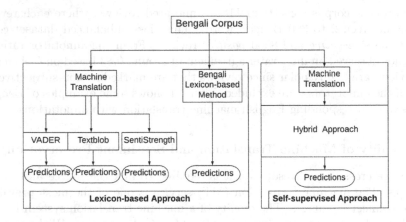

Fig. 2. Various approaches of sentiment analysis in Bengali review corpus

4.1   Lexicon-Based Methods

To find the efficacy of the lexicon-based methods for sentiment classification in the translated corpus, three popular lexicon-based tools, SentiStrength, VADER, and TextBlob are employed. A non-negative polarity score of SentiStrength and TextBlob refers to a positive class prediction; otherwise, we consider it as a negative prediction. For VADER, the compound score is used instead of the polarity score.

For Bengali, we utilize a publicly available Bengali sentiment lexicon [21] and a set of linguistic rules. This binary-weighted lexicon consists of around 700 opinion words, where positive and negative words have a weight of +1 and −1, respectively. Besides applying the word-level polarity, we employ a simple negation rule to address the shift of polarity. The class assignment based on the review polarity score is implemented similarly to English lexicon-based methods. In Bengali, only a few works employed the lexicon-based methods for sentiment classification due to a lack of standard language-specific resources (e.g., sentiment lexicon, POS tagger, dependency parser, etc.). Besides, their implementations are not publicly available.

4.2 Self-supervised Hybrid Methodology

Self-supervised learning utilizes data that is automatically labeled by learning patterns, exploiting the relationships between features, and employing rules. As the Bengali lexicon-based method [21] yields comparatively lower accuracy, we integrate an English lexicon-based method [23] in the self-supervised framework for automatically generating labels. The steps of the proposed methodology are shown in Fig. 3.

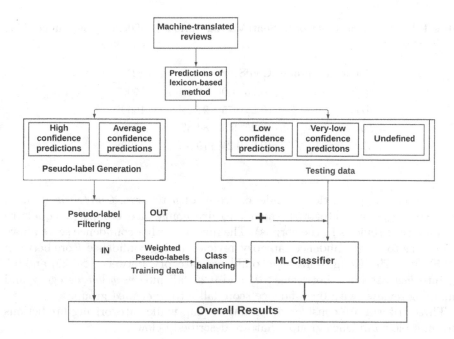

Fig. 3. Steps of the proposed self-supervised hybrid methodology

The proposed cross-lingual self-supervised method trains an ML classifier by following several steps. First, the Bengali reviews are translated to English utilizing Google Translate. Then we employ an English lexicon-based method [23] to generate highly accurate pseudo-labels, which are used as a training set for ML classifiers. Afterward, a filtering step is applied to remove some of the pseudo-labeled reviews from the training set. In the final step, weights are assigned to the filtered pseudo-labels to train a supervised ML classifier.

Pseudo-label Generation. To generate highly accurate pseudo-labels for the supervised ML classifier, we utilize a lexicon-based method, LRSentiA [23]. In addition to determining the semantic orientation of a review, LRSentiA provides the confidence score of the prediction. The prediction confidence score $ConfScore(r)$ of a review r is determined using the following equation-

$$ConfScore(r) = \frac{abs(P_{pos}(r) + P_{neg}(r))}{abs(P_{pos}(r)) + abs(P_{neg}(r))}$$

As the equation indicates, the confidence score of the review r, $ConfScore(r)$, depends on the positive terms, $P_{pos}(r)$ and negative terms, $P_{neg}(r)$ present in the review. A large presence of either positive or negative terms indicates a high confidence score. When the lexicon-based method predicts the class of a review with high confidence, then it is a highly polar review, as indicated by the above equation. These highly polarized reviews have a low chance for misclassification; thus can be used as pseudo-labeled training data.

Table 1. Prediction accuracy of LRSentiA across various confidence groups in machine translated corpus

Confidence group	ConfScore	Accuracy	#Review
High	(0.75, 1.0]	98.1%	7596
Average	(0.5, 0.75]	91.2%	1609
Low	(0.25, 0.5]	84.3%	633
Very-low	(0.0, 0.25]	79.5%	462

Based on the prediction confidence scores of n reviews, $ConfScore(r_1)$,...., $ConfScore(r_n)$, we categorize them into five confidence groups (n equals to the number of reviews in the corpus). The reviews with a confidence score above 0.75 belong to *high* confidence category, reviews having confidence score between (>0.5) and 0.75 belong to *average* confidence category, between (>0.25) and 0.5 fall into *low* category, between (>0) and 0.25 fall into *very-low* category and remaining reviews with 0 confidence score fall into *undefined* group.

Three criteria are considered, similar to [23], while categorizing predictions into multiple confidence groups that are described below.

[a]. Minimizing the inclusion of wrong predictions (i.e., inaccurate pseudo-labels) in the training set to restrict error propagation to the classifier.

[b]. Maximizing the number of reviews (i.e., a large training set) included in the training data.

[c]. Show the correlation between the prediction confidence score and the accuracy (i.e., prediction with a high confidence score implies correctness) to satisfy both criteria [a] and [b].

Both [a] and [b] assist in achieving better performance from the ML classifier. The highly accurate pseudo-labels ([a]) imply less error-propagation to the classifier, and a higher number of pseudo-labels ([b]) mean a large training set, which is required to get good performance from the ML model. [c] helps to determine which reviews should go to training data and which ones to be used as testing data.

We observe that discretizing the reviews into five categories satisfies all the criteria (i.e., [a], [b], and [c]) best; therefore, five confidence groups are used. Table 1 shows accuracies of different confidence groups. The results suggest that there exists a correlation between the prediction accuracy and confidence scores.

Pseudo-label Filtering. This step involves filtering out some of the pseudo-labels selected from *high* and *average* confidence groups. The goal is to improve the accuracy of pseudo-labels further that are used in the training process. We apply the consensus-based filtering based on the lexicon-based method and SVM classifier. We perform 10-fold cross-validation utilizing these pseudo-labeled data from *high* and *average* confidence groups. Based on the predictions of SVM, we only keep the reviews that are assigned to the same class by both SVM and the lexicon-based method. These reviews are utilized as training data for the proposed self-supervised method. The discarded reviews are added to the testing data along with the reviews from *low*, *very-low* and *undefined* categories. The default parameter settings of scikit-learn library [19] and unigram and bigram based tf-idf features are used for the SVM classifier.

After the filtering step, we find 7082 reviews belong to *high confidence* group with an accuracy of around 98.5%. Since the accuracy of this group is already high, the improvement is not significant. However, for the next confidence group, *average*, we observe improvement in the accuracy from 91.2% (1609 reviews) to 93.7% (1321 reviews).

Pseudo-label Weighting. In this step, we assign the weights of the pseudo-labels. We calculate the average confidence scores of *high* and *average* confidence groups. Based on the average confidence scores, we assign the weights of the pseudo-labels that are used as training data for the ML classifiers. The reviews belong to *high* confidence group have higher weight compared to *average* confidence group. The weights of the pseudo-labels (i.e., influence to the classifier) are set based on the group confidence score instead of its own confidence score, as it is a more flexible measure.

Model Training. As shown in Table 1, the *high* and *average* confidence categories of the lexicon-based method yield mostly accurate predictions and can be used as pseudo-labels for the supervised ML algorithms. However, we observe that the distributions of *high* and *average* confidence groups are biased toward *positive* class, contain a much higher number of *positive* samples (could also be attributed to the class distribution of the original dataset). The performance of a supervised ML algorithm can be affected by the presence of the class imbalance. To reduce the negative impact of class inequality, we apply a sampling algorithm, Synthetic Minority Over-sampling TEchnique (SMOTE) [15]. SMOTE is an oversampling method that creates synthetic minority class samples. However, this sampling technique was not able to eliminate the bias towards the *positive* class in our experiment.

Therefore, we use a balanced subset from the *high* and *average* prediction categories to train the supervised ML classifier. The number of instances of each class in the subset is determined by the minimum value of the positive class instance and negative class instance. The instances of the dominant class are randomly selected. The results reported here are the average of the results of 10 random selection.

The reviews belong to *low*, *very-low*, and *undefined* prediction categories in which the lexicon-based method yield low accuracy and the discarded reviews in pseudo-label filtering step are used as testing data for the supervised ML classifiers. We extract unigram and bigram word features from the reviews, calculate the tf-idf scores and feed the scores to the machine learning classifiers. We use the default parameters settings of scikit-learn library [19] for all the ML classifiers.

Overall Predictions. As described above, the overall predictions of the hybrid methodology is determined by the combined predictions of the lexicon-based method (i.e., for reviews belong to *high* and *average* confidence groups excluding filtered out reviews) and the ML classifier (i.e., *low*, *very-low*, and *undefined* confidence groups plus filtered out reviews). The lexicon-based method successfully classifies reviews that are highly polarized (i.e., belong to *high*, and *average* confidence categories), with an accuracy of above 90%. However, for less polar and hard-to-distinguish reviews, the lexicon-based method shows lower accuracy due to various reasons (e.g., the polarity of a review is not obvious or lexicon-coverage problems). Therefore, for reviews belong to these groups, we utilize an ML classifier that is more robust for classifying complicated cases.

5   Results and Discussion

We compare the performances of various classifiers in the Bengali corpus and its machine-translated English version utilizing accuracy, precision, recall, and macro F1 score. The results of ML classifiers are reported based on the default parameter settings of the scikit-learn library [19].

Table 2. The performances of lexicon-based classifiers in Bengali and machine-translated English corpus.

Language of corpus	Method	Precision	Recall	F1 Score	Accuracy
Translated English	VADER	0.846	0.707	0.771	82.56%
	TextBlob	0.863	0.705	0.776	82.79%
	SentiStrength	0.787	0.645	0.708	78.61%
Bengali	[21]	0.716	0.684	0.699	77.10%

Table 2 shows the performances of the lexicon-based methods in Bengali and translated English corpus. VADER and TextBlob exhibit similar F1 scores and accuracies, while SentiStrength performs relatively worse. VADER achieves an F1 score of 0.771 and an accuracy of 82.56%, while TextBlob obtains 0.776 and 82.79%, respectively. The Bengali lexicon-based method shows an F1 score of 0.699 and an accuracy of 77.10%.

Table 3. The performance of the proposed hybrid method in the machine-translated corpus integrating various ML classifiers

	Precision	Recall	F1 Score	Accuracy
Self-Supervised-Hybrid-SVM	0.891	0.903	0.897	91.5%
Self-Supervised-Hybrid-LR	0.876	0.919	0.897	90.8%
Self-Supervised-Hybrid-RF	0.888	0.858	0.872	90.0%
Self-Supervised-Hybrid-ET	0.888	0.872	0.880	90.5%

Table 3 shows the performance of the self-supervised hybrid approach in the machine-translated corpus. The best F1 score of 0.897 is achieved when either LR (Logistic Regression) or SVM (Support Vector Classifier) classifier is integrated into the hybrid method. SVM provides the best accuracy of 91.5%. The decision tree-based methods RF (Random Forest) and ET (Extra Trees Classifier) achieve relatively lower F1 scores.

Among the three English lexicon-based methods applied to the translated reviews, TextBlob and VADER perform similarly, while SentiStrength shows relatively lower efficacy. Compared to English lexicon-based methods, the Bengali lexicon-based method exhibits inferior performance. Sentiment analysis research in Bengali is still not matured; therefore, it lacks enough resources. For example, in Bengali, no sophisticated and comprehensive sentiment lexicon exists. The sentiment lexicon we use here is small in size, consists of around 700 opinion words; thus, it lacks coverage of sentiment words, which is reflected in the performance of the Bengali sentiment analysis tool.

The self-supervised hybrid methodology improves the performance of the sentiment classification. Substantial improvements in both the F1 scores and accuracies are observed compared to the lexicon-based methods. As seen by Table 1, the confidence score of the prediction of the lexicon-based method is highly correlated with the prediction accuracy. The *high* category has a prediction accuracy of above 95% and *average* category has prediction accuracy of over 90%. Therefore, predictions from these categories can be used as training data for supervised ML classifiers with minimal negative impact. As low-resource languages suffer from data annotation issues, the proposed approach can boost sentiment analysis research in resource-poor languages.

6   Summary and Conclusions

In this work, we present a cross-lingual self-supervised methodology for improving the performance of sentiment classification in Bengali by automatically generating pseudo-labels. The proposed approach has advantages over the existing supervised classification methods, as it does not require manual labeling of reviews. As annotated data are hardly available in Bengali, and no sophisticated tools are available for sentiment analysis in unlabeled Bengali text, we explore the adaptation of resources and tools from English. We show that the hybrid cross-lingual approach substantially improves the performance of sentiment classification in Bengali. The results imply that the proposed methodology can advance sentiment analysis research in resource-constraints languages such as Bengali.

References

1. Abdi, A., Shamsuddin, S.M., Hasan, S., Piran, J.: Deep learning-based sentiment classification of evaluative text based on multi-feature fusion. Inf. Process. Manag. **56**(4), 1245–1259 (2019)
2. Al-Amin, M., Islam, M.S., Uzzal, S.D.: Sentiment analysis of Bengali comments with word2vec and sentiment information of words. In: 2017 International Conference on Electrical, Computer and Communication Engineering (ECCE), pp. 186–190, February 2017. https://doi.org/10.1109/ECACE.2017.7912903
3. Balahur, A., Turchi, M.: Comparative experiments using supervised learning and machine translation for multilingual sentiment analysis. Comput. Speech Lang. **28**(1), 56–75 (2014). https://doi.org/10.1016/j.csl.2013.03.004
4. Balamurali, A., Joshi, A., Bhattacharyya, P.: Cross-lingual sentiment analysis for Indian languages using linked wordnets. In: COLING (2012)
5. Banea, C., Mihalcea, R., Wiebe, J., Hassan, S.: Multilingual subjectivity analysis using machine translation. In: 2008 Conference on Empirical Methods in Natural Language Processing, pp. 127–135 (2008)
6. Chen, X., Sun, Y., Athiwaratkun, B., Cardie, C., Weinberger, K.: Adversarial deep averaging networks for cross-lingual sentiment classification. Trans. Assoc. Comput. Linguist. **6**, 557–570 (2018)
7. Chowdhury, S., Chowdhury, W.: Performing sentiment analysis in Bangla microblog posts. In: 2014 International Conference on Informatics, Electronics Vision (ICIEV), pp. 1–6, May 2014

8. Das, A., Bandyopadhyay, S.: Sentiwordnet for Bangla. Knowl. Sharing Event-4: Task **2**, 1–8 (2010)
9. Das, A., Bandyopadhyay, S.: Topic-based Bengali opinion summarization. In: Proceedings of the 23rd International Conference on Computational Linguistics: Posters, pp. 232–240. Association for Computational Linguistics (2010)
10. Feng, Y., Wan, X.: Towards a unified end-to-end approach for fully unsupervised cross-lingual sentiment analysis. In: Proceedings of the 23rd Conference on Computational Natural Language Learning (CoNLL), pp. 1035–1044. Hong Kong, China, November 2019
11. Hassan, A., Amin, M.R., Al Azad, A.K., Mohammed, N.: Sentiment analysis on Bangla and romanized Bangla text using deep recurrent models. In: 2016 International Workshop on Computational Intelligence (IWCI), pp. 51–56. IEEE (2016)
12. He, Y., Zhou, D.: Self-training from labeled features for sentiment analysis. Inf. Process. Manag. **47**(4), 606–616 (2011)
13. Hutto, C.J., Gilbert, E.: Vader: a parsimonious rule-based model for sentiment analysis of social media text. In: Eighth International AAAI Conference on Weblogs and Social Media (2014)
14. Islam, M.S., Islam, M.A., Hossain, M.A., Dey, J.J.: Supervised approach of sentimentality extraction from Bengali Facebook status. In: 2016 19th International Conference on Computer and Information Technology (ICCIT), pp. 383–387, December 2016
15. Lusa, L., et al.: Smote for high-dimensional class-imbalanced data. BMC Bioinform. **14**(1), 106 (2013)
16. Meng, X., Wei, F., Liu, X., Zhou, M., Xu, G., Wang, H.: Cross-lingual mixture model for sentiment classification. In: Proceedings of the 50th Annual Meeting of the Association for Computational Linguistics: Long Papers - Volume 1, pp. 572–581 (2012)
17. Pang, B., Lee, L., Vaithyanathan, S.: Thumbs up?: Sentiment classification using machine learning techniques. In: Proceedings of the ACL-02 Conference on Empirical Methods in Natural Language Processing, vol. 10, pp. 79–86. Association for Computational Linguistics (2002)
18. Patra, B.G., Das, D., Das, A., Prasath, R.: Shared task on sentiment analysis in Indian languages (SAIL) tweets - an overview. In: Prasath, R., Vuppala, A.K., Kathirvalavakumar, T. (eds.) MIKE 2015. LNCS (LNAI), vol. 9468, pp. 650–655. Springer, Cham (2015). https://doi.org/10.1007/978-3-319-26832-3_61
19. Pedregosa, F., et al.: Scikit-learn: machine learning in Python. J. Mach. Learn. Res. **12**, 2825–2830 (2011)
20. Sazzed, S.: Cross-lingual sentiment classification in low-resource Bengali language. In: Proceedings of the Sixth Workshop on Noisy User-generated Text (W-NUT 2020), pp. 50–60 (2020)
21. Sazzed, S.: Development of sentiment lexicon in Bengali utilizing corpus and cross-lingual resources. In: 2020 IEEE 21st International Conference on Information Reuse and Integration for Data Science (IRI), pp. 237–244. IEEE (2020)
22. Sazzed, S., Jayarathna, S.: A sentiment classification in Bengali and machine translated English corpus. In: 2019 IEEE 20th International Conference on Information Reuse and Integration for Data Science (IRI), pp. 107–114 (2019)
23. Sazzed, S., Jayarathna, S.: Ssentia: a self-supervised sentiment analyzer for classification from unlabeled data. Mach. Learn. Appl. **4** (2021)
24. Thelwall, M., Buckley, K., Paltoglou, G., Cai, D., Kappas, A.: Sentiment strength detection in short informal text. J. Am. Soc. Inf. Sci. Technol. **61**(12), 2544–2558 (2010)

25. Tripto, N., Eunus Ali, M.: Detecting multilabel sentiment and emotions from Bangla Youtube comments. In: 2018 International Conference on Bangla Speech and Language Processing (ICBSLP), pp. 1–6 (2018)
26. Turney, P.D.: Thumbs up or thumbs down?: Semantic orientation applied to unsupervised classification of reviews. In: Proceedings of the 40th Annual Meeting on Association for Computational Linguistics, pp. 417–424. Association for Computational Linguistics (2002)
27. Xu, R., Yang, Y., Otani, N., Wu, Y.: Unsupervised cross-lingual transfer of word embedding spaces. In: Proceedings of the 2018 Conference on Empirical Methods in Natural Language Processing, pp. 2465–2474. Brussels, Belgium, October-November 2018
28. Zhang, L., Ghosh, R., Dekhil, M., Hsu, M., Liu, B.: Combining lexicon-based and learning-based methods for Twitter sentiment analysis. HP Laboratories, Technical Report HPL-2011 89 (2011)
29. Zhang, W., Zhao, K., Qiu, L., Hu, C.: Sess: a self-supervised and syntax-based method for sentiment classification. In: Proceedings of the 23rd Pacific Asia Conference on Language, Information and Computation, vol. 2, pp. 596–605 (2009)

Human Language Comprehension
in Aspect Phrase Extraction
with Importance Weighting

Joschka Kersting[1]([✉]) and Michaela Geierhos[2]

[1] Paderborn University, Warburger Street 100, Paderborn, Germany
`joschka.kersting@uni-paderborn.de`
[2] Bundeswehr University Munich, Research Institute CODE,
Carl-Wery-Straße 22, Munich, Germany
`michaela.geierhos@unibw.de`

Abstract. In this study, we describe a text processing pipeline that transforms user-generated text into structured data. To do this, we train neural and transformer-based models for aspect-based sentiment analysis. As most research deals with explicit aspects from product or service data, we extract and classify implicit and explicit aspect phrases from German-language physician review texts. Patients often rate on the basis of perceived friendliness or competence. The vocabulary is difficult, the topic sensitive, and the data user-generated. The aspect phrases come with various wordings using insertions and are not noun-based, which makes the presented case equally relevant and reality-based. To find complex, indirect aspect phrases, up-to-date deep learning approaches must be combined with supervised training data. We describe three aspect phrase datasets, one of them new, as well as a newly annotated aspect polarity dataset. Alongside this, we build an algorithm to rate the aspect phrase importance. All in all, we train eight transformers on the new raw data domain, compare 54 neural aspect extraction models and, based on this, create eight aspect polarity models for our pipeline. These models are evaluated by using Precision, Recall, and F-Score measures. Finally, we evaluate our aspect phrase importance measure algorithm.

Keywords: Aspect-based sentiment analysis · Aspect polarity model

1 Introduction

Sentiment Analysis (SA) is the process of automatically identifying and categorizing opinions expressed in a text, especially to determine whether the author's attitude towards a particular topic, product, etc. is positive, negative, or neutral. There are different approaches: Aspect-based Sentiment Analysis (ABSA) aims to identify expressed opinions about aspects of services or products. SA at the document or sentence-level does not address conflicting feelings, feelings expressed towards different aspects, and the granularity of human language in

© Springer Nature Switzerland AG 2021
E. Métais et al. (Eds.): NLDB 2021, LNCS 12801, pp. 231–242, 2021.
https://doi.org/10.1007/978-3-030-80599-9_21

general. ABSA is therefore an alternative method that allows fine-grained analysis, automatically extracting individual aspects and their scores. The development of ABSA has led to various studies and shared tasks [10,15,17].

Previous approaches have often failed to pursue a human-centered method by considering implicit or indirect mentions of aspects and ratings, as the studies focused on domains with common vocabulary in which nouns often explicitly indicate an aspect. These approaches treat nouns and noun phrases as the representation of aspects, or they consider them as sufficient [2,16,17], due to the commonly used review domains: Most reviews are written for products [6,17] or services [17]. Despite the available domains and their particularities, it is necessary to understand how users rate and why they do so in order to use the reviews available on the Internet. Hence, ABSA is a promising research topic.

However, to find complex indirect aspect phrases, current deep learning approaches need to be combined with supervised training data. Due to implicit mentions and the use of longer phrases, keyword spotting is not an option.

*Example 1. (Sentence from Physician Review). "Dr. Stallmann has **never once looked me in the eye**, but he **accurately described** the options and he also seemed to **know, and this is important to me, what he is doing**."*

In this example, some ratings are given for the aspects "*friendliness*", "*explanation*", and "*competence*" (printed in bold). As shown, these aspect phrases are rather complex, using insertions and different wording. They are not covered by previous machine learning models targeting ABSA, partly because they often appear in a different form and expression. For example, they do not directly mention that a physician has a "*good friendliness*" because this is a rather uncommon style in written physician reviews or everyday conversations.

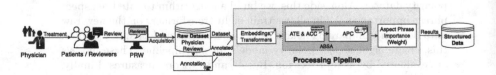

Fig. 1. Processing pipeline to structure and analyze unstructured text data.

Physician reviews can be found in various languages on physician review websites (PRWs) such as Ratemds[1] in English, or Jameda[2] in German. For example, users can rate a physician by assigning scores for rating classes and by writing a textual evaluation. Quantitative scores can be assigned to classes, such as the "*competence of the physician*". Assessed health services are strongly associated with trust; they are sensitive and personal.

As shown in Fig. 1, we build a fully functional text processing pipeline that takes raw text as input, vectorizes it, then extracts aspect phrases to finally add

---

[1] https://ratemds.com, accessed: 2020-12-17.
[2] https://jameda.de, accessed: 2020-12-17.

polarity scores. That is, we classify the extracted aspect phrases to determine whether the author evaluates a characteristic of the doctor negatively or positively. Then, our pipeline determines which of the phrase(s) has an increased importance weight. Overall, this implements a complete cycle from unstructured user-generated text to structured data.

Compared to related literature and our previous work [8–10], we here present a new aspect phrase dataset dealing with a physician practice team, an aspect polarity dataset and supervised learning algorithms, and a method for measuring aspect phrase's weight of importance. Furthermore, we train and test a number of machine learning approaches, including numerous domain-specific transformer models, and build a processing pipeline that converts unstructured physician reviews into structured data (cf. Fig. 1).

2 State of Research

Physician reviews are not like the standard data used for ABSA research. There is no standard service in the healthcare sector, as treatments from physicians and other healthcare providers heavily depend on the practitioner and the patient.

There are three core tasks in ABSA research: ATE, ACC, and APC (Aspect Term Extraction, Category Classification, and Polarity Classification) [2]. ATE means finding aspects in texts. This is important for performing subsequent steps, but as we discussed in a related study [8], much previous work relies on nouns, seed words, etc. For example, Pontiki et al. [18] write that "[a]n opinion target expression [...] is an explicit reference (mention) to the reviewed entity [...]. This reference can be a named entity, a common noun or a multi-word term". In their annotated datasets, they used common product or service domains (e.g., hotels) and achieved evaluation scores for ATE and ACC of about 50%. However, most studies use the data of the shared task by Pontiki et al. [17] or its predecessors [19,20], as survey studies show [24,25].

Previous ABSA approaches have neglected human-like language understanding without artificial constraints, thus limiting their methods and data domains, as we have previously described [8–10]. Therefore, most study designs cannot be applied to physician review data.

Recent approaches to ABSA use neural networks and deep learning methods, as surveys show [15,24,25]. They differ not only in the applied data (mostly from shared tasks [17,19,20]), but also in neural network architectures and do not perform ATE. Thus, they rather perform SA at the sentence or document level. However, it is clear that transformers such as BERT [7] have improved vector representations for use in other algorithms, while they can also be fine-tuned for downstream tasks such as tagging words and classifying texts. For example, our previous work [10] successfully applied transformer models to PRW data, but more traditional methods for language modeling such as FastText [1] are still competitive for physician reviews, as we have shown [8]. All in all, previous research has not explored and made the contributions described in Sect. 1, although researchers such as DeClercq et al. [6] built an ABSA pipeline for

Dutch social media data on retail, banking, and human resources. Nevertheless, the domain, approach, and data are entirely different. Based on our previous remarks and current studies, there is no alternative to supervised deep learning for ABSA with human-like aspect understanding [15,24,25].

Several datasets [6,14,17,20,23] have been created for ABSA so far. Here it can be seen that the polarity scales are usually threefold or twofold, i.e., they use either the positive, negative and neutral classes or only the first two. The importance weight of an aspect phrase is difficult to determine because aspect phrases are not very heterogeneous. There is no uniform vocabulary; so it is not sufficient to use rule-based or list-based approaches that determine importance with the infrequent preference of a word from a predefined list. In German, longer off-topic insertions are also common (*"He took a lot, and I want to add this after I clarify how I encountered my friend in the office, of time ..."*) and such cases are numerous, making it difficult to adapt ideas from the literature. Moreover, it is not known which rating scale should be applied here. However, from the ABSA datasets and their polarity scales, it can be inferred that a rather simple scale with two or three values is applicable. However, it is obvious that users assign different weights to aspect classes [13].

One of the many various approaches is to calculate the semantic information value, e.g., using the entropy or by measuring the cosine similarity between the embedding vectors of a phrase and the corresponding annotations for the class. However, this misses the point, because we have neural vectors with embedded semantics but do not see information scores or vector similarity as measures for importance. Another approach might be sentiment-intensity ranking: A study [21] uses words with the same meaning and ranks polar words by intensity, e.g., *"pleased, exhilarated"*. Such approaches do not fit because we do not have a traditional separation into sentiment and aspect words, and lexicon-based approaches are not flexible enough. Our phrases mostly cover both at once, e.g. in just one word like *"friendly"*, which indicates both friendliness as an aspect and a positive evaluation. The same applies to longer phrases (cf. Example 1). Therefore, a promising approach is to calculate the normalized frequency of aspect classes in the respective dataset. This provides a unique measure that also allows a comparison of the classes. A second possibility is to analyze linguistic structures which indicate a higher importance. Since adjectives are common in our data, intensified adjectives or additional adverbs could be a solid way to identifying important aspect phrases from physician reviews.

3   Data and Annotation Process

In our data and annotation process (aspect and polarity data), some of the data are based on our previous works which contain additional information, especially for the fkza and bavkbeg datasets [8–10].

Raw data were collected from three German-language PRWs[3] between March and July 2018 by using a spider to crawl all review and physician pages to reach a total of 400,000 physicians and over 2,000,000 review texts. The scales are based on the German/Austrian school grade system as well as star ratings. The number of quantitative rating classes varies greatly among the PRWs [5,8–10]. To train algorithms that extract and classify aspect phrases, we needed to find classes that could be annotated.

We considered all available quantitative rating classes from the three crawled websites and qualitatively merged classes, e.g., those related to the team's "*competence*" or to the "*waiting time for [an] appointment*", etc. The semantic merging of quantitative classes resulted in a larger set of rating classes. For ATE and ACC, we use three datasets in this study. The first two, fkza and bavkbeg, were taken from our previous studies [8–10], which present the dataset in detail and provide a tableau of examples. In short, fkza is an acronym of the German names of the classes translated into English as "*friendliness*", "*competence*", "*time taken*", and "*explanation*". The bavkbeg dataset covers the classes of "*treatment*", "*alternative healing methods*", "*relationship of trust*", "*child-friendliness*", "*care/commitment*", and "*overall/recommendation*". Fkza and bavkbeg apply to the physician as an aspect target. Bavkbeg has an overall rating class that applies equally to the physician, the practice, and the team. These three are the available aspect targets in the data. Like many systems, we perform ATE and ACC together [24], which is due to their mutual influence.

The third and newly annotated dataset is called bfkt, which aims at the physician's team as an aspect target. Since the target is different, some of the classes are similar to those in the fkza package. However, for human annotators identifying the aspect target clearly on the basis of the text is not an issue. To avoid annotation conflicts, certain rules can be established. The classes of bfkt are these: "*care/commitment*", "*friendliness*", "*competence*", and "*accessibility by telephone*"[4].

- "*Care/commitment*" refers to whether the practice team is (further) involved or interested in the patient's care and treatment: "***Such a demotivated*** assistant!"
- "*Friendliness*" deals with the friendliness, as in the package fkza, but aims at the team: "*Due to their **very nice manner**, there was no doubt about the team at any time.*"
- "*Competence*" describes the patient's perception of the team's expertise: "*The staff at the reception makes an **overstrained impression**.*"
- "*Accessibility by telephone*" indicates how easy it is to reach the team: "*You have to **try several times before you get someone on the phone**.*"

Since the PRWs focus on reviews of "doctors", this may explain why there are far fewer aspect phrases for bfkt. The annotation process began with one

---

[3] Jameda: https://jameda.de; Docfinder: https://docfinder.at; Medicosearch: https://medicosearch.ch; accessed 2021-01-11.

[4] Translated from German, with the team as the aspect target: "*Betreuung/Engagement*", "*Freundlichkeit*", "*Kompetenz*", and "*Telefonerreichbarkeit*".

person annotating the package, while we held ongoing discussions and reviews among a team of (computational) linguists. Active learning was performed once for all packages before annotations began, consistent with previous work [10]. Here, the goal was to find sentences that generally contain an evaluative statement. For this purpose, we used a neural network classifier. We then annotated several thousand sentences for bfkt. Since most of the sentences did not contain relevant statements, we again trained a sentence-level classifier using the existing annotations, ordering the sentences in the resulting file so that they contained at least one predicted class per line. This multi-label, multi-class classification problem at the sentence level helped us save time, which is consistent with what was done for bavkbeg [10]. Of more than 15,000 sentences in the bfkt dataset, about half contain an evaluative statement, and it was possible to annotate more than one mentioned aspect in a sentence. Most sentences tend to be short, and users generally write as they speak, indicating rating aspects in longer phrases like: *"It doesn't matter how many times you try, you will **never catch any of them over the wire**!"* During annotation, we also formulated rules and examples as guidelines for the annotators to follow, such as that phrases should be as short as possible but contain all important information, preferably without punctuation.

The annotation task was rather difficult due to the data and the direct and indirect long phrases it contained. We computed an IAA based on the tagged words, assigning a tag to each word indicating its class. All non-annotated words were tagged "no class". We used the annotations of the first annotator and randomly selected about 330–360 sentences (about 3% of fkza [8], bavkbeg [10]). The second annotator and another person then performed new annotations for the agreement. The values of all IAAs are shown in Table 1. All Cohen's Kappa [3] values can be considered as "substantial" agreement (0.61–0.80). One pair of annotators, "B&J", achieved an "almost perfect" agreement [12]. Krippendorff's Alpha [11] can be considered good as it leans to 1.0. However, the values are worse than for fkza and bavkbeg [10] with 0.654 (R&B) to 0.722 (R&B) for bfkt and fkza.

Table 1. Inter-annotator agreements for all used datasets (fkza & bavkbeg: [8,10]).[a]

	Dataset fkza			Dataset bavkbeg			Dataset bfkt			Polarity dataset		
	R& B	R& J	B& J	R& B	R& J	B& J	R& B	R& J	B& J	R& M	R& J	M& J
CK	0.722	0.857	0.730	0.731	0.719	0.710	0.654	0.673	0.806	0.917	0.923	0.918
KA	0.771			0.720			0.711			0.919		

[a] CK = Cohen's Kappa; KA = Krippendorff's Alpha.

The sentiment polarity annotations were conducted differently. As mentioned above, a distinction between aspect phrases and sentiment words is not possible. Since the aspect phrase and class annotations were difficult and several tens of thousands of sentences had to be annotated, the steps were separated and the polarity step was conducted later. For the polarity annotation, we randomly selected sentences containing aspect phrases from the datasets and deleted erroneous annotations from the file. We also included two newly annotated aspect datasets that were not yet complete, so we had the annotator also check the aspect phrases for errors. For each phrase, we needed to assign a positive or negative sentiment polarity. At first, we tried finer scales by using a neutral value. After testing and discussions we discovered that neither a finer granularity nor a neutral label are appropriate for our data, as the phrases do not have patterns that reveal finer nuances, and neutral evaluative statements are almost nonexistent in physician reviews. As for nuances such as a "highly positive", "midly positive" or "normal positive" polarity, it is difficult to distinguish between the phrases such as: *"very friendly"*, *"expressively friendly"*, *"always very friendly"* or *"always friendly as every time except once"*, *"indeed he was friendly today"*. These phrases show that nuances are hard to systematize, so adequate and consistent annotations for scales with increments are not possible.

After deleting the sentences that contained mistakes and the ones in which we experimentally annotated potentially neutral values, we have over 9,300 sentences with polarity annotations in general. For quality reasons, we computed the IAA shown in Table 1. As shown, the results are quite good. Since the Cohen's Kappa values are all above 0.90, the agreement is almost perfect [12]. However, this is not surprising for a human annotation of a binary phrase-sentiment polarity classification task. Krippendorff's Alpha can be considered as very good, with a value of 0.919, which is quite close to 1.0.

4   Method and Results

As our previous work has shown, supervised neural learning is the most promising path for ABSA in a serious data domain such as ours [8,10]. However, it was also shown that transformers perform well in ATE and ACC, especially when pre-trained on raw PRW data. Nevertheless, more traditional solutions such as FastText provided the best results, while a domain-trained BERT [7] performed slightly better or almost as well [10]. Due to this information, we want to further investigate using transformer models for our case, so we searched Huggingface for pre-trained transformer models for German.

For our experimental setup, we used IO tags (Inside, Outside) for ATE and ACC [8], e.g., *"I-friendliness_T"*. This step is critical because it is the most challenging and it starts the pipeline, so the other steps depend on the results (cf. Fig. 1). Therefore, we tested a large number of transformers and show these results in Table 2. First, we domain-trained the existing transformer models for German as well as the multilingual XLM-RoBERTa [4]. The domain-trained models are marked with a "+". As tests have shown, we do not have enough PRW data to train a transformer from scratch (no useful results), so we tested pre-trained transformers and domain-trained these further. In addition to fine-tuning, we built our own neural networks that used the word vectors generated by the transformer as input. We used XLM-RoBERTa for this purpose because the loss in domain training was extremely small. The loss was about 0.37 after

Table 2. Results[b] for the extraction and classification of aspect phrases (ATE, ACC) using broadly pre-trained and domain-trained ("+") transformers.

Model	bfkt			bavkbeg			fkza		
	P	R	F1	P	R	F1	P	R	F1
xlm-roberta-base+	0.81	0.70	0.75	0.83	0.82	**0.82**	0.86	0.80	**0.83**
∟ biLSTM-CRF+	0.78	0.76	**0.77**	0.81	0.81	0.81	0.86	0.79	**0.83**
∟ biLSTM-Attention+	0.82	0.70	0.75	0.78	0.81	0.79	0.83	0.80	0.82
xlm-roberta-base	0.80	0.70	0.74	0.83	0.81	0.81	0.85	0.80	0.82
MedBERT+	0.81	0.70	0.75	0.84	0.82	**0.82**	0.86	0.80	**0.83**
MedBERT	0.80	0.68	0.73	0.83	0.79	0.80	0.86	0.78	0.82
electra-base uncased+	0.16	0.20	0.18	0.10	0.14	0.12	0.15	0.20	0.17
electra-base uncased	0.79	0.70	0.74	0.82	0.81	0.81	0.85	0.80	0.82
distilbert-base cased+	0.80	0.69	0.74	0.83	0.80	0.80	0.85	0.80	0.82
distilbert-base cased	0.78	0.67	0.72	0.81	0.78	0.79	0.84	0.78	0.81
dbmdz bert-base uncased+	0.81	0.72	0.76	0.82	0.82	0.81	0.86	0.80	**0.83**
dbmdz bert-base uncased	0.80	0.70	0.74	0.83	0.80	0.81	0.86	0.80	0.82
dbmdz bert-base cased+	0.80	0.70	0.74	0.83	0.81	0.81	0.86	0.81	**0.83**
dbmdz bert-base cased	0.79	0.70	0.74	0.83	0.79	0.80	0.85	0.80	0.82
bert-base cased+	0.81	0.71	0.75	0.83	0.81	**0.82**	0.86	0.81	**0.83**
bert-base cased	0.79	0.68	0.73	0.82	0.80	0.80	0.86	0.78	0.82
FastText biLSTM-CRF+	0.77	0.70	0.73	0.80	0.76	0.78	0.83	0.79	0.81
FastText biLSTM-Attention+	0.74	0.71	0.72	0.81	0.74	0.77	0.82	0.77	0.79

[b] P = Precision, R = Recall, F1 = F1-score; all pre-trained transformer models are in German and can be found by their names on https://huggingface.co/models, accessed 2020-12-28. BiLSTM-CRF and Attention models are based on [10].

4 epochs compared to about 1.1–1.3 after 10 epochs for most German language models such as BERT (bert-base cased). This was different for Electra (a loss over 6.7). The parameters were tuned before the final runs. We used a train-test split of 90%/10% of the sentences extracted from the raw data (cf. Sect. 3).

XLM-RoBERTa achieves the best scores for the datasets bavkbeg (F1: 0.82) and fkza (F1: 0.82) with transformer fine-tuning and for bfkt (F1: 0.77) with a biLSTM-CRF model [10]. The train-test split was 80%/20% for transformers (epochs: 10) in most cases, and 90%/10% for the other neural networks (epochs: 6) after tuning the parameters. FastText was trained uncased, as we are using error-prone user-generated text data with medical terms. A general advantage cannot be seen (in contrast to previous work [8]), since cased transformers also perform well. This may be because the transformer approach computes embeddings ad-hoc, based on context, while FastText computes a fixed table in which each string is given a vector.

The other models in Table 2 that are not explicitly marked as (un-)cased are cased. While XLM-RoBERTa is well documented, which is another reason for its use, other German transformers were not. For MedBERT, a related paper was published after we had used it [22]. At least it was obvious that MedBERT was trained on data related to the medical domain. We use Precision, Recall and F1 as scores because accuracy is prone to error considering our high class imbalance. Most words in a sentence are not tagged as a specific class, but as "O" such as outside a phrase. Therefore, the accuracy values were all the same but did not reveal differences in the models. To reduce the imbalance and because the results were better, we used only the sentences containing aspect phrases.

The second step in the pipeline (cf. Fig. 1) involves sentiment polarity classification. We used the transformer architectures that performed best in the previous step, so only domain-tuned transformers and FastText embeddings were used. The results are shown in Table 3. Again, our own neural network performs best with an F1-score of 0.96, strengthened by XLM-RoBERTa embeddings. Again, we obtained the best results with a train-test split of 80%/20% for the transformer fine-tuning (epochs: 4) and 90%/10% for the other neural networks (epochs: 6). The task was performed as a binary text classification. We used two input layers that received the corresponding aspect phrase and its context. Our goal was to classify the aspect phrase; the context was represented by the sentence from which the phrase was extracted. The multilingual XLM-RoBERTa outperformed the transformers trained specifically for German.

Table 3. Sentiment polarity classification results.

Model	P	R	F1
xlm-roberta-base+	0.93	0.95	0.94
∟ biLSTM+	0.97	0.95	**0.96**
∟ CNN+	0.92	0.97	0.94
MedBERT+	0.90	0.95	0.92
dbmdz bert-base uncased+	0.92	0.93	0.92
bert-base cased+	0.92	0.91	0.91
FastText biLSTM+	0.92	0.95	0.93
FastText CNN+	0.91	0.93	0.92

The third step of the pipeline deals with measuring the importance of aspect phrases. After studying the available methods in Sect. 2, we concluded that three approaches are promising: First, importance can be derived from a normalized frequency of each aspect class. On this basis, the most frequent aspect classes are ranked as most important. Second, as suggested in Sect. 2, we set up a linguistic approach that uses part-of-speech (POS) tagging to identify adverbs and adjective superlatives. We suggested that the presence of an adverb increases importance, which is often the case: "*They were **very compassionate**.*", instead of just "*compassionate*", "*friendly*", etc. This also applies to longer phrases. The use of superlatives also shows a high importance: "*The woman at the front desk is the **worst listener** I have ever seen!*" We also included German indefinite pronouns: "*They had **many friendly** words.*" Third, we combined the two approaches and suggested that whenever either one of both suggests a high importance, this should be respected as an outcome. The exploration of possible methods also led to an investigation of which scale is appropriate. Consistent with our observations regarding the polarity scale (cf. Sect. 3), only a binary classification into higher and lower (normal) importance is possible. Finer gradations are not possible.

Table 4. Accuracy-agreement of humans with the importance weighting method.

Person	Statistic	Linguistic	Combined
J	0.51	**0.82**	0.62
R	0.54	**0.87**	0.65

To test our aspect-phrase-importance weighting, we had two human annotators label approximately 340 random phrases with higher or lower relative importance. Both knew the domain and were introduced to the task. During the initial annotations, they were allowed to see the results of the automatic approach. The evaluation results in Table 4 show the accuracy of the annotations with the automatic methods. As can be seen, the linguistic approach has the highest agreement: 0.82 for annotator J and 0.87 for annotator R. The high scores indicate the quality of the approach. The disadvantages of this method are that POS tagging sometimes fails, especially when distinguishing between adverbs and adjectives. Furthermore, POS tagging may fail for longer phrases and due to insertions that may contain superlatives that are not relevant to the corresponding aspect. The evaluation results may have a limited value because annotators may be biased on their linguistic knowledge or knowledge of the used methods.

5   Conclusion

We showed three datasets for ATE and ACC, one is new and deals with the team of a physician's office, and two deal with the physician as the aspect target. We also presented a new dataset for APC and calculated IAAs for all datasets, achieving good scores (cf. Table 1). To build a pipeline that converts user-generated, unstructured physician reviews into structured data, we trained a set of deep-learning models and developed a method for measuring the importance of aspect phrases. All of these were evaluated in detail in Tables 2, 3 and 4 and obtained good results. We tested 54 models for ATE and ACC, and another eight for APC. XLM-RoBERTa in its basic version emerged as the best model among all those tested. It is a multilingual model that also outperformed German-only models, which we consider a major finding, especially as we applied the models to long, complex, and user-generated phrases. Furthermore, due to resource constraints, we trained the base version of this pre-trained transformer instead of the large version. This large version is a promising tool for future experiments.

In all training steps, we applied human language comprehension to extract information in a human-like manner, conducting broad research by using and comparing a wide variety of neural models. In the future, we can build on these experiments to extract other aspect classes from data such as the accessibility and the opening hours. We see potential applications in domains with implicit aspect phrases. Since XLM-RoBERTa is capable of working with multiple languages, we plan to test our fine-tuned models on English physician reviews. The binary scales discussed and used to measure sentiment polarity and aspect phrase importance emerged as the only feasible solutions based on the data. Annotating the data based on context allows us and our models to treat irony accordingly. Parts of the pipeline methods presented here are in further development for a related study dealing with possible analyses based on it.

Acknowledgments. This work was partially supported by the German Research Foundation (DFG) within the Collaborative Research Center On-The-Fly Computing (SFB 901). We thank F. S. Bäumer, M. Cordes, and R. R. Mülfarth for their assistance with the data collection.

References

1. Bojanowski, P., Grave, E., Joulin, A., Mikolov, T.: Enriching word vectors with subword information. Trans. ACL **5**, 135–146 (2017)
2. Chinsha, T.C., Shibily, J.: A syntactic approach for aspect based opinion mining. In: Proceedings of the 9th IEEE International Conference on Semantic Computing, pp. 24–31. IEEE (2015)
3. Cohen, J.: A coefficient of agreement for nominal scales. Educ. Psychol. Measur. **20**(1), 37–46 (1960)
4. Conneau, A., et al.: Unsupervised cross-lingual representation learning at scale. In: Proceedings of the 58th Annual Meeting of the ACL, pp. 8440–8451. ACL, Online (2020)
5. Cordes, M.: Wie bewerten die anderen? Eine übergreifende Analyse von Arztbewertungsportalen in Europa. Master's thesis, Paderborn University (2018)
6. De Clercq, O., Lefever, E., Jacobs, G., Carpels, T., Hoste, V.: Towards an integrated pipeline for aspect-based sentiment analysis in various domains. In: Proceedings of the 8th ACL Workshop on Computational Approaches to Subjectivity, Sentiment and Social Media Analysis, pp. 136–142. ACL (2017)
7. Devlin, J., Chang, M.W., Lee, K., Toutanova, K.: BERT: pre-training of deep bidirectional transformers for language understanding. In: Proceedings of the 2019 Conference of the North American Chapter of the ACL: Human Language Technologies, Volume 1 (Long and Short Papers), pp. 4171–4186. ACL (2019)
8. Kersting, J., Geierhos, M.: Aspect phrase extraction in sentiment analysis with deep learning. In: Proceedings of the 12th International Conference on Agents and Artificial Intelligence: Special Session on Natural Language Processing in Artificial Intelligence, pp. 391–400. SCITEPRESS (2020)
9. Kersting, J., Geierhos, M.: Neural learning for aspect phrase extraction and classification in sentiment analysis. In: Proceedings of the 33rd International FLAIRS, pp. 282–285. AAAI (2020)

10. Kersting, J., Geierhos, M.: Towards aspect extraction and classification for opinion mining with deep sequence networks. In: Loukanova, R. (ed.) Natural Language Processing in Artificial Intelligence—NLPinAI 2020. SCI, vol. 939, pp. 163–189. Springer, Cham (2021). https://doi.org/10.1007/978-3-030-63787-3_6

11. Krippendorff, K.: Computing Krippendorff's Alpha-Reliability. Technical report 1–25-2011, University of Pennsylvania (2011)

12. Landis, J.R., Koch, G.G.: The measurement of observer agreement for categorical data. Biometrics **33**(1), 159–174 (1977)

13. Liu, Y., Bi, J.W., Fan, Z.P.: Ranking products through online reviews: a method based on sentiment analysis technique and intuitionistic fuzzy set theory. Inf. Fusion **36**, 149–161 (2017)

14. López, A., Detz, A., Ratanawongsa, N., Sarkar, U.: What patients say about their doctors online: a qualitative content analysis. J. General Internal Med. **27**(6), 685–692 (2012)

15. Nazir, A., Rao, Y., Wu, L., Sun, L.: Issues and challenges of aspect-based sentiment analysis: a comprehensive survey. IEEE Trans. Affective Comput. 1 (2020). https://doi.org/10.1109/TAFFC.2020.2970399

16. Nguyen, T.H., Shirai, K.: PhraseRNN: phrase recursive neural network for aspect-based sentiment analysis. In: Proceedings of the 2015 Conference on Empirical Methods in Natural Language Processing, pp. 2509–2514. ACL (2015)

17. Pontiki, M., Galanis, D., Papageorgiou, H., Manandhar, S., Androutsopoulos, I.: SemEval-2016 task 5: aspect based sentiment analysis. In: Proceedings of the 10th International Workshop on Semantic Evaluation, pp. 19–30. ACL (2016)

18. Pontiki, M., Galanis, D., Papageorgiou, H., Manandhar, S., Androutsopoulos, I.: SemEval-2016 Task 5: Aspect Based Sentiment Analysis (ABSA-16) Annotation Guidelines (2016)

19. Pontiki, M., Galanis, D., Pavlopoulos, J., Papageorgiou, H., Androutsopoulos, I., Manandhar, S.: SemEval-2014 task 4: aspect based sentiment analysis. In: Proceedings of the 8th International Workshop on Semantic Evaluation, pp. 27–35. ACL (2014)

20. Pontiki, M., Galanis, D., Papageorgiou, H., Manandhar, S., Androutsopoulos, I.: SemEval-2015 task 12: aspect based sentiment analysis. In: Proceedings of the 9th International Workshop on Semantic Evaluation, pp. 486–495. ACL (2015)

21. Sharma, R., Somani, A., Kumar, L., Bhattacharyya, P.: Sentiment intensity ranking among adjectives using sentiment bearing word embeddings. In: Proceedings of the 2017 Conference on Empirical Methods in Natural Language Processing, pp. 547–552. ACL (2017)

22. Shrestha, M.: Development of a language model for medical domain. Master's thesis, Rhine-Waal University of Applied Sciences (2021)

23. Wojatzki, M., Ruppert, E., Holschneider, S., Zesch, T., Biemann, C.: GermEval 2017: shared task on aspect-based sentiment in social media customer feedback. In: Proceedings of the GermEval 2017 - Shared Task on Aspect-based Sentiment in Social Media Customer Feedback, pp. 1–12. Springer (2017)

24. Zhang, L., Wang, S., Liu, B.: Deep learning for sentiment analysis: a survey. Wiley Interdisc. Rev.: Data Mining Knowl. Discov. **8**(4), 1–25 (2018)

25. Zhou, J., Huang, J.X., Chen, Q., Hu, Q.V., Wang, T., He, L.: Deep learning for aspect-level sentiment classification: survey, vision, and challenges. IEEE Access **7**, 78454–78483 (2019)

Exploring Summarization to Enhance Headline Stance Detection

Robiert Sepúlveda-Torres⁽✉⁾(iD), Marta Vicente(iD), Estela Saquete(iD),
Elena Lloret(iD), and Manuel Palomar(iD)

Department of Software and Computing Systems, University of Alicante,
Carretera de San Vicente s/n 03690, Alicante, Spain
{rsepulveda,mvicente,stela,elloret,mpalomar}@dlsi.ua.es

Abstract. The spread of fake news and misinformation is causing serious problems to society, partly due to the fact that more and more people only read headlines or highlights of news assuming that everything is reliable, instead of carefully analysing whether it can contain distorted or false information. Specifically, the headline of a correctly designed news item must correspond to a summary of the main information of that news item. Unfortunately, this is not always happening, since various interests, such as increasing the number of clicks as well as political interests can be behind of the generation of a headlines that does not meet its intended original purpose. This paper analyses the use of automatic news summaries to determine the stance (i.e., position) of a headline with respect to the body of text associated with it. To this end, we propose a two-stage approach that uses summary techniques as input for both classifiers instead of the full text of the news body, thus reducing the amount of information that must be processed while maintaining the important information. The experimentation has been carried out using the Fake News Challenge FNC-1 dataset, leading to a 94.13% accuracy, surpassing the state of the art. It is especially remarkable that the proposed approach, which uses only the relevant information provided by the automatic summaries instead of the full text, is able to classify the different stance categories with very competitive results, so it can be concluded that the use of the automatic extractive summaries has a positive impact for determining the stance of very short information (i.e., headline, sentence) with respect to its whole content.

Keywords: Natural language processing · Fake news · Misleading headlines · Stance detection

1 Introduction

Headlines are fundamental parts of news stories, summarizing the content and giving the reader a clear understanding of the article's content [9]. However, nowadays, the speed at which information spreads and the degree of information overload are considered by many to be reaching an unmanageable state [34].

© Springer Nature Switzerland AG 2021
E. Métais et al. (Eds.): NLDB 2021, LNCS 12801, pp. 243–254, 2021.
https://doi.org/10.1007/978-3-030-80599-9_22

Therefore, it is tempting to read only the headlines of news and share it without having read the entire story [15]. In this sense, a headline should be as effective as possible, without losing accuracy or being misleading [19], in order to maintain accuracy and veracity of the entire article.

Unfortunately, in practice, headlines tend to be more focused on attracting the reader's attention and going viral because of this, despite the lack of veracity within the information in the body text, thus leading to mis- or disinformation through erroneous/false facts or headline/body dissonance [6]. Headlines are considered misleading or incongruent when they significantly misrepresent the findings reported in the news article [7], by exaggerating or distorting the facts described in the news article. Some important nuances that are part of the news body text are missing in the headline, causing the reader to come to the wrong conclusion. Therefore, the reader cannot discover these inconsistencies if the news body text is not read [38].

In the research community, the task of automatically detecting misleading/incongruent headlines is addressed as a stance detection problem, which implies estimating the relative perspective, i.e., the stance of two pieces of text relative to a topic, claim or issue [14]. This is done through news body text analysis, determining the evidences from which the headline has been derived.

In this context, the main objective of our research is to propose a novel approach that automatically determines the stance of the headline with respect to its body text integrating summarization techniques in a two-stage classification problem, where both the news headline and its corresponding body text are given as input.

2   Related Work

Triggered by a greater demand for new technologies together with an increase in the availability of annotated corpora, headline stance detection task quickly emerged in the context of fake news analysis. In this context, research challenges and competitions, such The *Fake News Challenge*[1] (FNC-1) [2] were proposed.

FNC-1 was created using Emergent dataset [14] as a starting point [31] and it aimed to compile a gold standard to explore Artificial Intelligence technologies, especially ML and Natural Language Processing (NLP), applied to detection of fake news. The three best systems in this competition were Talos [3], Athene system [1] and UCLMR [30] in this order. Talos [3] applied a one-dimensional convolution neural networks (CNN) on the headline and body text, represented at the word level using Google News pretrained vectors. The output of this CNN is then sent to a multilayer perceptron (MLP) with 4-class output: *agree*, *disagree*, *discuss*, and *unrelated*, and trained end-to-end. Using this combination CNN-MLP, the system outperformed all the submissions and achieved the first position in the FNC-1 challenge. Outside the FNC-1 competition but using its dataset other work and experiments have been carried out. [40] addressed the

---

[1] http://www.fakenewschallenge.org/ (accessed online 18 March, 2021).

problem proposing a hierarchical representation of the classes, which combines agree, disagree and discuss in a new related class. A two-layer neural network is learning from this hierarchical representation of classes and a weighted accuracy of 88.15% is obtained with their proposal. Furthermore, [12] constructed a stance detection model by performing transfer learning on a RoBERTa deep bidirectional transformer language model by taking advantage of bidirectional cross-attention between claim-article pairs via pair encoding with self-attention. They reported a weighted accuracy of 90.01%. Outside the FNC-1 Challenge and dataset, there is other research that also addresses the stance detection tasks, determining the relation of a news headline with its body text by extracting key quotes [28] or claims [36].

Turning now into text summarization, its main potential is its ability to extract the most relevant information from a document, and synthesize its essential content. In this respect, one of the most outstanding areas in using summarization techniques is that of news, partly thanks to the development of appropriate corpora (e.g. DUC, Gigaword, CNN/DailyMail)[8], and the wide range of techniques and approaches to help digest this type of information [11,22,26,41]. Moreover, there is a significant amount of research on the task of headline generation using summarization techniques [4,10,39], and more recently using Deep Learning [16,18,33].

However, to the best of our knowledge, regarding disinformation, summarization for detecting fake news has only been proposed in [13], where an abstractive summarization model is applied. In this manner, the news article is first summarized, and the generated summary is used by the classification algorithm instead of the whole body text, which may be too long, or just the headline, which may be too short. Considering this aforementioned research results in which the accuracy is higher when using the summary compared to the full body text, our approach adopts this similar idea where the news article is reduced to its essential information, and exploits it further within a two-stage classifier to detect incongruities between headline and the body text of a news article.

3   Approach Architecture

Following the FNC-1 guidelines, the task of detecting misleading headlines tackled as a headline stance detection task involves classifying the stance of the body text with respect to the headline into one of the following four classes: a) *agrees*—agreement between body text and headline; b) *disagrees*—disagreement between body text and headline; c) *discusses*—same topic discussed in body text and headline, but no position taken; and, d) *unrelated*—different topic discussed in body text and headline.

To address this task, we propose an approach[2] that involves two-stages, thus addressing the task as a two-level classification problem: *Relatedness Stage*, and *Stance Stage*. Figure 1 illustrates the complete architecture. Next, a more

---

[2] Implementation available at https://github.com/rsepulveda911112/Headline-Stance-Detection.

detailed description of both stages and the different modules involved in performing the stance classification is provided.

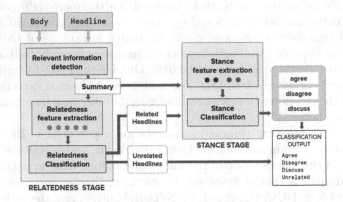

Fig. 1. Two-staged architecture devised to tackle the headline stance detection task

3.1   Relatedness Stage

The *Relatedness Stage* is in charge of determining whether or not the headline and the body of the news are related. The inputs of this stage are both the text body and the headline, resulting in a binary classification. The outputs of this stage are:

- The *headlines* classified as *related* or *unrelated*.
- The *summary* of the news content, obtained in a relevant information detection module.

For this, three modules are proposed: i) relevant information detection; ii) relatedness feature extraction; and, iii) relatedness classification.

Relevant Information Detection Module. This module aims to create a summary revealing the important information of the input news article in relation to its headline. Although different summarization approaches could be used for this purpose, we opt for the popular and effective TextRank extractive summarization algorithm [24], due to its good performance, execution time and implementation availability.[3] This algorithm represents the input text as a graph, where the vertices represent the sentences to be ranked, and the edges are the connections between them. Such connections are determined by the similarity between the text sentences measured with respect to their overlapping content. Then, a weight is computed for each of the graph edges indicating the strength of the connection between the sentences pairs/vertices linked by them. Once the

---

[3] https://pypi.org/project/sumy/.

graph is built, a weighted graph-based ranking is performed in order to score each of the text sentences. The sentences are then sorted in reversed order of their score. Finally, the top ranked sentences, in our case five, are selected to be included in the final summary.

Relatedness Feature Extraction. This module is focused on computing similarity metrics between the generated summary and the given headline. The computed features, which will be used in subsequent module, are:

- Cosine similarity between headline and summary TF-IDF vector without stop word [27].
- Overlap coefficient between headline and summary without stop words [23].
- BERT cosine similarity between headline and summary. We use sentences transformer [29].
- Positional Language Model (PLM) salience score between headline and summary, which has been shown to be effective for relevant content selection [35]
- Soft cosine similarity between headline and summary without stop words. We use word2vec vector [25].

Relatedness Classification. This module exploits the relatedness features previously computed, as well as the automatic summary to finally classify the headlines as *related* or *unrelated*. The proposed architecture is flexible to choose any model that allows classifiers to be improved.

In this case, the design of the relatedness classification module is based on fine-tuning the RoBERTa (Robustly optimized BERT approach) pre-trained model [21], applying a classifier to its output afterwards.

First, the headline and the summary are concatenated and processed with the RoBERTa model. The resulting vector is consecutively multiplied by the three features (Cosine similarity, Overlap coefficient, BERT cosine similarity, PLM salience score and Soft cosine similarity) to finally carry out the classification using a Softmax activation function in the output layer.

Specifically, we have chosen RoBERTa Large model (24 layer and 1024 hidden units) since it achieves state-of-the-art results in General Language Understanding Evaluation (GLUE) [37], Reading Comprehension Dataset From Examinations (RACE) [20] and Stanford Question Answering Dataset (SQuAD) benchmark. Similar to [12,21,32], in this work we fine-tune RoBERTa to efficiently address a task that involves comparing sentences.

In our model, the hyperparameter values are: maximum sequence length of 512; batch size of 4; training rate of 1e-5; and, training performed for 3 epochs. These values were established after successive evaluations, following previous experiments on this model [12,21,32].

3.2   Stance Stage

Once our approach has been able to identify the headlines that are related to their source text, the main goal of this stage is to determine their type considering the remaining stances: *agree*, *disagree* or *discuss*. Therefore, the claim made in the headline can be finally classified into one of three classes left.

The inputs of this stage are:

- The *headlines* classified as *related*.
- The *summary* of the news content.

These classified headlines together with the *unrelated* headlines determined before, will comprise the final output for the whole approach. To achieve this, this stage comprises the following modules:

Stance Feature Extraction. In this module, polarity features of the headline and the summary are computed using NLTK tool [5].

- Polarity positive and negative of the headline (Pol_head_pos, Pol_head_neg).
- Polarity positive and negative of the summary (Pol_sum_pos, Pol_sum_neg).

Stance Classification. Similar to the Relatedness classification module, this stage has been build using RoBERTa as foundation, selected as the model able to improve the classification. In this case, the four features of the stance feature extraction module are added, two dense layers are included to reduce dimensions and, finally, the Softmax classification layer. The hyperparameters of the model used in this classifier are the same as those of the Relatedness classification, except for the classification output which in this case is of three classes: *agree*, *disagree*, *discuss*. In all this classification process, the automatic summaries previously generated with TextRank are used.

4   Evaluation and Discussion

The evaluation of our proposed approach is conducted over the Fake News Challenge dataset (FNC-1) whose instances are labeled as *agree*, *disagree*, *discuss* and *unrelated*. The dataset contains 1,683 news with their headlines and was split into a training set (66.3%) and a testing set (33.7%), where neither the headlines nor the body text overlapped.

To measure our approach's performance, a set of incremental experiments were conducted, where each of the two stages of the proposed architecture were first evaluated independently, and then, the whole approach was validated. By this means, we can first measure the effectiveness of this stage in isolation, also conducting an ablation study to verify whether or not the features used in each of the stages of the classifier make a positive contribution.

In addition to the average accuracy and *Relative Score* metric originally proposed in the FNC-1 challenge,[4] we also take into account the F_1 class-wise,

---

[4] This metric assigns higher weight to examples correctly classified, as long as they belonged to a different class from the *unrelated* one.

and a macro-averaged $F_1{}^5$ ($F_1 m$) metrics [17]. The advantage of these additional metrics is that it is not affected by the size of the majority class.

Table 1 shows the performance obtained in *Relatedness Stage* (first classifier).

Table 1. Relatedness classification results using automatic summaries

System	F_1 Score		$F_1 m$
	Related	Unrelated	
Relatedness Stage FNC-1-Summary	98.22	99.31	98.77

The ablation study for this stage consisted on performing five different experiments removing each time one specific feature with the aim of gain better insights on how each of these features contribute to the proposal. Results are shown in Table 2 and indicate that the most influential feature for the classification is the Cosine similarity since the experiment that does not use this feature obtains the worst results, although the classification results are still very high.

Table 2. Ablation study results for the features used in the *Relatedness Stage*

Removed feature	F_1 Score		$F_1 m$
	Related	Unrelated	
Cosine similarity	97.52	99.04	98.28
BERT cosine similarity	97.66	99.11	98.38
PLM salience score	97.91	99.19	98.55
Overlap coefficient	98.04	99.24	98.64
Soft cosine similarity	98.05	99.26	98.66

Concerning the validation of the *Stance Stage* in isolation, only the examples tagged as *related* from the FNC-1 Gold-Standard are used. Table 3 shows the performance results obtained in the *Stance Stage* (second classifier).

Table 3. *Stance Stage* results

Removed feature	F_1 Score			$F_1 m$
	Agree	Disagree	Discuss	
Stance Stage FNC-1	74.54	64.54	87.69	75.59

As we did with the *Relatedness Stage*, an ablation study (Table 4) was carried out, where the *Stance Stage* classifier was tested removing each of the proposed

---

[5] This is computed as the mean of those per-class F scores.

features (Pol_head_pos, Pol_head_neg, Pol_sum_pos, Pol_sum_neg). The included features clearly show their positive influence in the performance of the classifier. In this case the most influential feature for the classification is the *Pol_head_pos*.

Table 4. Ablation study results for the features used in the *Stance Stage*

Removed feature	F_1 Score			F_1m
	Agree	Disagree	Discuss	
Pol_head_pos	71.64	56.99	87.10	71.91
Pol_head_neg	72.19	58.84	88.12	73.05
Pol_sum_neg	71.68	61.31	88.11	73.70
Pol_sum_pos	73.08	59.94	88.26	73.76

Finally, the results of the whole approach, which integrates the Relatedness and Stance classifiers together with the sole use of automatic summaries for these two classifiers are shown in Table 5. This table contains the performance for the class-wise F_1, macro-average F_1m, accuracy (Acc.) and the relative score (Rel. Score). Moreover, it also provides the results obtained by competitive state-of-the-art systems together with additional configurations that were also tested.

Table 5. Complete approach performance and comparison with state-of-the-art systems

System	F_1 Score				F_1m	Acc.	Rel. Score
	Agree	Disagree	Discuss	Unrelated			
Talos [3]	53.90	3.54	76.00	99.40	58.21	89.08	82.02
Athene [1]	48.70	15.12	78.00	99.60	60.40	89.48	82.00
UCLMR [30]	47.94	11.44	74.70	98.90	58.30	88.46	81.72
Human Upper Bound [1]	58.80	66.70	76.50	99.70	75.40	–	85.90
Dulhanty et al. [12]	73.76	55.26	85.53	99.12	78.42	93.71	90.00
Zhang et al. [40]	67.47	**81.30**	83.90	**99.73**	**83.10**	93.77	89.30
OurApproach-1stage	71.64	53.31	85.25	99.29	77.37	93.58	89.92
OurApproach-2stages	**74.22**	64.29	**86.00**	99.31	80.95	**94.13**	**90.73**

The 3 first rows are the top-3 best systems that participated in the FNC-1 challenge, calculated using the confusion matrices and results published [30] or made available by the authors.[6],[7]

The fourth row corresponds to the *Human Upper Bound* [1], and is the result of conducting the FNC-1 stance detection task manually.

---

[6] https://github.com/hanselowski/athene_system/ (accessed online 15 March, 2021).
[7] https://github.com/Cisco-Talos/fnc-1 (accessed online 15 March, 2021).

Next, the fifth and sixth rows include the results of recent approaches [12, 40] that also addressed the headline stance detection task using the FNC-1 dataset, but did not participate in this challenge. Since there was no public code available, these results were also calculated from the confusion matrices provided in their respective papers.

The seventh row indicates the results for our approach but configured only as a single classifier (*OurApproach-1stage*). Finally, the last row belongs to our approach, using our proposed two-stage classification (*OurApproach-2stages*). Regardless whether the classification is conducted in 1 or 2 stages, both approaches use for the whole process the features extracted and the summaries created from the full body text.

As can be seen in Table 5, *OurApproach-2stages* is competitive enough with respect to the other systems, given that it only uses short summaries for the classification process, and not the full body text as the other systems use, so the information reduction does not imply a high loss in the results obtained, being better than the FNC-1 participants, and the human upper bound. Furthermore, the results also validate the fact that dividing the classification into two stages is beneficial and yields better performance with respect to using our proposed model with a single classifier (rows 7th and 8th), especially for detecting disagreement between the headline and the news article. At this point, it is worth noting that the results previously obtained with the independent evaluation of the *Stance Stage* are slightly better the ones of whole approach (see Table 3). This was already expected since errors derived from the *Relatedness Stage* were avoided in the former, simulating an ideal environment.

Whereas our approach outperforms the other automatic systems in terms of *agree* and *discuss* classes, accuracy, and relative score, it was outperformed in the *disagree* class by [40] and in the *unrelated* class by top-3 best systems that participated in the FNC-1 challenge and [40]. When the results obtained by the participants in the FNC-1 competition are analyzed independently for each of the classes, it can be seen that except for the classification of *unrelated* headlines —whose results are close to 100% in F1 measure, and this happens also for the remaining approaches as well— for the remaining classes, the results are very limited. The systems that participated in the FNC-1 competition have a very reduced performance especially in detecting the *disagree* stance, whereas the detection of *agree* is around 50% in F1 measure and for *discuss* around 75% for the best approach. Outside the FNC-1 competition, the performance increases in all categories, being the *disagree* category one of the most challenging to classify, in which only the approach proposed in [40] obtains surprisingly high results for this category compared to the remaining methods.

5   Conclusions and Future Work

This paper presented an approach for stance detection, i.t., for automatically determining the relation between a news headline and its body. Its novelty relies on two key premises: i) the definition of a two-stages architecture to tackle the stance classification problem; and ii) the use of summarization instead of the whole news body. To show the appropriateness of the approach, different experiments were carried out in the context of an existing task —Fake News Challenge FNC-1—, where the stance of a headline had to be classified into one of the following classes: *unrelated, agree, disagree*, and *discuss*. The experiments involved validating each of the proposed classification stages in isolation together with the whole approach, as well as a comparison with respect the state of the art in this task.

The results obtained by our system were very competitive compared to other systems obtaining 94.13% accuracy, as well as the highest result in FNC-1 relative score compared with the state of the art (90.73%). Given that the use of summaries provided good results in this preliminary research, as a future goal, we would like to study more in-depth the impact of the summarization techniques in the stance detection process, by using other summarization approaches, or analysing how the length of the summaries affect the performance of the approach, among other issues to be researched. Moreover, we also plan to include in our stance detection approach, new learning strategies and discourse aware techniques, with the final aim to help to combat online fake news, a societal problem that requires concerted action.

Acknowledgements. This research work has been partially funded by Generalitat Valenciana through project "SIIA: Tecnologias del lenguaje humano para una sociedad inclusiva, igualitaria, y accesible" (PROMETEU/2018/089), by the Spanish Government through project "Modelang: Modeling the behavior of digital entities by Human Language Technologies" (RTI2018-094653-B-C22), and project "INTEGER - Intelligent Text Generation" (RTI2018-094649-B-I00). Also, this paper is also based upon work from COST Action CA18231 "Multi3Generation: Multi-task, Multilingual, Multimodal Language Generation".

References

1. Hanselowski, A., Avinesh, P.V.S., Schiller, B., Caspelherr, F.: Description of the system developed by team athene in the FNC-1 (2017). https://github.com/hanselowski/athene_system/blob/master/system_description_athene.pdf
2. Babakar, M., et al.: Fake News Challenge - I (2016). http://www.fakenewschallenge.org/. Accessed 29 May 2020
3. Baird, S., Sibley, D., Pan, Y.: Talos targets disinformation with fake news challenge victory (2017). blog.talosintelligence.com/2017/06/talos-fake-news-challenge.html. Accessed 29 May 2020
4. Banko, M., Mittal, V.O., Witbrock, M.J.: Headline generation based on statistical translation. In: Proceedings of the 38th Annual Meeting on Association for Computational Linguistics, pp. 318–325. Association for Computational Linguistics (2000)
5. Bird, S., Klein, E., Loper, E.: Natural Language Processing with Python: Analyzing Text with the Natural Language Toolkit. O'Reilly Media, Inc. (2009)
6. Chen, Y., Conroy, N.J., Rubin, V.L.: News in an online world: the need for an "automatic crap detector". In: Proceedings of the 78th ASIS&T Annual Meeting: Information Science with Impact: Research in and for the Community. American Society for Information Science (2015)

7. Chesney, S., Liakata, M., Poesio, M., Purver, M.: Incongruent headlines: yet another way to mislead your readers. Proc. Nat. Lang. Process. Meets J. **2017**, 56–61 (2017)
8. Dernoncourt, F., Ghassemi, M., Chang, W.: A repository of corpora for summarization. In: Proceedings of the Eleventh International Conference on Language Resources and Evaluation. European Language Resources Association (2018)
9. van Dijk, T.A.: News as Discourse. L. Erlbaum Associates, Communication Series (1988)
10. Dorr, B., Zajic, D., Schwartz, R.: Hedge trimmer: a parse-and-trim approach to headline generation. In: Proceedings of the North American of the Association for Computational Linguistics, Text Summarization Workshop, pp. 1–8 (2003)
11. Duan, Y., Jatowt, A.: Across-time comparative summarization of news articles. In: Proceedings of the 12th ACM International Conference on Web Search and Data Mining, pp. 735–743. Association for Computing Machinery, New York (2019)
12. Dulhanty, C., Deglint, J.L., Daya, I.B., Wong, A.: Taking a stance on fake news: Towards automatic disinformation assessment via deep bidirectional transformer language models for stance detection. arXiv preprint arXiv:1911.11951 (2019)
13. Esmaeilzadeh, S., Peh, G.X., Xu, A.: Neural abstractive text summarization and fake news detection. Computing Research Repository abs/1904.00788 (2019)
14. Ferreira, W., Vlachos, A.: Emergent: a novel data-set for stance classification. In: Proceedings of the Conference of the North American Chapter of the Association for Computational Linguistics, pp. 1163–1168. Association for Computational Linguistics (2016)
15. Gabielkov, M., Ramachandran, A., Chaintreau, A., Legout, A.: Social clicks: what and who gets read on Twitter? ACM SIGMETRICS Performance Eval. Rev. **44**, 179–192 (2016)
16. Gavrilov, D., Kalaidin, P., Malykh, V.: Self-attentive model for headline generation. In: Azzopardi, L., Stein, B., Fuhr, N., Mayr, P., Hauff, C., Hiemstra, D. (eds.) ECIR 2019. LNCS, vol. 11438, pp. 87–93. Springer, Cham (2019). https://doi.org/10.1007/978-3-030-15719-7_11
17. Hanselowski, A., et al.: A retrospective analysis of the fake news challenge stance-detection task. In: Proceedings of the 27th International Conference on Computational Linguistics, pp. 1859–1874. Association for Computational Linguistics (2018)
18. Iwama, K., Kano, Y.: Multiple news headlines generation using page metadata. In: Proceedings of the 12th International Conference on Natural Language Generation, pp. 101–105. Association for Computational Linguistics (2019)
19. Kuiken, J., Schuth, A., Spitters, M., Marx, M.: Effective headlines of newspaper articles in a digital environment. Digit. J. **5**(10), 1300–1314 (2017)
20. Lai, G., Xie, Q., Liu, H., Yang, Y., Hovy, E.: RACE: large-scale reading comprehension dataset from examinations. In: Proceedings of the 2017 Conference on Empirical Methods in Natural Language Processing, pp. 785–794. Association for Computational Linguistics (2017)
21. Liu, Y., et al.: RoBERTa: A robustly optimized BERT pretraining approach. arXiv preprint arXiv:1907.11692 (2019)
22. Mackie, S., McCreadie, R., Macdonald, C., Ounis, I.: Experiments in newswire summarisation. In: Ferro, N., et al. (eds.) ECIR 2016. LNCS, vol. 9626, pp. 421–435. Springer, Cham (2016). https://doi.org/10.1007/978-3-319-30671-1_31
23. Metcalf, L., Casey, W.: Metrics, similarity, and sets. In: Cybersecurity and Applied Mathematics, pp. 3–22. Elsevier (2016)

24. Mihalcea, R., Tarau, P.: TextRank: bringing order into text. In: Proceedings of the 2004 Conference on Empirical Methods in Natural Language Processing, pp. 404–411. Association for Computational Linguistics (2004)
25. Mikolov, T., Chen, K., Corrado, G., Dean, J.: Efficient estimation of word representations in vector space. arXiv preprint arXiv:1301.3781 (2013)
26. Nenkova, A.: Automatic text summarization of newswire: lessons learned from the document understanding conference. In: Proceedings of the 20th National Conference on Artificial Intelligence, vol. 3, pp. 1436–1441. AAAI Press (2005)
27. Passalis, N., Tefas, A.: Learning bag-of-embedded-words representations for textual information retrieval. Pattern Recogn. **81**, 254–267 (2018)
28. Pouliquen, B., Steinberger, R., Best, C.: Automatic detection of quotations in multilingual news. Proc. Recent Adv. Nat. Lang. Process. **2007**, 487–492 (2007)
29. Reimers, N., Gurevych, I.: Sentence-bert: Sentence embeddings using siamese bert-networks. arXiv preprint arXiv:1908.10084 (2019)
30. Riedel, B., Augenstein, I., Spithourakis, G.P., Riedel, S.: A simple but tough-to-beat baseline for the Fake News Challenge stance detection task. Computing Research Repository, CoRR abs/1707.03264 (2017)
31. Silverman, C.: Lies, damn lies and viral content (2019). http://towcenter.org/research/lies-damn-lies-and-viral-content/. Accessed 29 May 2020
32. Slovikovskaya, V.: Transfer learning from transformers to fake news challenge stance detection (FNC-1) task. arXiv preprint arXiv:1910.14353 (2019)
33. Tan, J., Wan, X., Xiao, J.: From neural sentence summarization to headline generation: a coarse-to-fine approach. In: Proceedings of the 26th International Joint Conference on Artificial Intelligence, pp. 4109–4115. AAAI Press (2017)
34. Tsipursky, G., Votta, F., Roose, K.M.: Fighting fake news and post-truth politics with behavioral science: the pro-truth pledge. Behav.Soc. Issues **27**(1), 47–70 (2018). https://doi.org/10.5210/bsi.v27i0.9127
35. Vicente, M.E., Pastor, E.L.: Relevant content selection through positional language models: an exploratory analysis. Proces. del Leng. Nat. **65**, 75–82 (2020)
36. Vlachos, A., Riedel, S.: Identification and verification of simple claims about statistical properties. Proc. Conf. Empirical Methods Nat. Lang. Process. **2015**, 2596–2601 (2015)
37. Wang, A., Singh, A., Michael, J., Hill, F., Levy, O., Bowman, S.: GLUE: a multi-task benchmark and analysis platform for natural language understanding. In: Proceedings of the 2018 EMNLP Workshop BlackboxNLP: Analyzing and Interpreting Neural Networks for NLP, pp. 353–355. Association for Computational Linguistics (2018)
38. Wei, W., Wan, X.: Learning to identify ambiguous and misleading news headlines. In: Proceedings of the 26th International Joint Conference on Artificial Intelligence, pp. 4172–4178. AAAI Press (2017)
39. Zajic, D., Dorr, B., Schwartz, R.: Automatic headline generation for newspaper stories. In: Proceedings of the Workshop on Automatic Summarization 2002, pp. 78–85 (2002)
40. Zhang, Q., Liang, S., Lipani, A., Ren, Z., Yilmaz, E.: From stances' imbalance to their hierarchical representation and detection. In: The World Wide Web Conference, pp. 2323–2332. ACM (2019)
41. Zhu, C., Yang, Z., Gmyr, R., Zeng, M., Huang, X.: Make lead bias in your favor: A simple and effective method for news summarization. arXiv preprint arXiv:1912.11602 (2019)

Predicting Vaccine Hesitancy and Vaccine Sentiment Using Topic Modeling and Evolutionary Optimization

Gokul S. Krishnan[1,2]([✉]), S. Sowmya Kamath[1], and Vijayan Sugumaran[3]

[1] Healthcare Analytics and Language Engineering (HALE) Lab,
Department of Information Technology, National Institute of Technology Karnataka,
Surathkal, India
sowmyakamath@nitk.edu.in

[2] Robert Bosch Centre for Data Science and Artificial Intelligence,
Indian Institute of Technology Madras, Chennai, India
010917@imail.iitm.ac.in

[3] Department of Decision and Information Sciences, School of Business
Administration, Oakland University, Rochester, USA
sugumara@oakland.edu

Abstract. The ongoing COVID-19 pandemic has posed serious threats to the world population, affecting over 219 countries with a staggering impact of over 162 million cases and 3.36 million casualties. With the availability of multiple vaccines across the globe, framing vaccination policies for effectively inoculating a country's population against such diseases is currently a crucial task for public health agencies. Social network users post their views and opinions on vaccines publicly and these posts can be put to good use in identifying vaccine hesitancy. In this paper, a vaccine hesitancy identification approach is proposed, built on novel text feature modeling based on evolutionary computation and topic modeling. The proposed approach was experimentally validated on two standard tweet datasets – the flu vaccine dataset and UK COVID-19 vaccine tweets. On the first dataset, the proposed approach outperformed the state-of-the-art in terms of standard metrics. The proposed model was also evaluated on the UKCOVID dataset and the results are presented in this paper, as our work is the first to benchmark a vaccine hesitancy model on this dataset.

Keywords: Evolutionary computation · Machine learning · Natural language processing · Population health analytics · Topic modeling

1 Introduction

In the last few decades, the world has faced several epidemics and contagious viral diseases such as SARS, MERS, H1N1, Zika, Ebola etc., currently superceded by

---

G. S. Krishnan—Work done as part of doctoral research work at HALE Lab, NITK Surathkal.

the "once-in-a-century pandemic", COVID-19 [17]. Organizations such as WHO and health governing bodies of most countries have a huge task of keeping their population healthy. During critical situations like pandemics like the ongoing COVID-19 crisis, the world community has allocated huge amount of financial, research and human resources towards developing effective vaccines for managing disease outbreaks through structured vaccination of vulnerable population groups. Although the success of vaccines in disease control have been proven time and again, making them the obvious and successful measure for managing contagious disease outbreaks, it is quite unfortunate that a growing number of people deem it unnecessary, "against the natural order" and unsafe [4].

Public opinion on vaccinations can be diverse - e.g., majority of the population may be voluntarily ready to submit to the vaccination shots, whereas a significant number may be skeptical about it, despite strong recommendations from the medical community. Vaccine hesitancy is one of the most critical factors that affect effective vaccination policies, owing to the lack of confidence, disinclination or negative opinion towards a vaccine [9]. Vaccine hestinacy has resulted in reduced vaccination coverage and increased risk of epidemics and disease outbreaks that are often easily preventable via mass vaccination [4]. With the widespread adoption of Open Social Network (OSN) platforms such as Twitter and Facebook, such negative opinions can have a negative impact on the efforts of governments and public health organizations. Therefore, it is very important for public health governing national bodies to understand the prevalence of vaccine hesitancy in populations and public sentiment towards vaccination programmes. In normal cases, the public opinions are recorded through surveys and interactive programmes, which are not only difficult to organize and time-consuming, but also tend to under-represent all kinds of citizens, and hence may not be generalizable [6,13].

Automated computational population health surveillance systems that can be modeled to identify vaccine hesitancy is a potentially advantageous solution to these challenges as mining OSN data can provide essential insights to health governing bodies for making informed and possibly better decisions. In this paper, we present a vaccine hesitancy prediction model that can effectively detect vaccine hesitancy in public based on OSN media posts. The proposed approach leverages the concepts of evolutionary computation (Particle Swarm Optimization (PSO)), topic modeling (Latent Dirichlet Allocation (LDA)) and neural networks like Convolutional Neural Networks (CNN) to achieve this objective. The key contributions of this work are as follows:

1. Design of a PSO based topic modeling approach that can dynamically determine the optimal number of latent topic clusters, for OSN data.
2. Design of a PSO-CNN wrapper for dynamically determining the optimal number of topics for LDA topic modeling and for effectively identifying any vaccine hesitancy.
3. Benchmarking the proposed vaccine hesitancy identification approach on open standard datasets.

The rest of the paper is organized as follows. Section 2 provides an overview on existing related works in the domains of interest. Section 3 discusses in detail the system architecture of the proposed model. In Sect. 4, we present the experimental results along with a discussion on the performance of the approach, followed by conclusion and potential scope for future work.

2   Related Work

The research community has shown significant interest in modeling OSN data for a wide variety of tasks. We discuss some relevant works in each of these categories, in the context of the chosen tasks. Computational techniques like Natural Language Processing (NLP) and ML have great potential in performing predictive analytics based tasks on OSN data. Several works in the areas of influenza or flu monitoring/detection [1,2,16], adverse drug event detection [3,14], vaccine sentiment [6], vaccine behaviour/vaccine shot detection (whether vaccine shot was received or not) [6,9], vaccine hesitancy/vaccine intent (whether vaccine is intended to be taken or not) [6], etc., have been proposed over the past decade. Huang et al. [6] presented a study that made use of several natural language classifiers to analyze Twitter users' behavior towards influenza vaccination. They performed prediction tasks such as vaccine relevance, vaccine shot detection, vaccine intent detection and vaccine sentiment.

Moslehi and Haeri [12] proposed a hybrid method based on PSO, Genetic Algorithm (GA) and gain ratio index to select optimal feature subsets. Gomez et al. [5] proposed a GA based evolutionary approach for learning some meta-rules which can help further optimize text classification. While these approaches showed capabilities of evolutionary computation being applied towards text classification, these approaches fail to extract effective feature representations for a specific prediction task.

Li et al. [11] proposed an auxiliary word embedding based topic modeling approach for text classification. Steinskog et al. [15] proposed a topic modeling and pooling techniques based approach for aggregation of tweet texts. While these approaches showed the effectiveness of topic modeling in NLP tasks, other approaches (by Zhao et al. [18] and Ignatenko et al. [8]) were put forward by to determine the number of topic clusters, a known research problem in topic modeling techniques. Though these approaches could determine a certain number of topics for topic modeling, the choice is not based on the prediction task to be performed. In this paper, we propose the use of PSO, an evolutionary optimization algorithm, and ensemble it with a wrapper technique based on CNN to determine an optimal number of topic clusters for LDA topic modeling technique and use it to effectively model features for training a vaccine hesitancy identification model.

3   Proposed Approach

The overall workflow of the proposed approach is depicted in Fig. 1. We used standard datasets consisting of OSN data for the experiments. A preprocessing pipeline involving several basic NLP techniques were used to clean and pre-process the corpus. All special characters except white spaces were removed. Tokenization was performed on the corpus to break down the text into units called tokens; stemming and lemmatization were applied to bring the words to root form and finally, stopping was also performed to filter out frequent unimportant words. The next set of processes that involve the feature modeling and prediction modeling are explained in subsequent sub sections.

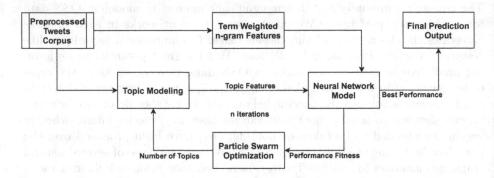

Fig. 1. Workflow of Proposed Vaccine Hesitancy Identification Approach

Term Weighted n-gram Feature Generation. The preprocessed tokens obtained from the tweets corpus were modeled into a vector representation using a Term Frequency - Inverse Document Frequency (TF-IDF) vectorizer to create term weighted n-gram features. TF-IDF, a statistical measure that signifies the weightage or importance of words within a document, is often considered as textual features in text mining based prediction modeling. In the proposed approach, TF-IDF weights for n-grams (i.e., n = 1, 2, 3) were extracted and the top 2000 weights were considered to be Feature Set 1 (hereafter referred to as FS1).

Latent Dirichlet Allocation. Topic modeling is an approach that clusters documents into a set of topics that most represent the documents in an unsupervised manner. Latent Dirichlet Allocation (LDA) is a popular probabilistic topic modeling approach that assigns a given set of documents to topic clusters. LDA theorizes that for a set of words appearing in a given document, it belongs to a certain number of topics with certain probabilities. In this work, LDA is applied to the preprocessed tweets corpus, and vectorized probabilities of each topic for a document are considered as features for the proposed prediction model, i.e., Feature Set 2 is hereafter referred to as FS2.

Similar to unsupervised clustering techniques, determining the number of topics while performing the LDA topic modeling approach is a challenging and critical task. Determining the optimal number of LDA topic clusters also pertains to the process of deriving the optimal number of features in the topic feature vector generated by the topic model. The solution subspace to search for the optimal number of topics is quite large and therefore, evolutionary optimization approaches are an apt choice. We adopted the usage of PSO algorithm for this and the adaptation is explained in detail next.

PSO Based LDA Topic Modeling. We utilize PSO, an evolutionary optimization algorithm, for dynamically determining the optimal number of topics for various prediction tasks. Towards this objective, a wrapper, named PSO-CNN, is proposed. The feature sets – FS1 and FS2 are combined and fed into a PSO-CNN wrapper, in which the neural network model was adopted from the popular TextCNN model [10] for effective text classification. The performance of the TextCNN model in terms of F-score was considered as the fitness performance for the proposed PSO based topic modeling approach.

Initially, a swarm of particles, along with particle positions were initialized as a set of number of topic clusters for the LDA model. For each position, say i, the best classification performance of the TextCNN model in terms of F-score is considered as the *local_best$_i$* score, and the same of the entire swarm is considered as the *global_best* score. The new next positions, x_{i+1}, and next velocities, v_{i+1}, of the initalized particles are calculated and updated based on PSO equations (Eq. 1 and 2), where, $c1$ and $c2$ are constants, whose values were empirically found to be 0.5 and 0.2 respectively and $r1$ and $r2$ are random real numbers.

$$v_{i+1} = w * v_i + c_1 * r_1 * (local\_best_i - x_i) + c_2 * r_2 * (global\_best - x_i) \quad (1)$$

$$x_{i+1} = x_i + v_{i+1} \quad (2)$$

In our work, a set of eight particles were used and number of iterations was set to 50. The position at which the best performance was observed, i.e., the position of *global_best*, was considered to be the optimal number of topic clusters for LDA topic modeling for the task of vaccine hesitancy identification.

The TextCNN model adopted in the proposed PSO-CNN wrapper model consists of three 1D convolution layers with 512 filters which were of sizes 5,6 and 7 respectively. The number of nodes in the input layer indicates the optimal number of topics as determined by the PSO based topic modeling approach. Further, 1D Maxpool layers and a Rectified Linear Unit (ReLU) activation function were used along with each convolution layer, followed by concatenating and flattening layers. Additionally, a 50% dropout was also introduced to reduce chances of overfitting. Finally, the output layer consisted of a sigmoid activation function. The optimizer used for training was rmsprop and the loss function used was binary cross-entropy. The performance for vaccine hesitancy prediction task was extensively tested, the details of which are presented in Sect. 4.

4   Experimental Results and Discussion

The performance of the proposed vaccine hesitancy identification approach was benchmarked on two standard tweet datasets, created specifically for vaccine hesitancy tasks. The performance was measured using standard classification metrics – precision, recall and F-score. The proposed model was also benchmarked against state-of-the-art approach for one dataset.

Datasets. We used two datasets to benchmark the performance of the proposed vaccine hesitancy identification model. First, the flu vaccine dataset (FVD) provided by Huang et al. [6] was used for this prediction task which consists of around 10,000 tweets related to influenza vaccine. It is to be noted that, the irrelevant tweets labelled in the dataset and any rows with missing labels were dropped, after which a total of 9,513 instances were available for the analysis. Second, the UK COVID-19 Vaccine tweets dataset (hereafter referred as UKCOVID) collected and released by Hussain et al. [7] was also used for the experiments. The dataset originally consists of 40,268 tweets from the United Kingdom with respect to the context of vaccination for COVID-19 pandemic. The original dataset released by the authors consisted of only tweet IDs as per policy of Twitter. However, only 24,309 tweets could be retrieved due to issues such as deleted tweets or private accounts. The vaccine hesitancy is indicated as sentiment labels for tweets in three categories – positive, negative and neutral. The characteristics of the two datasets are as shown in Tables 1a and 1b.

Table 1. Dataset statistics

(a) Dataset Statistics of FVD

Feature	Frequency
Unique tweets	9,513
Users	9,334
Words	1,54,204
Intend/Receive *(Positive)*	3,148
Hesitancy *(Negative)*	6,365

(b) Dataset Statistics of UKCOVID

Feature	Frequency
Unique tweets	24,309
Words	63,863
Positive	10,230
Negative	8,245
Neutral	5,834

Results. When applied on FVD dataset, the PSO-CNN wrapper based topic modeling technique determined the optimal number of LDA topic clusters to be 634. Along with 2,000 top n-gram features, total number of textual features came to 2,634. The performance of the proposed approach was compared to that of the state-of-the-art approach by Huang et al. [6]. Similar to their approach, the performance of the proposed approach was also measured after the 5-fold cross validation. The comparison of performance is as shown in Table 2, from which, it can be observed that the proposed approach outperformed Huang et al.'s approach in terms of Recall and F-score by 5% and 2% respectively. Higher values of recall and F-score indicate that the proposed approach was able to reduce the number of False Negatives (FNs).

The proposed approach was also applied on the UKCOVID dataset, and the optimal number of LDA topic clusters were determined to be 600. Using 2,000 top n-gram features, total number of textual features used were about 2,600. The performances in terms of precision, recall and F-score has been benchmarked for this dataset and the results are shown in Table 2. The dataset is quite recent and therefore, no other works have benchmarked any performance on this dataset yet due to which we did not perform any comparison. The classification performance in terms of Recall and F-score shows that there is scope for improvement. This is due to the misclassification of true neutral sentiment as either positive or negative. This is one of the limitations of the current model, towards which we plan to design techniques as part of future work.

Table 2. Flu vaccine hesitancy: performance of proposed approach

Approach	Dataset	Precision	Recall	F-Score
Huang et al. [6]	FVD	0.84	0.80	0.82
LDA+PSO+TextCNN *(Proposed)*	FVD	0.84	0.84	0.84
LDA+PSO+TextCNN *(Proposed)*	UKCOVID	0.75	0.51	0.60

Discussion. From Table 2, it can be observed that the proposed vaccine hesitancy identification approach outperforms the existing approach by Huang et al. Huang et al. [6] by 2% in terms of F-score. As the proposed approach is an entirely text-dependent model, it is able to 'understand' the natural language text and figure out the vaccine hesitancy sentiment of the user, which demonstrates its suitability for quantifying vaccine hesitancy sentiment. The performance benchmarking on the UKCOVID dataset not only ensures future research promotion, but also highlights an approach that can be put to use in the current real world scenario of the COVID-19 pandemic.

5   Conclusion and Future Work

In this paper, a novel approach leveraging topic modeling and evolutionary optimization for predicting vaccine hesitancy and identifying negative sentiments towards vaccination using OSN data has been proposed. The proposed approach is built on effective usage of the PSO algorithm and LDA topic modeling approach, along with CNN based prediction model to identify vaccine hesitancy in tweets by users. Experimental validation revealed that the proposed approach outperformed state-of-the-art approaches on the FVD dataset. In addition, the performance of the same on the newly released COVID-19 based UKCOVID dataset was also benchmarked. The purely natural language text dependent model proved to be effective in identifying vaccine hesitancy and can be considered as a tool to identify population sentiment towards COVID-19 vaccines in the current pandemic scenario. As part of future work, we plan to explore further improvements for effective identification of neutral sentiment towards vaccine policies. We further intend to benchmark the proposed approach on more diverse datasets. Moreover, we also plan to explore the use of other topic modeling approaches and also word embedding approaches as part of textual feature modeling.

References

1. Alshammari, S.M., Nielsen, R.D.: Less is more: with a 280-character limit, Twitter provides a valuable source for detecting self-reported flu cases. In: Proceedings of the 2018 International Conference on Computing and Big Data, pp. 1–6. ACM (2018)
2. Byrd, K., Mansurov, A., Baysal, O.: Mining Twitter data for influenza detection and surveillance. In: Proceedings of the International Workshop on Software Engineering in Healthcare Systems, pp. 43–49. ACM (2016)
3. Cocos, A., Fiks, A.G., Masino, A.J.: Deep learning for pharmacovigilance: recurrent neural network architectures for labeling adverse drug reactions in Twitter posts. JAMIA **24**(4), 813–821 (2017)
4. Dubé, E., Laberge, C., Guay, M., Bramadat, P., Roy, R., Bettinger, J.A.: Vaccine hesitancy: an overview. Hum. Vaccines Immunotherapeutics **9**(8), 1763–1773 (2013)
5. Gomez, J.C., Hoskens, S., Moens, M.F.: Evolutionary learning of meta-rules for text classification. In: Proceedings of the Genetic and Evolutionary Computation Conference Companion, pp. 131–132 (2017)
6. Huang, X., Smith, M.C., Paul, M.J., et al.: Examining patterns of influenza vaccination in social media. In: Workshops at 31st AAAI Conference on Artificial Intelligence (2017)
7. Hussain, A., Tahir, A., Hussain, Z., Sheikh, Z., Gogate, M., et al.: Artificial intelligence-enabled analysis of UK and US public attitudes on Facebook and twitter towards Covid-19 vaccinations (2020)
8. Ignatenko, V., Koltcov, S., Staab, S., Boukhers, Z.: Fractal approach for determining the optimal number of topics in the field of topic modeling. J. Phys.: Conf. Ser. **1163** (2019)
9. Joshi, A., Dai, X., Karimi, S., Sparks, R., Paris, C., MacIntyre, C.R.: Shot or not: comparison of NLP approaches for vaccination behaviour detection. In: Proceedings of the 2018 EMNLP Workshop, pp. 43–47 (2018)
10. Kim, Y.: Convolutional neural networks for sentence classification. arXiv preprint arXiv:1408.5882 (2014)
11. Li, C., Wang, H., Zhang, Z., Sun, A., Ma, Z.: Topic modeling for short texts with auxiliary word embeddings. In: ACM SIGIR Conference on Research and Development in Information Retrieval, pp. 165–174 (2016)
12. Moslehi, F., Haeri, A.: An evolutionary computation-based approach for feature selection. J. Ambient Intell. Hum. Comput. 1–13 (2019)
13. Parker, A.M., Vardavas, R., Marcum, C.S., Gidengil, C.A.: Conscious consideration of herd immunity in influenza vaccination decisions. Am. J. Prev. Med. **45**(1), 118–121 (2013)
14. Sarker, A., et al.: Utilizing social media data for pharmacovigilance: a review. J. Biomed. Inform. **54**, 202–212 (2015)

15. Steinskog, A., Therkelsen, J., Gambäck, B.: Twitter topic modeling by tweet aggregation. In: Proceedings of the 21st Nordic Conference on Computational Linguistics, pp. 77–86 (2017)
16. Wakamiya, S., Kawai, Y., Aramaki, E.: Twitter-based influenza detection afterflu peak via tweets with indirect information: text mining study. JMIR Public Health Surveillance **4**(3), e65 (2018)
17. Yang, H., Ma, J.: How an epidemic outbreak impacts happiness: factors that worsen (vs. protect) emotional well-being during the coronavirus pandemic. Psychiatry Res. **289**, 113045 (2020)
18. Zhao, W., et al.: A heuristic approach to determine an appropriate number of topics in topic modeling. BMC Bioinform. **16**, S8 (2015). https://doi.org/10.1186/1471-2105-16-S13-S8

Sentiment Progression Based Searching and Indexing of Literary Textual Artefacts

Hrishikesh Kulkarni[1](✉) and Bradly Alicea[2]

[1] Georgetown University, Washington, DC, USA
[2] Orthogonal Research and Educational Laboratory, Champaign- Urbana, USA

Abstract. Literary artefacts are generally indexed and searched based on titles, meta data and keywords over the years. This searching and indexing works well when user/reader already knows about that particular creative textual artefact or document. This indexing and search hardly takes into account interest and emotional makeup of readers and its mapping to books. In case of literary artefacts, progression of emotions across the key events could prove to be the key for indexing and searching. In this paper, we establish clusters among literary artefacts based on computational relationships among sentiment progressions using intelligent text analysis. We have created a database of 1076 English titles + 20 Marathi titles and also used database http://www.cs.cmu.edu/~dbamman/booksummaries.html with 16559 titles and their summaries. We have proposed Sentiment Progression based Search and Indexing (SPbSI) for locating and recommending books. This can be used to create personalized clusters of book titles of interest to readers. The analysis clearly suggests better searching and indexing when we are targeting book lovers looking for a particular type of books or creative artefacts.

Keywords: Literature · Creative artefacts · Searching · NLP · Text analysis · Machine learning · Information retrieval · Sentiment mining

1 Introduction and Related Work

1.1 Searching and Indexing Literature

Searching, recommending and indexing literary artefacts is generally driven by names of authors, topics, and keywords. This is very effective but very primitive method and cannot cope up with uncertainty and variations associated with user interests. It comes with its own advantages and challenges. But when we take into account millions of unknown titles; does this indexing in true sense gives us book titles best suited to our interest and emotional makeup? Actually, these primitive indexing mechanisms create bias while making some titles best seller and do injustice to many classic literary creations by unknown champions. This even refrains newcomers from creating novel literary experiments. This makes many worthy literary creations even sometimes vanish behind the curtains of brands created by this name and author-based system. While digital libraries are taking care of availability of books, we here propose an algorithm for fair

© Springer Nature Switzerland AG 2021
E. Métais et al. (Eds.): NLDB 2021, LNCS 12801, pp. 264–271, 2021.
https://doi.org/10.1007/978-3-030-80599-9_24

indexing and recommending literary artefacts. The joy and satisfaction of reading has more to do with plot, theme, progression of emotional upheavals than title or name of author. Indexing based on sentiment progression and thematic changes could help in dealing with bias of indexing and making justice to many titles those could not reach to book lovers. Availability, reachability and personalized indexing can help in solving this problem. This will reduce the bias, nurture creativity and bust the monopolies in the literary world.

1.2 Related Work

Indexing, listing and searching of books help readers in selecting the title of their interest. When we were focused on physical books – book catalogues are maintained alphabetically. These catalogues in digital platform were very useful for searching and locating a book if the title is known to a reader. These catalogues were extended with same paradigm of indexing for digital books. You can search even contents in these digital artefacts. But this basic paradigm of catalogues, indexing and searching has many limitations. You need to know a book if you are searching one. Keyword based search works effectively for scientific books but for fictions it fails miserably. The concepts of decoding character relationships for indexing and recommending is the core idea proposed in this paper. There are many attempts to decode relationships among characters. Decoding relationships among characters in narratives [1] can be considered as one of the core aspects while analyzing it. Relationship among characters at various places in a narrative is indicative of sentiment progression. These relationships can be modelled in different ways using semi-supervised machine learning [1]. Narrative structure is core to this analysis [2]. The role of sentiment in this pattern mapping is crucial to such relationship models [3]. Topic transition is generally determined using consistency analysis and coherence [4].

Linguistic perspective and contextual event analysis can play a vital role in narrative assessment. Surprises bring unexpected changes in relationships and event progressions, differentiating adorable events [5]. The overall narrative can be viewed as an emotional journey with variations in interestedness. This journey progresses as various relationships in the given narrative unfold. In the concept journey, different concepts are battling for existence and key concept prove to be ultimate winner. Emotional aspects blended in very personalized culture are at the helm of this journey. These emotional aspects are associated with part of stories or creative textual artefacts depicted through different impacting sentences [6]. In any of such scenarios decoding personality and culture with personality vector analysis prove to be effective for mapping [7, 8]. Researchers also used text-based analysis for clustering books [9]. The progression of relationships among characters in a narrative can be used for searching and indexing of books. This can take book catalogues beyond traditional limitations and hence searching can be possible based on progression of emotions, and interestedness of readers. This paper proposes 'Sentiment Progression based Search and Indexing' (SPbSI) to overcome limitations of traditional indexing approaches.

2   Sentiment Progression Based Search and Indexing (SPbSI)

The proposed method is divided into four important phases:

1. Keyword-based core character identification and selection of pivot points
2. Sentiment progression analysis across pivot points
3. Derive similarity using 'Sentiment Progression Similarity Indicator' (SPSI)
4. Indexing and preparing catalogue for sentiment progression search.

3   Data Analysis

A database of (1076 English + 20 Marathi) book titles from different genre is prepared and used for testing and learning. Another database used is http://www.cs.cmu.edu/~dba mman/booksummaries.html. One of the sample books we used as an example - titled *"Rage of Angels"* is a part of both of these datasets. The analysis on this data is preformed using SPbSI to determine relevance for indexing.

4   Mathematical Model

4.1   Core Character Identification and Pivot Point Selection

Core characters are crucial to narrative and the story cannot progress without them. They are identified based on their frequency and relationships with other characters. To explain the concept, we have chosen two interesting fictions those were read by 50 out of 150 + book lovers from the BDB book club[1]: First one is '*Rage of Angels*' (https:// en.wikipedia.org/wiki/Rage_of_Angels) published by Sidney Sheldon in 1980 titled & the second one is Marathi Classic *KraunchVadh* by V.S. Khandekar (https://en.wikipe dia.org/wiki/Vishnu_Sakharam_Khandekar).

The algorithm to identify core characters is developed around core words and pivot points. Here core word is defined as a word that belongs to keyword set and has highest frequency of occurrence across the text space of interest. This word acts as a reference while creating cluster of words. Similarly, Core Character (CC) is one of the prime characters in narrative and is defined based on its presence and association with other prime characters. Equation 1 gives mathematical definition of CC.

$$\forall c \in c | C \in [CC] \ and \ c \to [CC] \ where \ [CC] \neq \Phi \tag{1}$$

Going through narrative in an iterative fashion, the core characters are identified. The characters Jenifer, Michael and Adam are identified as core characters in *Rage of Angels*, while Sulu, Dilip, and Bhagvantrao in *Kraunch-Vadh*. Pivot point is a location in a narrative marked by intense interaction where we perform sentiment analysis.

---

[1] BDB Book Club is a major book club run by BDB India Pvt Ltd in Pune https://bdbipl.com/index.php/bdb-book-club/.

4.2   Sentiment Progression Analysis across Pivot Points (PP)

The sentiment and emotional index at a pivot point using expressive word distribution is used to derive sentiment index. The progression of sentiment across these pivot points represents the behavior and nature of narrative. PP detection algorithm has identified 10 pivot points across the novel for characters Sulu and Dilip. Similarly, there are 8 pivot points for Sulu and Bhagwantrao. The detail algorithm SPbSI for indexing based on pivot point determination and sentiment association is given in algorithm 1.

Algorithm 1. Indexing for effective sentiment progression search

---

Task 1: Initialization (Determine Pivot Points)

1. $CC_{NAR} = \forall c \in c | C \in [CC]$ and $c \rightarrow [CC]$ where $[CC] \neq \Phi$

Task 2: Identify Pivot Point based on frequency, presence and Sentiment

2. $for\ \forall\ SJ_i,\ CC_i$ where $CC\ ] \neq \Phi\ \&\&\ j \leq PP\_max\ \ do$

3. $SI_i[SJ_iCC_i] = Sentiment(CC_i) + \alpha$

4. $With\ gradient\ decent\ determine\ local\ sentiment\ maxima$

5. $PP_j = Local\ Maxima$

6. $j++$

7. end

Task 3: Determine Sentiment Progression (based on Sentiment Value at Pivot Point (PP))

8. $for\ \forall\ PP\ do$

7. $if([SV@PP] \neq \Phi)$

8. $Insert\ SV\ in\ to\ series$

Task 4: Cluster formation of Sentiment Progression

9. $Sentiment\ Distanc\ SD = \frac{\sum_{i=1}^{n} CF(i)^2 \times N(i)}{\sum_{i=1}^{n} N(i)}$

10. $Sentiment\ Progression\ Similarity\ Indicator\ SPSI = \frac{1}{(1+ln\ (1+SD))}$

11. $for\ \forall\ Series\ IF\ [(Series_{Length} <> M)]$

12. $Make\ length\ of\ series = M$

13. $for\ \forall\ Series\ where\ [Series] \neq \Phi\ \&\&\ Series\ length = M$

14. $Create\ SPSI\ Matrix$

15. $for\ \forall\ Series\ where\ [Series] \neq \Phi\ \&\&\ \exists(SPSI) > DT$

16. $IF\ [(SPSI(i,j) > Dynamic\ Threshold\ DT)]$

17. $Combine\ Seires\ (i,J)$

19. $endif$

20. end

End of Algorithm

The extraction of sentiment (emotional positivity and negativity in this case) with reference to context of story includes pivot point identification, extracting sentiment at a particular event. These sentiments are progressed from one pivot point to next one. Thus, Model' (SPbSI) identifies sentiment progression from one pivot point to another.

4.3   Derive Similarity Using Sentiment Progression Similarity Indicator

Statistically Sentiment Progression Similarity Indicator (SPSI) gives behavioral similarity between two sentiment progression patterns. Every pivot point has a sentiment value. Hence every book has a sentiment progression series and can be represented as a data series. Let's take two creative textual artefacts at a time and get corresponding two sentiment progression series. It is highly likely that these two series will have different number of pivot points. Hence, we add supporting points so that both series have equal number of elements. Hence series will look like:

$$RS \ni S(i) = S_1(i) + S_2(i) \tag{2}$$

Here, RS is a series derived by summing corresponding pivot point sentiment values. This is used to calculate the probable value for corresponding series. The probable sentiment value (PS) in accordance with sentiment progression is determined using Eq. 3.

$$PS = \frac{\sum_{i=1}^{n} S_1(i)}{\sum_{i=1}^{n} S(i)} \tag{3}$$

PS is used to derive expected sentiment value with assumption that sentiment progression in second series is same. It us used to calculate correction factor CF.

$$CF(i) = \frac{PS \times RS(i) - S_1(i)}{\sqrt{RS(i) \times PS \times (1 - PS)}} \tag{4}$$

The sentiment distance SD between two text artefacts is given by Eq. 5.

$$SD = \frac{\sum_{i=1}^{n} CF(i)^2 \times N(i)}{\sum_{i=1}^{n} N(i)} \tag{5}$$

Here N is normalization factor $N(i) = \sqrt{R(i)}$. The SPSI is calculated using Eq. 6

$$SPSI = \frac{1}{(1 + ln(1 + SD))} \tag{6}$$

SPSI will drop slowly with increase in sentiment distance. It is close to 1 for patterns those look alike & approaches to zero for completely different patterns.

4.4 Indexing and Preparing Catalogue for Effective Sentiment Progression Search

Iteratively similarity between sentiment progression of every pair of creative textual artifacts is calculated. This leads to SPSI matrix. The diagonal of this matrix is always 1. Two series with maximum similarity are combined to reduce the $(n \times n)$ matrix to $(n-1 \times n-1)$ and so on. This process continues till the similarity between all representative patterns is less than dynamic threshold. This pcess results in getting clusters with representative sentiment progression patterns. A Progression Similarity Matrix for nine books is depicted in Fig. 1.

Here out of these 9 series during first iteration series (7, 9) are combined. Further the resultant series is combined with series 3, then with series 6 and later with series 5. Thus a representative series is formed for cluster made up of series (3, 5, 6, 7, 9). Similarly, series (1, 4) are combined and that series is combined with series 8. Thus, a cluster is formed of series (1, 4, 8). Thus, at the end of iteration 1 the matrix will be of size 3 × 3, with members representing cluster (3, 5, 6, 7, 9), cluster (1, 4, 8) and (2).

Thus, each representative cluster pattern is converted in to an index point.

Series	1	2	3	4	5	6	7	8	9
1	1	0.32	0.11	0.76	0.36	0.16	0.15	0.62	0.11
2	0.32	1	0.18	0.22	0.31	0.28	0.14	1	0.23
3	0.11	0.18	1	0.16	0.58	0.54	0.73	0.25	0.41
4	0.76	0.22	0.16	1	0.37	0.26	0.39	0.57	0.25
5	0.36	0.31	0.58	0.37	1	0.49	0.52	0.16	0.53
6	0.16	0.28	0.54	0.26	0.49	1	0.66	0.11	0.50
7	0.15	0.32	0.73	0.39	0.52	0.66	1	0.29	0.86
8	0.62	0.14	0.25	0.57	0.16	0.11	0.29	1	0.15
9	0.11	0.23	0.41	0.25	0.53	0.50	0.86	0.15	1

Fig. 1. Progression similarity matrix.

4.5   Handling Unequal Length Pivot Point Data Sets

Handling unequal length data series is the most challenging aspect of this method. To deal with this, we distributed pivot points based on its distribution across the book for the shorter length data series. The biggest gap is filled with interpolation first. This process is continued till the length of two data series becomes same. The 30% length difference can be handled with this method.

5   Experimentation

5.1   Baselines

Creative artifacts are generally indexed and catalogued using titles or author names. In some very special cases support is provided using metadata and keywords. Thus, indexing in the past performed using two different approaches. In the first approach it is based on metadata, author names, genre and titles. In the second approach textual similarity is used across the complete text or on the summary. The first approach is developed as the baseline-1 while the second one is developed as baseline-2.

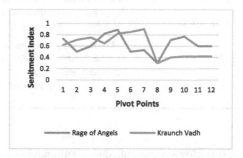

Fig. 2. Comparison of sentiment progression

5.2   Results

Table 1 gives sentiment indices (SI) at normalized pivot points for fictions *Rage of Angels* and *KraunchVadh*. The Sentiment Progression Similarity Indicator (SPSI) between these two fictions is 0.649149. Figure 2 depicts sentiment progression.

Table 1. Pivot point mapping and ranking across the fiction (normalized values)

Pivot points	SI RageofAngels	SI KraunchVadh	Pivot Points	SI RageofAngels	SI KraunchVadh
1	0.73	0.62	7	0.53	0.9
2	0.5	0.71	8	0.3	0.3
3	0.6	0.75	9	0.71	0.4
4	0.82	0.65	10	0.77	0.42
5	0.89	0.82	11	0.6	0.42
6	0.5	0.85	12	0.6	0.42

Table 2. Sentiment progression based indexing

Sr. No.	Representative Sentiment Progression	Example Books	Remarks
1		*Rage of Angels, KraunchVadh,* etc	Middle portion creates higher sustained positive emotions with peak at pivot point 11
2		*Nothing lasts forever, Amrutvel,* etc	Slowly leads to lower point and there are surprises and excitement towards end
3		*Alchemist, Rikama Devhara,* etc.	Distributed surprises excitement leading to multiple spikes across the narrative
4		*Kite Runner, Five Point Someone,* etc.	Begins with excitements and multiple spikes and followed by sudden negativity with some positive resolve towards end

Books from database (http://www.cs.cmu.edu/~dbamman/booksummaries.html) are used for experimentation. We needed complete text of the book, hence the number of samples used for experimentation are kept limited. The response of 25 book lovers is compiled for analysis of outcome. Total 100 top books are indexed and catalogued using SPbSI. This outcome is compared with results from baseline algorithm where book lovers look for a book of a particular type. The representative behavioral patterns are depicted in Table 2. 100 books are used for Indexing. This indexing based on clusters is verified using inputs from book lovers. Out of these 100 books for 92 all the book lovers were in agreement of indexing. On the other side for baseline-1 only 55 booklovers endorsed the outcome. Baseline-2 based on text similarity could find 66 books indexed as per expectations of the book lovers. The behavioral pattern and indexing are depicted in Table 2. Around 38% improvement could be obtained for the given set of data using SPbSI over the baseline-1 and 26% over baseline-2. Thus, readers look for sentiment progression rather than metadata related to book. Though the sample size is small one, it is representative of overall similarity.

6 Conclusion

Indexing and searching narratives and creative textual artefacts using author names and titles comes with its own challenges. While searching narrative based on features and behaviors, sentiment progression could prove to be a valid alternative to traditional way of indexing. Book similarity in terms of reader preferences depends on progression of sentiment. This paper proposed an approach of indexing and searching of books based on 'Sentiment Progression based Searching and Indexing' (SPbSI). The results are analyzed with reference to data collected from 25 book lovers, but the method may be scaled to the analysis of thousands of candidates. The proposed algorithm gives around 26% improvement over the base line algorithm. The algorithm SPbSI can further be improved using moving window-based similarity approach which can make possible even to recommend certain part of a narrative or creative artifact to readers.

References

1. Iyyer, M., Guha, A., Chaturvedi, S., Boyd-Graber, J., Daume, H.: Feuding families and former friends: unsupervised learning for dynamic fictional relationships. In: ACL, San Diego, California (2016)
2. Chaturvedi, S., Srivastava, S., Daume, H., Dyer, C.: Modeling evolving relationships between characters in literary novels. In: AAAI, Phoenix, Arizona (2016)
3. Li, J., Jia, R., He, H., Liang, P.: Delete, retrieve, generate: a simple approach to sentiment and style transfer. In: ACL, New Orleans, Louisiana (2018)
4. Lund, J.: Fine-grained Topic Models Using Anchor Words. Dissertation. Brigham Young University, Provo, Utah (2018)
5. Oard, D.W., Carpuat, M., et al.: Surprise languages: rapid-response cross-language IR. In: ACM NTCIR-14 Conference, 10 June 2019 Tokyo Japan (2019)
6. Quan, C., Ren, F.: Selecting clause emotion for sentence emotion recognition. In: International Conference on Natural Language Processing and Knowledge Engineering, Tokushima, Japan (2011)
7. Kulkarni, H., Alicea, B.: Cultural association based on machine learning for team formation. arXiv, 1908.00234 (2019)
8. Kulkarni, H., Marathe, M.: Machine learning based cultural suitability index (CSI) for right task allocation. In: IEEE International Conference on Electrical, Computer and Communication Technologies (IEEE ICECCT), Coimbatore, India (2019)
9. Spasojevic, N., Poncin, G.: Large scale page-based book similarity clustering. In: 2011 International Conference on Document Analysis and Recognition, Beijing, pp. 119–125 (2011). https://doi.org/10.1109/ICDAR.2011.33

Social Media

Argument Mining in Tweets: Comparing Crowd and Expert Annotations for Automated Claim and Evidence Detection

Neslihan Iskender[1]([✉]), Robin Schaefer[2], Tim Polzehl[1,3],
and Sebastian Möller[1,3]

[1] Quality and Usability Lab, Technische Universität Berlin, Berlin, Germany
{neslihan.iskender,sebastian.moeller}@tu-berlin.de
[2] Applied Computational Linguistics, University of Potsdam, Potsdam, Germany
robin.schaefer@uni-potsdam.de
[3] German Research Center for Artificial Intelligence (DFKI), Berlin, Germany
tim.polzehl@dfki.de

Abstract. One of the main challenges in the development of argument mining tools is the availability of annotated data of adequate size and quality. However, generating data sets using experts is expensive from both organizational and financial perspectives, which is also the case for tools developed for identifying argumentative content in informal social media texts like tweets. As a solution, we propose using crowdsourcing as a fast, scalable, and cost-effective alternative to linguistic experts. To investigate the crowd workers' performance, we compare crowd and expert annotations of argumentative content, dividing it into claim and evidence, for 300 German tweet pairs from the domain of climate change. As being the first work comparing crowd and expert annotations for argument mining in tweets, we show that crowd workers can achieve similar results to experts when annotating claims; however, identifying evidence is a more challenging task both for naive crowds and experts. Further, we train supervised classification and sequence labeling models for claim and evidence detection, showing that crowdsourced data delivers promising results when comparing to experts.

Keywords: Argument mining · Crowdsourcing · Corpus annotation

1 Introduction

With the rapid development of social media sites, especially Twitter, have begun to serve as a primary media for argument and debate, leading to increasing interest in automatic argument mining tools [15]. However, they require considerable amounts of annotated data for the given topic to achieve acceptable performance, increasing the cost and organizational efforts of data set annotation by linguistic experts enormously [10]. As a result, crowdsourcing has become an attractive

© Springer Nature Switzerland AG 2021
E. Métais et al. (Eds.): NLDB 2021, LNCS 12801, pp. 275–288, 2021.
https://doi.org/10.1007/978-3-030-80599-9_25

alternative to expert annotation, helping researchers generate data sets quickly and in a cost-effective way [7]. Although some researchers have applied crowd-sourcing to argument annotation [7,12,16], they did not focus on social media text which has character limitations and tends to be written informally without following specific rules for debate or opinion expression. So, focusing on social media increases the subjectivity and complexity of the argument annotation task [1,17]. Therefore, the appropriateness of crowdsourcing for it should be investigated.

This paper addresses this gap by conducting crowd and expert experiments on a German tweet data set[1], comparing annotations quantitatively, and investigating their performance for training argument mining tools. By placing a strong focus on the comparison of the crowd and expert annotations, we extend our previous study on tweet-based argument mining [13], which presents the first results for training performance of the expert annotations also used in this work. Like in our previous work, we apply a claim-evidence model, where *claim* is defined as a controversial opinion and *evidence* as a supportive statement related to a claim. Both components are further referred to as *Argumentative Discourse Units* (ADU) [11].

2   Related Work

Related work has investigated argument mining in tweets primarily from the viewpoint of corpus annotation and argument component detection. In an early work from 2016, the *Dataset of Arguments and their Relations on Twitter* (DART) was presented [2]. 4000 English tweets were annotated by three experts on the full tweet level for general argumentative content (stating high consistency as Krippendorff's α: 0.74 for inter-annotator agreement (IAA)), thereby refraining from further separating between claim and evidence. Also, topics were heterogeneous, including, for instance, tweets on product releases, which may contain different argumentation frequency, density and clarity. This may have facilitated individual annotation tasks. An applied logistic regression model yielded an F1 score of 0.78 on argument detection.

Another line of research approached argument mining on Twitter by focusing on evidence detection [1]. In contrast to our work, tweets were annotated for specific evidence types, e.g., news or expert opinion, and the annotators' level of expertise was not reported in the paper. Also, the full tweet was the unit of annotation, which reduced the task's complexity and might be reflected in their high Cohen's κ score of 0.79. An SVM classifier achieved an F1 score of 0.79 for the evidence detection task.

More recently, argument annotation work on Swedish social media was presented [9]. Annotators (one expert and seven *"trained annotators with linguistic backgrounds"*) labeled argumentative spans in posts from discussion forums (Cohen's κ: 0.48). While this research did not focus on tweets, it still shows the difficulty of creating high-quality consistent argument annotations in social

---

[1] Corpus repository: https://github.com/RobinSchaefer/climate-tweet-corpus.

media data. Work on argument mining on data from various Greek social media sources, including tweets, was presented by [4]. The study included data annotation, however IAA was not presented, which hinders comparison. Moreover supervised classification and sequence labeling models were trained (F1: 0.77 and 0.42), which we adopt in our work.

As previous research on argument annotation of social media text reveals, the annotators were either experts [2,9], or their level of expertise was not reported or questioned [1,4]. Our research extends these studies by investigating the effect of annotator's expertise on the ADU annotation, focusing on claim detection in addition to general argument detection [2,4,9] and evidence detection [1] on the domain of highly controversial *climate change* tweets on Twitter.

3   Experiments

In our experiments, we used a data set with 300 German tweet pairs extracted from the Twitter API on the climate change debate. Each pair in the data set consists of a *context* tweet and a *reply* tweet as a response to the context tweet. The average word count of context tweets is 26.64, the shortest one with one word and the longest one with 49 words; the average word count of reply tweets is 27.44, the shortest one with one word, the longest one with 52 words.

3.1   Crowdsourcing Study

We collected crowd annotations using the Crowdee[2] Platform. We designed a task specific pre-qualification test for crowd worker selection. All crowd workers who passed Crowdee's German language test with a score of 0.9 or above were admitted for the pre-qualification test. In the pre-qualification test, we explained at first the general task characteristics and provided definitions and examples for the argumentative content dividing it into its two components *claim* and *evidence*. We defined *claim* as "the author's personal opinion, position or presumption" and *evidence* as "content intended to support a claim". In line with previous research, we decided on using relatively broad ADU definitions due to the rather informal nature of argumentation in tweets, which is hard to capture with more narrow definitions. Further, we provided text annotation guidelines such as only to annotate the smallest understandable part in a reply tweet as claim, only to annotate evidence if it relates to a claim from the tweets shown, and to ignore personal political beliefs, as well as the spelling or grammatical errors.

After reading the instructions, crowd workers were asked to annotate claim and evidence in tweet pairs. The first question "Is there any claim in the reply tweet?" was displayed with the two answer options "yes" and "no". The second question "Is there evidence in the reply tweet?" was displayed with the four answer options "yes, evidence in the reply tweet relates to a claim in the reply

---

[2] https://www.crowdee.com/.

tweet.", "yes, evidence in the reply tweet relates to a claim in the context tweet.", "yes, evidence in the reply tweet relates to a claim in both tweets.", and "no, there is no evidence.". We refer to these questions as voting questions in Sect. 4. If crowd workers selected an answer option with "yes" in any of the voting questions, they were asked to label the text part containing claim or evidence, which we refer to as text annotation in Sect. 4.

Each question was displayed on a separate page, and the pre-qualification task included the annotation of three different tweet pairs. Crowd workers could achieve a maximum of 12 points for answering each of the voting questions correctly, and we kept crowd workers exceeding 8 points. Additionally, the author's team evaluated manually crowd workers' answers for three text annotation questions and eliminated crowd workers who labeled the non-argumentative content in tweets as claim or evidence. Overall, 101 crowd workers participated in the pre-qualification test completing the task in 15 h with an average work duration of 546 s. Based on our selection criteria, 54 crowd workers were accepted for the main task.

Out of 54 admitted crowd workers, 42 crowd workers participated in the main task. Further, five unique crowd workers per tweet pair annotated claim and evidence using the same task design as in the pre-qualification test, resulting in 1500 crowd answers. We published a total of 1500 tasks in batches, and each batch was completed within a maximum of five days, with an average work duration of 394 s. Here, we observed that the main task's average task completion duration was lower than for the pre-qualification task, although the main task included the annotation of two more tweet pairs. The reason for this is probably the following: after doing the task a couple of times, crowd workers did not need to read the definitions and instructions at the beginning of the task, which led to a lower task completion duration.

3.2   Expert Evaluation

Two experts, one of them a Ph.D. student at a linguistics department and co-author of this paper, and the other one a student in linguistics, annotated the same 300 tweet pairs using the same task design as the crowdsourcing study. At first, they annotated the tweet pairs separately using the Crowdee platform. After the first separate evaluation round, the IAA scores, Cohen's κ, showed that the experts often diverted in their assessment. To reach consensus among experts, we arranged physical follow-up meetings with the two experts, which we refer to as mediation meetings. In these meetings, experts discussed the reasons and backgrounds of their annotations for tweet pairs in case of substantial disagreement and eventually aligned them if consensus was obtained. Eventually, acceptable IAA scores were reached for the voting questions of claim and evidence. This procedure also led to several suggestions regarding the refinement of annotation guidelines which will be discussed in Sect. 6.

4 Comparing Crowd with Expert

Results are presented for the two voting questions (claim and evidence) and the text annotations from the crowdsourcing and expert evaluation. We analyzed 1500 crowd answers using majority vote as the aggregation method for the voting questions, leading to 300 majority voted crowd answers and 600 expert answers for 300 tweet pairs. Further, we investigate the general annotation of argumentative content by combining claim and evidence annotation under the label *argument*.

4.1 Comparing Voting

Before comparing expert votings for argument, claim and evidence with the crowd, we calculated Cohen's κ and Krippendorff's α scores to measure the IAA between two experts and the raw agreement scores in %. We analyzed both the voting with four answer options and binary evidence voting deducted from four answer options.

Table 1. Raw agreement in %, Cohen's κ and Krippendorff's α scores between two experts for argument, claim and evidence votings before mediation and after mediation

	Before mediation			After mediation		
	Agr. in %	κ	α	Agr. in %	κ	α
Argument	87.7	0.47	0.47	90.7	0.62	0.62
Claim	85.7	0.45	0.45	90	0.62	0.62
Evidence (binary)	65.7	0.34	0.31	71.7	0.44	0.43
Evidence (4 options)	61.7	0.32	0.31	67.7	0.41	0.41

Looking at Table 1, we see that the mediation meetings increase all of the agreement scores, and the Cohen's κ score for argument and claim reaches a substantial level (0.6–0.8] [6]. However, the mediation meetings increase the Cohen's κ scores for evidence only from fair (0.20–0.40] to moderate (0.40–0.60]. Also, we calculated Krippendorff's α, which is technically a measure of evaluator disagreement rather than agreement. Although the mediation meetings increase the Krippendorff's α scores, still they leave room for improvement ($\alpha < 0.667$) [5]. This result shows that identifying argumentative content, especially evidence, is even for experts a subjective and ambiguous task, which is also reflected by the raw agreement scores in % for evidence.

Next, we calculated raw agreement in % between crowd and experts, and between the two experts before and after mediation as shown in Fig. 1. Here, we observe that both before and after mediation, crowd workers reach comparable results as experts in terms of the raw agreement in %, achieving an agreement above 85 % for argument and claim. However, crowd-expert agreements for evidence is lower than the expert-agreement, especially when using the scale with four answer options. It shows that evidence identification by determining to

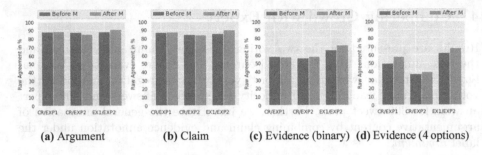

Fig. 1. Barplots of raw agreement in percentage for argument, claim, evidence (binary), and evidence (4 options) between crowd and experts, and between experts before and after mediation (M = Mediation, CR = Crowd, EXP = Expert)

which tweet evidence relates is a complex and subjective task, notably for crowd workers. Therefore, we use the results from the binary evidence votings in our further analysis.

To investigate the differences between crowd and expert for voting questions, we calculated the non-parametric T-Test, Mann-Whitney U Test. The test results revealed significant differences for argument and claim between crowd and experts both before and after mediation. The median values of crowd and experts clearly showed that the crowd workers identified arguments and claims in more tweets than the experts (argument: $N_{cr} = 282$, $N_{exp1} = 273$, $N_{exp2} = 255$; claim: $N_{cr} = 261$, $N_{exp1} = 255$, $N_{exp2} = 251$). Moreover, the Mann-Whitney U test results for evidence also revealed significant differences between crowd and expert 2 and between two experts both before and after mediation. Looking at the median values, we observed that expert 2 identified more evidence in tweets than expert 1 and crowd workers ($N_{cr} = 162$, $N_{exp1} = 166$, $N_{exp2} = 175$). The significant difference between the two experts for evidence is in line with our previous expert IAA analysis.

Analyzing the Spearman correlation coefficients between the crowd and expert, we saw that crowd-expert correlation for argument ($r_{cr/exp1} = 0.35, r_{cr/exp2} = 0.31, p < 0$) was at a weak level, where experts reached a moderate correlation before mediation ($r = 0.47, p < 0$). On the contrary, crowd correlation with expert 1 for claim ($r_{cr/exp1} = 0.42, r_{cr/exp2} = 0.31, p < 0$) achieved a similar level of correlation as the correlation between two experts ($r = 0.45, p < 0$). After expert mediation, the correlation between two experts increased to 0.62 both for argument and claim, while correlation between crowd and expert remained at the same level for argument ($r_{cr/exp1} = 0.36, r_{cr/exp2} = 0.25, p < 0$) and claim ($r_{cr/exp1} = 0.42, r_{cr/exp2} = 0.30, p < 0$). Note, the crowd and expert correlations for evidence were of an overall weak level regardless of the mediation (before mediation: $r_{cr/exp1} = 0.15, r_{cr/exp2} = 0.15, r_{exp1/exp2} = 0.36$, after mediation: $r_{cr/exp1} = 0.14, r_{cr/exp2} = 0.17, r_{exp1/exp2} = 0.46, p < 0$). These weak/moderate correlations before mediation demonstrate again the subjectivity of the task, especially for evidence.

As our last analysis on the voting consistency, we calculated Fleiss' κ scores between crowd and two experts. Before mediation, they reached a Fleiss' κ score of 0.36 for argument and 0.37 for claim, which is also at a similar level of expert-agreement before mediation. After mediation, the Fleiss' κ score increased to 0.40 for argument and 0.43 for claim. This shows that mediation meetings contribute to the robustness of expert votings, indicating that a similar approach between crowd workers could increase the crowd votings' robustness as well. Similarly, the Fleiss' κ score for evidence increased from 0.20 to 0.24 after mediation, however, still remaining at a weak level.

4.2   Comparing Text Annotations

In this section, we compare the text annotations for claim and evidence given by crowd and experts. As explained in Sect. 4.1, the mediation meetings did not affect the relationship between crowd and expert votings remarkably, therefore we only focus on the annotations after mediation in this section. To compare the text annotations with each other, we follow a similar logic to ROUGE-1, which describes the overlap of unigrams (each word) between the system and reference summaries [8]. In our case, we compare the location of labeled text characters by crowd and expert, computing the precision

$$Precision = \frac{\text{location of crowd labeled characters} \cap \text{location of expert labeled characters}}{\text{location of crowd labeled characters}}$$ and

recall

$$Recall = \frac{\text{location of crowd labeled characters} \cap \text{location of expert labeled characters}}{\text{location of expert labeled characters}}$$

to calculate the F1 score ($F1\ score = 2 \times \frac{\text{Precision} \times \text{Recall}}{\text{Precision} + \text{Recall}}$).

We applied three different methods for comparing text annotations: *mean*, *majority vote* and *similarity*. In the first approach, we considered all five different crowd annotations for each tweet pair and computed the F1 score between each of five crowd workers and experts, calculating the mean of the five F1 scores as a final result. In our second approach, we followed a similar strategy to voting and calculated the majority vote for each annotated character location by comparing annotations from five different crowd workers. The resulting majority-voted character locations were used to calculate F1 scores between crowd and experts. In the last approach, we calculated F1 scores between each of five crowd workers for each tweet pair and selected the individual crowd worker whose text annotation has the highest average F1 score with other crowd workers. Then, we used this crowd worker's answer for calculating the F1 score between crowd and expert. It should be noted that we calculated the F1 scores only in case of positive claim or evidence voting from both naive and expert annotators.

Figure 2 shows the cumulative histograms and density estimation plots of argument, claim and evidence F1 scores for all three approaches. As the density plot for all text annotations between the two experts shows, the experts either do not agree on the text annotations or they agree 100 %. However, crowd's and experts' text annotations F1 score is distributed equally centered around the score 0.5 for argument and claim (see Fig. 2a and Fig. 2d) and around the score 0.3 for evidence (see Fig. 2g) using the mean approach. For the majority vote approach, we observe that argument and claim annotations get close to the score

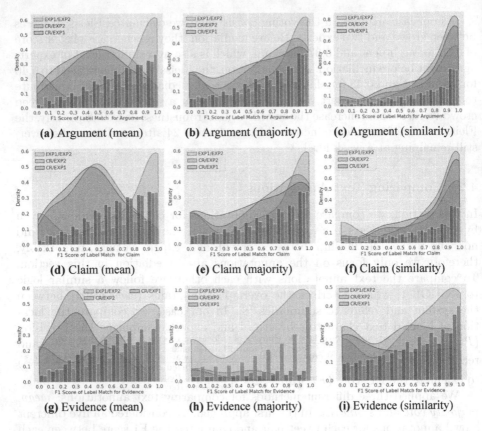

(a) Argument (mean)   **(b)** Argument (majority)   **(c)** Argument (similarity)

(d) Claim (mean)   **(e)** Claim (majority)   **(f)** Claim (similarity)

(g) Evidence (mean)   **(h)** Evidence (majority)   **(i)** Evidence (similarity)

Fig. 2. Cumulative histograms and density estimation plots for the annotation match for argument (first row), claim (second row) and evidence (third row) between crowd and experts, and between two experts (CR = Crowd, EXP = Expert)

1, but still, its density is not at the level of the experts' F1 score (see Fig. 2b and Fig. 2e); and the crowd workers cannot agree on the text annotations for evidence (see Fig. 2h). As the Figs. 2c, 2f and 2i demonstrate, the similarity approach produce results most similar to experts' F1 score, especially for argument and claim. Therefore, we recommend using this approach when collecting data from multiple crowd workers.

5   Training Argument Mining Models on Annotated Tweets

In this section, we present experimental results from training supervised classification and sequence labeling models on full tweet and ADU annotations of crowds and experts, respectively. As features BERT [3] embeddings were created by using deepset.ai's pretrained *bert-base-german-cased* model[3].

We compare different annotation sets (crowd vs expert) and layers (argument vs claim vs evidence). Models are trained both on individual expert and crowd annotations and on combinations of these. Models are tested either with test sets obtained from a train-test split (Tables 2 and 4) or by using expert annotations as gold standard (Tables 3 and 5). As shown in Sect. 4.1, all argument classes form the respective majority class, which is why we report weighted F1 scores. For comparison we also show unweighted macro scores in Tables 2 and 4. All scores are 10-fold cross-validated.

5.1   Supervised Classification

We trained supervised classification models on full tweet annotations derived from the ADU annotations (voting questions in experiments). Thus, a classifier's task is to separate tweets containing an ADU from non-argumentative tweets. Results (Tables 2 and 3) are obtained using eXtreme Gradient Boosting. Models trained on non-mediated expert annotations mostly yield promising weighted F1 scores (0.71–0.91). Unweighted F1 scores are comparatively low. This indicates the models' problems with identifying minority classes, which is intensified by the small corpus size. Notably, the reduction appears especially to be caused by low recalls. Models trained on mediated expert data show less variance between annotators. Also, training on combined expert annotation sets yields substantially better results than training on individual expert annotation sets.

Results obtained by crowd annotations show an interesting pattern. While models trained on all crowd annotations can generally compete with expert models, weighted F1 scores derived from crowd majority annotations are reduced (F1: 0.57/0.58) with the exception of evidence targets. For argument and claim targets the difference between weighted and unweighted F1 scores is less severe than for expert annotations. Also, utilizing combined crowd and expert annotations yields acceptable results. Testing models trained on mediated expert data with gold annotations (see Table 3) yields mainly similar results to the scores shown in Table 2. However, testing all crowd annotation sets with expert annotations does not perform well. Adding expert annotations to the training set notably improves results with the exception of evidence annotations.

5.2   Sequence Labeling

Sequence labeling models were trained on the ADU annotations in order to build a system that can extract argumentative spans from tweets (text annotations from crowd and experts). We applied Conditional Random Fields for this task. Here, we use the similarity method instead of majority for deriving a single set from the crowd annotations, as this showed best results during text annotation

---

[3] https://huggingface.co/bert-base-german-cased.

Table 2. Supervised classification results (M = mediation; CS = corpus size; p = partial (i.e. only experts are mediated); w = weighted).

Annotator	M	CS	Argument				Claim				Evidence			
			F1 (w)	F1	P	R	F1 (w)	F1	P	R	F1 (w)	F1	P	R
Expert 1	−	300	0.84	0.60	0.68	0.59	0.81	0.57	0.66	0.57	0.57	0.54	0.56	0.55
Expert 2	−	300	**0.91**	0.77	0.93	0.72	0.86	0.71	0.85	0.67	**0.71**	0.70	0.73	0.70
Expert (both)	−	600	0.90	0.77	0.79	0.78	**0.88**	0.78	0.78	0.78	0.69	0.69	0.71	0.69
Expert 1	+	300	0.87	0.66	0.79	0.64	0.84	0.63	0.72	0.61	0.62	0.60	0.63	0.61
Expert 2	+	300	0.90	0.76	0.91	0.74	0.87	0.72	0.87	0.69	0.69	0.68	0.70	0.68
Expert (both)	+	600	**0.95**	0.89	0.93	0.87	**0.93**	0.86	0.90	0.84	**0.75**	**0.75**	0.77	0.75
Crowd (majority)	−	300	0.57	0.53	0.55	0.53	0.58	0.53	0.54	0.54	**0.81**	0.46	0.43	0.50
Crowd (all)	−	1,500	**0.87**	0.86	0.87	0.86	**0.81**	0.79	0.80	0.79	0.78	0.61	0.65	0.60
Crowd + Expert	p	2,100	0.80	0.80	0.81	0.80	0.78	0.78	0.79	0.78	0.76	0.69	0.73	0.67

Table 3. Supervised classification results, tested with gold annotations (Expert 1 or Expert 2). Expert annotations are mediated. Only weighted F1 scores are reported.

Annotator	Argument		Claim		Evidence	
	Expert 1	Expert 2	Expert 1	Expert 2	Expert 1	Expert 2
Expert 1	−	0.86	−	0.83	−	0.61
Expert 2	0.90	−	0.88	−	0.59	−
Expert (both)	0.89	0.88	0.86	0.87	0.61	0.66
Crowd (majority)	0.47	0.49	0.48	0.47	0.55	0.38
Crowd (all)	0.21	0.21	0.11	0.14	0.43	0.28
Crowd + Expert	0.81	0.84	0.74	0.72	0.49	0.33

analysis (see Sect. 4.2). Looking at Table 4, models trained on non-mediated data yields promising results for argument (0.83) and evidence detection (0.70). Weighted F1 scores for claim detection are comparatively low. However, training on both expert sets results in a notable improvement on this task. Compared to classification, unweighted precision and recall show less divergence. Training sequence labeling models on mediated expert data hardly changes results. However, improvements are achieved by utilizing both expert annotation sets.

Using all crowd annotations results in reduced scores for argument labels, and comparable results for claim and evidence labels in comparison to experts. Combining crowd and expert annotations improves the results. Testing models with gold annotations (see Table 5) shows patterns similar to previously discussed results. Importantly, crowd similarity annotations yield results comparable to expert annotations or better when tested with gold annotations.

Table 4. Sequence labeling results. (M = mediation; CS = corpus size, p = partial (i.e. only experts are mediated); w = weighted).

Annotator	M	CS	Argument				Claim				Evidence			
			F1(w)	F1	P	R	F1(w)	F1	P	R	F1(w)	F1	P	R
Expert 1	–	300	0.72	0.62	0.62	0.62	0.57	0.58	0.59	0.58	**0.70**	0.60	0.60	0.61
Expert 2	–	300	**0.83**	0.68	0.69	0.68	0.57	0.60	0.61	0.60	0.62	0.62	0.63	0.62
Expert (both)	–	600	0.80	0.70	0.72	0.71	**0.65**	0.67	0.69	0.66	**0.70**	0.67	0.71	0.68
Expert 1	+	300	0.72	0.61	0.61	0.62	0.57	0.59	0.60	0.59	0.71	0.61	0.61	0.61
Expert 2	+	300	0.81	0.67	0.67	0.68	0.57	0.60	0.61	0.60	0.62	0.61	0.62	0.61
Expert (both)	+	300	**0.86**	0.78	0.80	0.78	**0.69**	0.71	0.73	0.70	**0.76**	0.72	0.75	0.72
Crowd (similarity)	–	300	0.53	0.54	0.55	0.53	**0.55**	0.56	0.58	0.55	**0.81**	0.64	0.64	0.64
Crowd (all)	–	1500	**0.64**	0.60	0.64	0.58	0.54	0.59	0.63	0.57	0.64	0.61	0.65	0.59
Crowd + Expert	p	2100	0.72	0.65	0.71	0.62	0.59	0.63	0.67	0.61	0.68	0.63	0.69	0.60

Table 5. Sequence labeling results, tested with gold annotations (Expert 1 or Expert 2). Expert annotations are mediated. Only weighted F1 scores are reported.

Annotator	Argument		Claim		Evidence	
	Expert 1	Expert 2	Expert 1	Expert 2	Expert 1	Expert 2
Expert 1	–	0.77	–	0.57	–	0.62
Expert 2	0.74	–	0.56	–	0.66	–
Expert (both)	0.74	0.79	0.58	0.59	0.67	0.63
Crowd (similarity)	0.73	0.73	0.66	0.65	0.75	0.62
Crowd (all)	0.75	0.83	0.57	0.61	0.74	0.64
Crowd + Expert	0.75	0.80	0.57	0.61	0.74	0.64

6 Discussion and Outlook

Our extensive empirical comparison of crowd and expert ADU annotations in Sect. 4 showed that this task has a high level of subjectivity and ambiguity, even for experts. Even after mediation, experts only reached moderate IAA scores for evidence, indicating that distinguishing between claim and evidence is even harder than claim identification. We observed similar results when comparing crowd and expert annotations, where crowd workers could reach a comparable level of raw agreement in % as experts for argument and claim, while crowd-expert agreement for evidence remained at moderate level for both expert and crowd assessment. Also, the results from Sect. 4.2 confirmed this finding. Here, we also demonstrated a method for determining the "reliable" crowd worker for text annotation who can achieve similar results as experts.

Despite the annotation differences, the results from Sect. 5.1 showed that training with all crowd annotations delivers similar results as experts. However, when using gold annotations for testing classification models, the crowd could not achieve comparable results to experts. For sequence labeling (see Sect. 5.2), training with crowd annotations produced close results to single experts for claim and evidence, but combining both experts led to better results than for the crowd. Also, when using an expert data set as the test set for sequence labeling, crowd text annotations achieved expert-level F1 scores. As results of models trained on crowd worker annotations derived by the similarity method and on expert annotations are comparable when tested with gold annotations, we argue that the similarity annotations are reliable.

The reasons for different annotations between crowd and experts, especially for evidence, may be due to the text structure of tweets, which are characterized by a certain degree of implicitness, thereby entailing substantial subjectivity for the annotation task. Further, subjectivity also complicates the decision on the exact boundary between claim and evidence units. As evidence is defined as occurring only in relation to a claim, determining claim-evidence boundaries is of particular importance. So, one may consider separating evidence annotation from claim annotation. Annotating claims in a first step, followed by subsequent evidence labeling, would reduce annotators' degrees of freedom and thereby possibly increase the IAA. Limiting the allowed number of ADU annotations per tweet could positively affect IAA scores as fewer boundaries between claim and evidence have to be drawn.

In future work, we suggest adjustments to the definitions of argument components based on the results from expert mediation sessions. Given the peculiarities of tweets, we consider it appropriate to utilize a relatively broad interpretation, especially of the concept *claim*. Still, it may be fruitful to define more narrow claim and evidence definitions resulted from expert mediation sessions. For example, one could focus on *major claims* [14], which could be defined as a tweet's single main position or opinion, i.e., the argumentative reason why it was created. This may decrease the task subjectivity. Additionally, evidence might relate to a tweet outside the presented tweet pairs, so showing more than one context tweet may help the evidence annotation process. Another helpful approach may be arranging mediation sessions between crowd workers since the mediation between experts increased their agreement.

Despite the limitations, this paper makes an important contribution to human annotation research of argument mining in tweets. The organizational efforts and the cost of expert annotation at scale can be enormous, which is a great challenge in a fast-moving field like argument mining. Therefore, finding reliable ways of using crowdsourcing can be a promising solution, and we hope to see more research in this field.

References

1. Addawood, A., Bashir, M.: What is your evidence? A study of controversial topics on social media. In: Proceedings of the Third Workshop on Argument Mining (ArgMining2016), August 2016, pp. 1–11. Association for Computational Linguistics, Berlin, Germany (2016). https://doi.org/10.18653/v1/W16-2801
2. Bosc, T., Cabrio, E., Villata, S.: DART: a dataset of arguments and their relations on Twitter. In: Proceedings of the Tenth International Conference on Language Resources and Evaluation (LREC 2016), May 2016, pp. 1258–1263. European Language Resources Association (ELRA), Portorož, Slovenia (2016)

3. Devlin, J., Chang, M.W., Lee, K., Toutanova, K.: BERT: pre-training of deep bidirectional transformers for language understanding. In: Proceedings of the 2019 Conference of the North American Chapter of the Association for Computational Linguistics: Human Language Technologies, vol. 1 (Long and Short Papers), June 2019, pp. 4171–4186. Association for Computational Linguistics, Minneapolis, Minnesota (2019). https://doi.org/10.18653/v1/N19-1423

4. Goudas, T., Louizos, C., Petasis, G., Karkaletsis, V.: Argument extraction from news, blogs, and social media. In: Likas, A., Blekas, K., Kalles, D. (eds.) Artificial Intelligence: Methods and Applications, pp. 287–299. Springer International Publishing, Cham (2014)

5. Krippendorff, K.: Content Analysis: An Introduction to Its Methodology, Sage publications, Thousand Oaks (1980)

6. Landis, J.R., Koch, G.G.: The measurement of observer agreement for categorical data. Biometrics **33**(1), 159–174 (1977)

7. Lavee, T., et al.: Crowd-sourcing annotation of complex NLU tasks: a case study of argumentative content annotation. In: Proceedings of the First Workshop on Aggregating and Analysing Crowdsourced Annotations for NLP, November 2019, pp. 29–38. Association for Computational Linguistics, Hong Kong (2019). https://doi.org/10.18653/v1/D19-5905

8. Lin, C.Y.: ROUGE: A package for automatic evaluation of summaries, pp. 74–81 (July 2004)

9. Lindahl, A.: Annotating argumentation in Swedish social media. In: Proceedings of the 7th Workshop on Argument Mining, December 2020, pp. 100–105. Association for Computational Linguistics, Online (2020)

10. Miller, T., Sukhareva, M., Gurevych, I.: A streamlined method for sourcing discourse-level argumentation annotations from the crowd. In: Proceedings of the 2019 Conference of the North American Chapter of the Association for Computational Linguistics: Human Language Technologies, vol. 1 (Long and Short Papers), June 2019, pp. 1790–1796. Association for Computational Linguistics, Minneapolis, Minnesota (2019). https://doi.org/10.18653/v1/N19-1177

11. Peldszus, A., Stede, M.: From argument diagrams to argumentation mining in texts: a survey. Int. J. Cogn. Inform. Nat. Intell. **7**(1), 1–31 (2013). https://doi.org/10.4018/jcini.2013010101

12. Reisert, P., Vallejo, G., Inoue, N., Gurevych, I., Inui, K.: An annotation protocol for collecting user-generated counter-arguments using crowdsourcing. In: Isotani, S., Millán, E., Ogan, A., Hastings, P., McLaren, B., Luckin, R. (eds.) Artificial Intelligence in Education, pp. 232–236. Springer International Publishing, Cham (2019)

13. Schaefer, R., Stede, M.: Annotation and detection of arguments in tweets. In: Proceedings of the 7th Workshop on Argument Mining, December 2020, pp. 53–58. Association for Computational Linguistics, Online (2020)

14. Stab, C., Gurevych, I.: Identifying argumentative discourse structures in persuasive essays. In: Proceedings of the 2014 Conference on Empirical Methods in Natural Language Processing (EMNLP), October 2014, pp. 46–56. Association for Computational Linguistics, Doha, Qatar (2014). https://doi.org/10.3115/v1/D14-1006

15. Stede, M., Schneider, J.: Argumentation Mining, Synthesis Lectures in Human Language Technology, vol. 40. Morgan & Claypool (2018)

16. Toledo-Ronen, O., Orbach, M., Bilu, Y., Spector, A., Slonim, N.: Multilingual argument mining: Datasets and analysis. In: Findings of the Association for Computational Linguistics: EMNLP 2020, November 2020, pp. 303–317. Association for Computational Linguistics, Online (2020). https://doi.org/10.18653/v1/2020. findings-emnlp.29

17. Šnajder, J.: Social media argumentation mining: The quest for deliberateness in raucousness (2016)

Authorship Attribution Using Capsule-Based Fusion Approach

Chanchal Suman[1]([✉]), Rohit Kumar[1], Sriparna Saha[1],
and Pushpak Bhattacharyya[2]

[1] Department of Computer Science and Engineering,
Indian Institute of Technology Patna, Patna, India
{1821cs11,rohit.cs17,sriparna}@iitp.ac.in
[2] Department of Computer Science and Engineering,
Indian Institute of Technology Bombay, Mumbai, India

Abstract. Authorship attribution is an important task, as it identifies the author of a written text from a set of suspect authors. Different methodologies of anonymous writing, have been discovered with the rising usage of social media. Authorship attribution helps to find the writer of a suspect text from a set of suspects. Different social media platforms such as Twitter, Facebook, Instagram, etc. are used regularly by the users for sharing their daily life activities. Finding the writer of micro-texts is considered the toughest task, due to the shorter length of the suspect piece of text. We present a fusion based convolutional Neural Network model, which works in two parts i) feature extraction, and ii) classification. Firstly, three different types of features are extracted from the input tweet samples. Three different deep-learning based techniques, namely capsule, LSTM, and GRU are used to extract different sets of features. These learnt features are combined together to represent the latent features for the authorship attribution task. Finally the softmax is used for predicting the class labels. Heat-maps for different models, illustrate the relevant text fragments for the prediction task. This enhances the explain-ability of the developed system. A standard Twitter dataset is used for evaluating the performance of the developed systems. The experimental evaluation shows that proposed fusion based network is able to outperform previous methods. The source codes are available at https://github.com/chanchalIITP/AuthorIdentificationFusion.

Keywords: Authorship identification · Capsule · Fusion · LSTM · GRU

1 Introduction

Forensic authorship analysis is the process of examining the characteristics of a questioned text in order to draw conclusions on its authorship. Its application involves analyzing long fraud document, terrorist conspiracy texts, short letters, blog posts, emails, SMS, Twitter streams or Facebook status updates to check

© Springer Nature Switzerland AG 2021
E. Métais et al. (Eds.): NLDB 2021, LNCS 12801, pp. 289–300, 2021.
https://doi.org/10.1007/978-3-030-80599-9_26

the authenticity and identify fraudulence. Authorship analysis can be carried out in different ways: 1) authorship attribution, 2) authorship verification, and 3) authorship profiling [7]. In authorship profiling, characteristics (e.g., age, gender, native language, race, personality) of an author are determined after analyzing different texts written by the author [3]. Authorship verification is the task of assessing whether a specific individual writes a suspicious text [7]. In authorship attribution (AA), given the examples of the writings of a number of authors, for an anonymous text, the author is determined. These days, social media play a vital role in our life. People write about their daily life activities via different social media platforms like Twitter, Facebook etc. Most of the data created on these social media applications are micro text. A micro or short text message could be a tweet or a comment which is around 140 characters or less. The authorship analysis of a micro-text is challenging due to the smaller length of text [14].

The traditional strategy of developing the authorship attribution (AA) model, deals with extraction of different features from the text data, and then feeding the generated vector to different available machine learning classifiers (mainly Support Vector Machine) [8,22]. Convolutional neural networks (CNNs) also perform well in this area [4,5,10,23]. The recently proposed n-gram based CNN model with fastText word embeddings has set the state-of-the-art results on the Twitter dataset released by [4]. Pooling layer of CNN reduces the computational complexity of convolution operations. It captures the invariance of local features. But, pooling operations loose information regarding spatial relationships, which causes the mis-classification of objects based on their orientation or proportion. Capsules consider the spatial relationships between entities and learn these relationships via dynamic routing [16]. This has motivated us to use capsule networks for the attribution task. Some recent works have shown the efficacy of a deep learning system using a combination of features learnt from different deep learning models [2]. Following these concepts, in addition to the capsules, we have fused features extracted from different modules for better representation of text features.

Our developed system mainly consists of two parts, i) feature extraction, and ii) classification. The feature extraction part is based on an ensemble technique. Different hidden representations are learnt from three different deep learning models, i.e., Convolutional neural network with capsule, ii) Long Short-Term Memory (LSTM), and iii) Gated Recurrent Unit (GRU). These representations learn the higher-level features from the given text sample. We subsequently fuse these features for representing the latent feature for the AA task. Finally, the fused vector is fed to the softmax layer for the final classification. In this way, the proposed system extracts the authorship information using different features learnt from multiple networks.

In order to show the effectiveness of our developed system, the twitter dataset released by [21] is used for experimentation. An accuracy of 85.35% is achieved for 1000 tweets per author (for 50 authors), using the character unigram representation. The results show that, our developed systems outperform the previous state-of-the-art models. Using heat maps for different models, the working of our developed system is shown. Below, we have listed the contributions of this work.

1. To the best of our knowledge, fusion of different features extracted from capsule, LSTM, and GRU is carried out for the first time for solving the task of authorship attribution.
2. We set new state-of-the-art values, by outperforming the previous ones.
3. A detailed ablation study of different features extracted from the tweet samples has also been performed.
4. The workings of different models are shown using heat-map on the test data, which helps in finding the relevant text-fragments of the suspect text. This gives explain-ability to our developed system.

The paper is organized as follows: Previous works are discussed in Sect. 2. In Sect. 3, we discussed our proposed approach. The dataset used for implementation is discussed in Sect. 4. In Sect. 5, we have discussed the results achieved using our model. The performance comparison of our developed model with other previous approaches is presented in Sect. 6. A short analysis of heat-maps generated from different system is presented in Sect. 7. Finally, we conclude this work in Sect. 8.

2   Related Work

Traditional methods for dealing with authorship attribution task, involve extraction of features related to content and style. Term-Frequency Inverse-Document Frequency at character or word n-gram level, and Bag-of-Words are mostly used as content features. Whereas usage of punctuation, capital letters, POS tags, digits represents the stylistic features of an author [24]. Different ML classifiers are trained on these features, for the final classification task. Mainly logistics regression, and support vector machines are used as the classifier [1,17]. Some of the recent works are using different deep learning techniques like convolutional neural networks, and siamese networks too [4,13,14,19,20,23]. A comprehensive literature review on the AA task, is presented in [15], focusing on dark sides of AA.

Table 1. Some recent works on Twitter dataset

Approach	Dataset
Character n-gram (n = 1,2) on a CNN architecture [23]	Standard Twitter Dataset [21]
Combination of character and word n-grams with flexible pattern, and sub-word embedding above MLP [14]	Standard Twitter Dataset [21]
Text representation and tag representation (posting style) fed to CNN [12]	Standard Twitter Dataset [21]
Pre-trained word embedding and character bigram on a multi channel CNN [4]	Standard Twitter Dataset [21]
User representation is learnt from siamese network [13]	Standard Twitter Dataset [21]

Word and character level CNNs are explored for AA in [20]. It is noticed that character level CNNs outperform the traditional simple approaches based on support vector machines and logistic regression. In [19], character n-grams (n = 3,4,5) are fed to a multi-layer CNN approach, with max-pooling. Character level n-grams have been used in [23], for authorship attribution in short texts. The results show that CNN based prediction outperforms the techniques based on LSTM and hand-crafted feature extraction. For better visualisation of model learning capacity, saliency score has been used to highlight the text modules responsible for the classification. Authors in [14], utilized character n-grams, flexible patterns, word n-grams, and sub-word embedding on a multi-layer perceptron network to observe its effect on the Twitter dataset [21]. In [4], authors have showed that the combination of embedding layers captures different stylometric features. Authors in [13], have shown that siamese networks are useful in learning the user tweets, with small amount of data only. We have also tabulated some of the important works on the Twitter authorship data in Table 1.

The previous studies show that mainly CNN is working effectively on short texts like tweets, messages etc. This has motivated us to use CNN as our basic framework, on top of which different features extracted from various modules are combined. We have discussed the developed framework below.

3   The Proposed Approach

The Convolution operator in a CNN is represented by the weighted sum of lower layers, thus it is difficult to carry out these features into upper layers in case of complex objects. In this way, it can be said that CNNs do not consider hierarchical relationships [16]. To overcome these shortcomings, pooling layers are introduced. Pooling can reduce the computational complexity of convolution operations and capture the invariance of local features. However, pooling operations loose information regarding spatial relationships and are likely to misclassify objects based on their orientation or proportion. The capsule network is a structured model, which solves the problems of CNNs. To learn the existence of visual entities and encoding them into vectors, there are locally invariant groups which are known as capsules. Capsule networks use a non-linear function called as squashing for grouping of neurons [16].

Long Short Term Memory (LSTM) [11] and Gated Recurrent Unit (GRU) [9] are variants of recurrent neural network (RNN) designed to solve the issue of learning long-term dependencies. For capturing the style of an user, we need information from the past and next part of the writing/future context. Thus, for capturing the contexts from both the past and the future, we have considered bidirectional LSTM and GRU for the purpose of feature extraction.

Our developed system predicts the author of the written sample in two steps, i) feature extraction, and ii) author identification. Below, the complete system is discussed in detail.

3.1   Feature Extraction

The proposed system contains three sister networks for learning three different sets of features, namely capsule network extracted features, LSTM network extracted features, and GRU network extracted features. The three networks are called as sister networks, because the error is back propagated in all of them.

1. Capsule features: The capsule network learns temporal as well as spatial features from the convolutional layer. This hybrid combination of features helps in learning the feature maps of lower and higher level representations. At first, the tweets are fed to convolutional layers for learning lower-level text features.

$$t_{c1} = tweet(conv_1) \qquad (1)$$

$$t_{c11} = t_{c1}(conv_1) \qquad (2)$$

The convoluted vector is then fed to the capsule layer for learning the hybrid features representing the higher-level text features.

$$t_{Capsule} = t_{c11}(Capsule) \qquad (3)$$

The output vector $t_{Capsule}$ represents the capsule features obtained from the text sample.

2. LSTM features: Similar to the above network, the tweet samples are fed to an LSTM network. LSTM helps in learning the long-term dependency of text samples.

$$t_{c2} = tweet(conv_2) \qquad (4)$$

$$t_{c12} = t_{c2}(conv_2) \qquad (5)$$

The tweet samples are first fed to two convolutional networks for representing texts into vector forms. The convoluted vector is then passed to LSTM network for getting LSTM features.

$$t_{LSTM} = t_{c12}(LSTM) \qquad (6)$$

3. GRU features: GRU and LSTM are used for solving the long-term dependency problem of RNN. Thus, we have used GRU as well for learning the text features.

$$t_{c3} = tweet(conv_3) \qquad (7)$$

$$t_{c13} = t_{c3}(conv_3) \qquad (8)$$

$$t_{GRU} = t_{c13}(GRU) \qquad (9)$$

In each of the networks, tweets are fed to a separate convolutional layer via input layer for the convolution operation. The convoluted output vector is then passed through another convolution layer for learning higher-level features. The learnt features are then passed to capsule layer. Similarly, the convoluted features are also passed to the LSTM and GRU layers for learning three different categories of features. Now, these three extracted features are fused and passed to classification layer for final prediction.

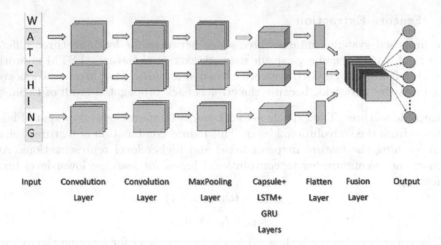

Fig. 1. The proposed fusion-based model

3.2 Classification

Finally all these three features are flattened together. The fused feature vector F_{tweet} is the final vector representing the input tweet. In this way, the final text vector represents all the three features extracted from the feature extraction layers.

$$F_{tweet} = [t_{capsule}; t_{LSTM}; t_{GRU}] \tag{10}$$

This tweet representation is then fed to the dense layer for the prediction of author labels.

$$Labels = F_{tweet}(Dense) \tag{11}$$

The proposed model is depicted in Fig. 1. The kernel size is different in each of the sister networks, so those are treated as three different networks. Thus three different types of features are learnt.

4   Dataset Used and Implementation Details

We have used the same dataset released by [21], which is used by other previous approaches too [4,14,23]. The dataset contains a total of 7000 authors, out of which we selected 50 authors at random, where each author has 1000 tweets. The total number of words present in dataset are $101, 180, 659$, with $14, 454.38$ average words per author. There are a maximum of 97 words present in the tweet whereas the minimum sample size is 2, and the average sample size is 14.40.

The implementation details for the developed model are shown in Table 2.

Table 2. Hyperparameters for neural network architecture

Layer	Number of Layers	Hyperparameters
Convolutional-capsule	1	filters:[500], kernel size:[3, 4]
Convolutional-LSTM	1	filters:[500], kernel size:[1, 3]
Convolutional-GRU	1	filters:[500], kernel size:[1, 3]
Capsules	1	no. of capsule:1, dim. of capsule:72
Dense	1	No of authors

Table 3. Performance of different input types on selected methods

Input type	Capsule-1	Capsule-2	LSTM	GRU	Max-Pooling
Unigram	84.03	84.35	67.40	68.98	62.52
Bigram	77.06	77.35	70.80	66.41	69.17
Trigram	76.98	75.94	62.53	61.93	69.04

5   Results and Discussion

Our developed system considers the fusion of capsule, LSTM and GRU features as input. Character n-grams have been used as the input in many of the previous AA tasks [4,14,23], thus we have also developed our model over character n-grams.

Firstly, the single models using capsule, LSTM, and GRU are developed over character n-grams. Accuracy values of 84.03%, 84.35%, 67.40%, and 68.98% are achieved using character uni-grams on capsule-1, capsule-2, LSTM, and GRU, respectively. Using character bi-grams and character tri-grams, the performances are lower than uni-grams. These results are shown in Table 3.

The existing approaches use max-pooling, so the results using max pooling are also shown. An accuracy value of 62.52% is achieved using character uni-grams over max-pooling settings. From the results, it is noticed that character uni-grams with capsule-2 is giving best results, followed by GRU and LSTM. Capsule-2 and capsule-1 are similar architectures, except the filter sizes. The filter size for capsule-2 is 4 and filter size for capsule-1 is 3. Thus, character uni-grams are considered for developing the fusion model. Bi-grams and tri-grams have also been explored in the fusion model, but the performance is not

Table 4. Performance of the fusion models on test data

Method	Uni-gram	Bi-gram
LSTM+GRU	73.17	66.13
LSTM+Capsule-2	85.29	77.34
GRU+Capsule-2	85.33	77.84
Capsule-2+LSTM+GRU	85.35	78.22

upto the mark, and results over character uni-grams are better. We have fused the capsule-2, LSTM, GRU in four different ways. They are, i) LSTM+GRU, ii) LSTM+Capsule-2, iii) GRU+Capsule-2, and iv) Capsule-2+LSTM+GRU. Accuracy values of 73.17%, 85.29%, 85.33%, and 85.35% are achieved using uni-grams on the above mentioned methods in the respective order. Using bi-grams, the respective accuracies are 66.13%, 77.34%, 77.84%, and 78.22%. The results are shown in Table 4. Thus, it can be concluded that the fusion of all the three features over character uni-grams performs better than others.

Table 5. Comparison of our proposed approach with other works, N: No. of Tweets per user

N	Capsule-4	Fusion	Char-word-CNN	CNN-II	CNN-I	LSTM-W	CNN-W
50	58.63	47.18	47	42	31.2	39.6	45
100	67.23	57.18	57	46.8	31.8	45.4	50
200	73.86	69.69	64	53.9	37	49.9	53.5
500	80.06	80.54	73	63.0	45.7	53.9	62.2
1000	84.64	85.35	79	68.1	51.5	65.06	66.2

We have randomly selected 50 authors for performing the experiments as done in [4, 21, 23]. In order to analyse the behavior of our developed systems on different sets of authors, we have created 10 different sets of 50 authors having 100 tweets for each. These sets are created after randomly choosing authors from the dataset. The experiments are performed using our developed approach, Char-Word-CNN, CNN-II, and CNN-I, CNN-W, and LSTM-W models on these sets. From the results, it is noticed that, the performances are different for different sets of authors. Number of human-like authors and number of bot-like authors are also shown in the Table 6. These results clearly demonstrate the effects of bot-like and human-like authors. In [23], it is reported that nearly 30% of authors behave like automated bots. Bots are automated programs which pose as humans with the aim at influencing users with commercial, political or ideological purposes [18]. The set of authors, having less number of bot-like authors is having less accuracy in comparison to the set having more number of bot-like authors. Still in all the generated sets of authors, the performances of our developed model are better than the current SOTA ([4]). This analysis also shows the effectiveness of capsule based architectures.

6   Comparison with Other Works

We have compared our results with the previous state-of-the-art methods. The approaches proposed in [12,14] are mainly based on stylometric feature learning. Since, our main idea is to use deep learning methodologies directly, without any feature engineering thus we have not compared our results with those approaches. The architecture proposed in [4] is the current SOTA for the Twitter dataset [21]. All the methods used for comparison are described below.

Table 6. Accuracy (in %) values for the different developed models on varying set of 50 authors {H: No. of Human-like Authors, B: No. of Bot-like Authors}

Set	H	B	Our Approach	Char-Word-CNN	CNN-II	CNN-I
A	42	8	47.19	46.4	16.4	16.4
B	40	10	51.80	31.6	25.6	20
C	39	11	53.07	34.8	38.8	30.4
D	37	13	55.06	35.6	21.2	17.6
E	37	13	47.46	32.8	27.6	18.4
F	37	13	58.67	42	27.2	25.6
G	37	13	50.54	24	25	24.2
H	36	14	55.60	39.2	36	29.6
I	32	18	51.80	30	30.4	20.4
J	30	20	61.03	29.6	46.8	31.8

1. Char-Word-CNN: Character bi-grams and word embeddings generated from fastText ([6]) are used as inputs in a multi-channel CNN architecture. This multi-channel learns different stylometric features of an author [4]. This architecture is the current state-of-the-art system for the authorship attribution on the used dataset.
2. CNN-II: Character bi-grams are fed to a CNN architecture, for the classification task [23]. This architecture is the base model and is the first work on short texts using character n-grams over CNN architecture.
3. CNN-I: In this system, the character uni-grams are fed to a CNN architecture [23].

In Table 5, the performance of our fusion model is compared with the performances of other methods as listed above. The accuracy values achieved by our approach are better than the current state-of-the-art (Char-Word-CNN). Out of 7000 authors, only random 50 authors are chosen, and the results are reported in [4,23]. Thus, it is not fair to compare the reported accuracy values with those attained by our developed system. As the source code is also not available, thus, we have implemented their approaches, i.e., CNN-II, and Char-Word-CNN and executed them on our randomly selected authors. From the results, it is clear that, the performances of our model is better than the SOTA system.

7   Analysis

A rigorous analysis of heat-maps generated through our developed system is performed. Heat-maps are helpful in identifying the words, which are important for prediction. As discussed in the earlier section, there are two types of authors (bot-like, and human-like). Thus, we have drawn heat-maps for one bot-like user, and one human-like user.

The time in Hollywood, CA, is 10:19:33 pm\n'

(a) Heat-map of CNN-II for bot-like author-1: The model assigns more weight to the word "Hollywood" rather than focusing on writing style of author.

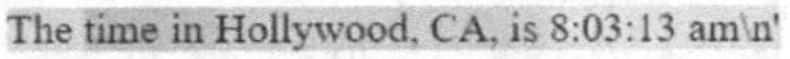

The time in Hollywood, CA, is 8:03:13 am\n'

(b) Heat-map of CNN-I for bot-like author-1: As all the tweets are almost similar for this author, the model predicts on the basis of different words in the tweet. Rather than assigning focus on repeating words, this model focuses on the number at the last. This model assigns weights to all words that are common in the tweets.

The time in Hollywood, CA, is 10:19:33 pm\n'

(c) Heat-map of fusion-model for bot-like author-1: Our model assigns more weight to the time pattern than the repeating words. This shows that, the developed system learns the writing style of the user more efficiently.

Fig. 2. Heat-maps generated from different models for bot-like author

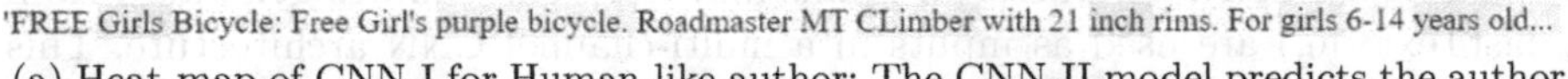

'FREE Girls Bicycle: Free Girl's purple bicycle. Roadmaster MT CLimber with 21 inch rims. For girls 6-14 years old...

(a) Heat-map of CNN-I for Human-like author: The CNN-II model predicts the author on the basis of the words he/she uses frequently; here the word "Girl" is assigned more weight as it is used by the author frequently but as mentioned above the author has repeated the word "girl" three number of times in his tweets but all the words are not given equal weight-age by the model.

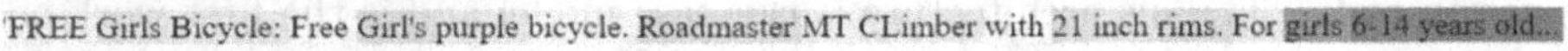

'FREE Girls Bicycle: Free Girl's purple bicycle. Roadmaster MT CLimber with 21 inch rims. For girls 6-14 years old...

(b) Heat-map of CNN-I for Human-like author: Similar to the CNN-I, this model also assigns equal weight-ages to all the occurrences of the same word while predicting the author and this also means the model learns the style of writing of the author.

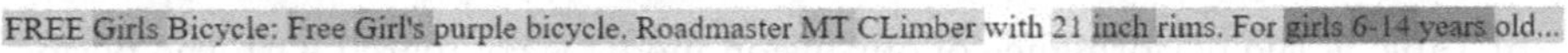

FREE Girls Bicycle: Free Girl's purple bicycle. Roadmaster MT CLimber with 21 inch rims. For girls 6-14 years old...

(c) Heat-map of fusion-model for Human-like author: The fusion-model correctly finds the unique characteristic of the author in repeating similar words in the tweet; in this tweet the "Girl" word is used thrice and the model has assigned equal weight-age to all the occurrences of the word.

Fig. 3. Heat-maps for Human-like author

In Figs. 2 and 3 the heat-maps generated for a sample of bot-like author, and human-like author are shown, respectively. The heat-maps are drawn for CNN-II, CNN-I, and fusion-based model. It can be seen that the previous models CNN-II, and CNN-I assign more priority to the word 'Hollywood'. On the other hand, our proposed fusion-based model assigns more priority to the time ('10:19:33'). Hollywood and time both are present in most of the samples. But the numerical value for the time is different in different samples, only the pattern of writing time is similar. Similarly, in Fig. 3 the heat-map drawn for a sample of human-like user is shown. From the above examples, it can be concluded that our developed system is capable of capturing the writing pattern of the user more effectively.

8   Conclusion and Future Work

The application area of the authorship attribution task includes analyzing long fraud document, terrorist conspiracy texts, short letters, blog posts, emails, SMS, Twitter streams or Facebook status updates to check the authenticity and identify fraudulence. In this work, a fusion-based capsule network is developed for solving the authorship attribution task. Character uni-grams are fed to a convolution layer for learning text representations. The convoluted vector is fed to different components such as LSTM, GRU, and capsule. Different features extracted from these components are fused together and then used for the final classification. Our work illustrates the effect of fusing these three sets of features, by achieving gain in performance. An accuracy of 85.35% is achieved using the fused model, and it outperforms the previous developed systems. With the help of heat-maps, we have also shown the relevant fragments of text sample, for solving the AA task.

In future, we would like to use different style based features with the neural network settings. Different discourse features can also be added to the developed system, which can lead to performance improvement.

References

1. Aborisade, O., Anwar, M.: Classification for authorship of tweets by comparing logistic regression and naive bayes classifiers. In: 2018 IEEE International Conference on Information Reuse and Integration (IRI), pp. 269–276. IEEE (2018)
2. Akhtar, S., Ghosal, D., Ekbal, A., Bhattacharyya, P., Kurohashi, S.: All-in-one: emotion, sentiment and intensity prediction using a multi-task ensemble framework. In: IEEE Transactions on Affective Computing (2019)
3. Al Marouf, A., Hasan, M.K., Mahmud, H.: Comparative analysis of feature selection algorithms for computational personality prediction from social media. IEEE Trans. Comput. Soc. Syst. **7**(3), 587–599 (2020)
4. Aykent, S., Dozier, G.: Author identification of micro-messages via multi-channel convolutional neural networks. In: 2020 IEEE International Conference on Systems, Man, and Cybernetics (SMC), pp. 675–681. IEEE (2020)
5. Bagnall, D.: Author identification using multi-headed recurrent neural networks. arXiv preprint arXiv:1506.04891 (2015)
6. Bojanowski, P., Grave, E., Joulin, A., Mikolov, T.: Enriching word vectors with subword information. Trans. Assoc. Comput. Linguist. **5**, 135–146 (2017)
7. Brocardo, M.L., Traore, I., Saad, S., Woungang, I.: Authorship verification for short messages using stylometry. In: 2013 International Conference on Computer, Information and Telecommunication Systems (CITS), pp. 1–6. IEEE (2013)
8. Castillo, E., Vilarino, D., Cervantes, O., Pinto, D.: Author attribution using a graph based representation. In: 2015 International Conference on Electronics, Communications and Computers (CONIELECOMP), pp. 135–142. IEEE (2015)

9. Chung, J., Gulcehre, C., Cho, K., Bengio, Y.: Empirical evaluation of gated recurrent neural networks on sequence modeling. arXiv preprint arXiv:1412.3555 (2014)

10. Fourkioti, O., Symeonidis, S., Arampatzis, A.: Language models and fusion for authorship attribution. Inf. Process. Manag. **56**(6), 102061 (2019)

11. Hochreiter, S., Schmidhuber, J.: Long short-term memory. Neural Comput. **9**(8), 1735–1780 (1997)

12. Huang, W., Su, R., Iwaihara, M.: Contribution of improved character embedding and latent posting styles to authorship attribution of short texts. In: Wang, X., Zhang, R., Lee, Y.-K., Sun, L., Moon, Y.-S. (eds.) APWeb-WAIM 2020. LNCS, vol. 12318, pp. 261–269. Springer, Cham (2020). https://doi.org/10.1007/978-3-030-60290-1_20

13. Joshi, M., Singh, P., Zincir-Heywood, N.: Compromised tweet detection using siamese networks and fasttext representations. In: 2019 15th International Conference on Network and Service Management (CNSM), pp. 1–5. IEEE (2019)

14. Joshi, M., Zincir-Heywood, N.: Classification of micro-texts using subword embeddings. In: RANLP (2019)

15. Juola, P.: Authorship studies and the dark side of social media analytics. J. UCS **26**(1), 156–170 (2020)

16. Kim, J., Jang, S., Park, E., Choi, S.: Text classification using capsules. Neurocomputing **376**, 214–221 (2020)

17. Madigan, D., et al.: Author Identification on the Large Scale p. 20 (2005)

18. Rangel, F., Rosso, P.: Overview of the 7th author profiling task at pan 2019: bots and gender profiling in twitter. In: Proceedings of the CEUR Workshop, Lugano, Switzerland, pp. 1–36 (2019)

19. Rhodes, D.: Author Attribution with CNN's p. 8 (2015)

20. Ruder, S., Ghaffari, P., Breslin, J.G.: Character-level and Multi-channel Convolutional Neural Networks for Large-scale Authorship Attribution. arXiv:1609.06686 [cs] (September 2016). http://arxiv.org/abs/1609.06686, arXiv: 1609.06686

21. Schwartz, R., Tsur, O., Rappoport, A., Koppel, M.: Authorship attribution of micro-messages. In: Proceedings of the 2013 Conference on Empirical Methods in Natural Language Processing, pp. 1880–1891 (2013)

22. Segarra, S., Eisen, M., Ribeiro, A.: Authorship attribution through function word adjacency networks. IEEE Trans. Signal Process. **63**(20), 5464–5478 (2015)

23. Shrestha, P., Sierra, S., González, F.A., Montes, M., Rosso, P., Solorio, T.: Convolutional neural networks for authorship attribution of short texts. In: Proceedings of the 15th Conference of the European Chapter of the Association for Computational Linguistics: vol. 2, Short Papers, pp. 669–674 (2017)

24. Stamatatos, E.: A survey of modern authorship attribution methods. J. Am. Soc. Inf. Sci. Technol. **60**(3), 538–556 (2009)

On the Explainability of Automatic Predictions of Mental Disorders from Social Media Data

Ana Sabina Uban$^{(\boxtimes)}$, Berta Chulvi, and Paolo Rosso

Pattern Recognition and Human Language Technology (PRHLT),
Universitat Politècnica de València, Valencia, Spain
ana.uban+acad@gmail.com, berta.chulvi@upv.es, prosso@dsic.upv.es

Abstract. Mental disorders are an important public health issue, and computational methods have the potential to aid with detection of risky behaviors online, through extracting information from social media in order to retrieve users at risk of developing mental disorders. At the same time, state-of-the-art machine learning models are based on neural networks, which are notoriously difficult to interpret. Exploring the explainability of neural network models for mental disorder detection can make their decisions more reliable and easier to trust, and can help identify specific patterns in the data which are indicative of mental disorders. We aim to provide interpretations for the manifestations of mental disorder symptoms in language, as well as explain the decisions of deep learning models from multiple perspectives, going beyond classical techniques such as attention analysis, and including activation patterns in hidden layers, and error analysis focused on particular features such as the emotions and topics found in texts, from a technical as well as psycholinguistic perspective, for different social media datasets (sourced from Reddit and Twitter), annotated for four mental disorders: depression, anorexia, PTSD and self-harm tendencies.

1 Introduction and Previous Work

Mental disorders are a serious public health issue, and many mental disorders are underdiagnosed and undertreated. The early detection of signs of mental disorders is important, since, undetected, mental disorders can develop into more serious consequences, constituting a major predictive factor of suicide [33]. Computational methods have a great potential to assist with early detection of mental disorders of social media users, based on their online activity.

There is an extensive body of research related to automatic mental disorder detection from social media data. The majority of research has focused on the study of depression [1,6,7,35], but other mental illnesses have also been studied, including generalized anxiety disorder [27], schizophrenia [16], post-traumatic stress disorder [3,4], risks of suicide [19], anorexia [13] and self-harm [13,34]. The majority of studies provide either quantitative analyses, or predictors built

© Springer Nature Switzerland AG 2021
E. Métais et al. (Eds.): NLDB 2021, LNCS 12801, pp. 301–314, 2021.
https://doi.org/10.1007/978-3-030-80599-9_27

using simple machine learning models, such as SVMs and logistic regression [5, 6], with few studies using more complex deep learning methods [23, 26, 29–31]. As features, most previous works use traditional bag of words n-grams [3], as well as some domain-specific representations, such as lexicons [5, 28], or Latent Semantic Analysis [22, 28]. There are few studies which compare multiple different aspects of the language, such as topics and emotions [26, 27, 30].

Quantitative analyses in existing research on mental disorders have found that people suffering from depression manifest changes in their language, such as greater negative emotion and high self-attentional focus [5, 29], or an increased prevalence of certain topics, such as medications or bodily issues such as lack of sleep, expressing hopelesness or sadness [24, 28]. Nevertheless, correlation studies are limited in discovering more complex connections between features of the text and mental health disorder risks. Moreover, research on mental health disorders from a computational perspective has been generally disconnected from mental health research in psychology, with few computational studies providing interpretations from a psychological perspective [15].

In practice, models based on neural networks are vastly successful for most NLP applications. Nevertheless, neural networks are notoriously difficult to interpret. Recently, there is increasing interest in the field of explainability methods in machine learning including in NLP [8], which aim for providing interpretations of the decisions of neural networks. If any system for mental disorder detection is to be developed into a tool to assist social media users, it is essential that its decision-making process is understandable in the name of transparency. Especially in the medical domain, using black-box systems can be dangerous for patients and is not a realistic solution [10, 36]. Moreover, recently, the need of explanatory systems is required by regulations like the General Data Protection Regulation (GDPR) adopted by the European Union. Additionally, the behavior of powerful classifiers modelling complex patterns in the data has the potential to help uncover manifestations of the disease that are potentially difficult to observe with the naked eye, and thus assist clinicians in the diagnosis process.

In the field of mental disorder detection, there are not many studies attempting to explain the behavior of models. We note one such example [2], where the authors analyze attention weights of a neural network trained for automatic anorexia detection. Nevertheless, recent studies have shown the limitations of using attention analysis for interpretability [25, 32]. In our study, we aim to go beyond explainability techniques based on the analysis of attention weights.

We intend to explore the explainability of mental disorder prediction models from different perspectives. We center our analysis around neural network models trained to identify signs of mental disorders from social media data for the four different mental health disorders, using various features to extract information reflecting different levels of the language, and through performing various complementary analyses of the behavior of the model and features used. In this way, we aim to discover the most relevant features that indicate mental disorder symptoms based on text data, analyze the way they manifest in text, as well as provide interpretations of our quantitative findings from a social psychology perspective.

2   Classification Experiments

2.1   Datasets

In order to obtain a wider picture on how mental disorders manifest in social media, we include in our analysis datasets from different sources, containing social media data labelled for several disorders and manifestations thereof: depression, anorexia, self-harm, and PTSD, and gathered from two different social media platforms: Reddit and Twitter.

ERisk Reddit Datasets on Depression, Anorexia and Self-harm. The eRisk CLEF lab[1] is focused on the early prediction of mental disorder risk from social media data, focused on disorders such as depression [12], anorexia and self-harm tendencies [13,14]. Data is collected from Reddit posts and comments selected from specific relevant sub-reddits. Users suffering from a mental disorder are annotated by automatically detecting self-stated diagnoses. Healthy users are selected from participants in the same sub-reddits (having similar interests), thus making sure the gap between healthy and diagnosed users is not trivially detectable. A long history of posts are collected for the users included in the dataset, up to years prior to the diagnosis.

CLPsych Twitter Dataset on Depression and PTSD. CLPsych (Computational Linguistics and Clinical Psychology) is a workshop and shared task organized each year around a different topic concerning computational approaches for mental health. In 2015 [4], the shared task challenged participants to detect Twitter users suffering from depression and PTSD. Labelling of the data was done semi-automatically, through an initial selection based on self-stated diagnoses, followed by human curation. For each user, their most recent public tweets were included in the dataset.

Twitter Dataset on Depression. To complement the CLPsych dataset, we include a second Twitter dataset labelled for depression. This dataset was collected and introduced in [26], following a similar methodology, based on self-stated diagnoses. Tweets published within a month of the diagnosis statement were included for each positive user. This short time frame is an exception compared to the other datasets considered. Non-depressed users were selected among Twitter users never having posted any tweet containing the character string "depress". In all datasets, the posts containing the mention of a diagnosis were excluded. Table 1 contains statistics describing all datasets considered.

2.2   Experimental Setup

We center our analyses on training deep learning models to predict mental disorders in social media data, which we will try to analyze in the following sections in order to explain their behavior.

---

[1] https://early.irlab.org/.

Table 1. Datasets statistics.

Dataset	Users	Positive%	Posts	Words
ERisk depression	1304	16.4%	811,586	25 M
ERisk anorexia	1287	10.4%	823,754	23 M
ERisk self-harm	763	19%	274,534	6 M
CLPsych depression	822	64.1%	1,919,353	26 M
CLPsych PTSD	1078	72.6%	2,541,214	19 M
Twitter depression [26]	519	50.2%	52,080	500 K

First, we train and test our model for classifying between healthy users and those suffering from a disorder, for each of the datasets and disorders independently. Secondly, we perform similar experiments for cross-disorder classification: we try to automatically distinguish between users suffering from different disorders, in an attempt to understand not only on linguistic patterns used by people diagnosed with an disorder, but also compare how these patterns differ (or coincide) across different disorders.

For the task of identifying users on social media suffering from a mental disorder, we model the problem as a binary classification task, training a deep learning model separately for each of the disorders and datasets considered. In the case of cross-disorder classification, we consider separately the two data sources: Reddit and Twitter, and perform experiments to distinguish between disorders present in each of the datasets: depression vs PTSD for the CLPsych (Twitter) datasets, and depression vs anorexia vs self-harm for the eRisk (Reddit) datasets. In this setup, we ignore the healthy users, and only focus on identifying the particular disorder that users are suffering from. We consider these as multi-label classification tasks (using a sigmoid activation for the final layer of our deep learning model for both tasks, instead of softmax), taking into account the fact that some users might be suffering from multiple disorders, given the known incidence of co-morbidity of mental disorders [11].

2.3   Model and Features

We choose a hierarchical attention network (HAN) as our model: a deep neural network with a hierarchical structure, including multiple features encoded with LSTM layers and two levels of attention. The HAN is made up of two components: a *post-level encoder*, which produces a representation of a post, and a *user-level encoder*, which generates a representation of a user's post history. The post-level encoder and the user-level encoder are modelled as LSTMs. The word sequences encoded using pre-trained GloVe embeddings and passed to the LSTM are then concatenated with the other features to form the hierarchical post encoding. The obtained representation is passed to the user-encoder LSTM, which is connected to the output layer. Posts are truncated or padded to sequences of 256 words. The post-level encoder LSTM has 128 units, and the user-level LSTM has 32 units. The dense layers for encoding the lexicon features and the stopwords feature have 20 units each. We use the train/test split provided by the shared task organizers, done at the user level, making sure users occurring in one subset don't occur in the other. Since individual posts are too short to be accurately classified, we construct our datapoints by concatenating groups of 50 posts, sorted chronologically. We publish all the code used for experiments reported in this paper in a public repository, which includes more details on the network's architecture[2].

We represent social media texts using features that capture different levels of the language (semantic, stylistic, emotions etc.) and train the model to predict mental disorder risk for each user.

Content Features. We include a general representation of text content by transforming each text into word sequences.

Style Features. The usage pattern of function words is known to be reflective of an author's style, at an unconscious level [18]. As stylistic features, we extract from each text a numerical vector representing function words frequencies as bag-of-words, which are passed through an additional dense layer of 20 units. We complement function word distribution features with other syntactical features extracted from the LIWC lexicon, as described below.

LIWC Features. The LIWC lexicon [20] has been widely used in computational linguistics as well as some clinical studies for analysing how suffering from mental disorders manifests in an author's writings. LIWC is a lexicon mapping words of the English vocabulary to 64 lexico-syntactic features of different kinds, with high quality associations curated by human experts, capturing different levels of language: including style (through syntactic categories), emotions (through affect categories) and topics (such as money, health or religion).

Emotions and Sentiment. We dedicate a few features to representing emotional content in our texts, since the emotional state of a user is known to be highly correlated with her mental health. Aside from the sentiment and emotion categories in the LIWC lexicon, we include a second lexicon: the NRC emotion lexicon [17], which is dedicated exclusively to emotion representation, with categories corresponding to a wider and a more fine-grained selection of emotions, containing the 8 Plutchik's emotions [21], as well as *positive/negative* sentiment categories: *anger, anticipation, disgust, fear, joy, sadness, surprise, trust*. We represent LIWC and NRC features by computing for each category the proportion of words in the input text which are associated with that category.

Our choice of model is motivated both by its hierarchical attention mechanism, and by the multiple features used, which allow for interpretability from different perspectives.

---

[2] https://github.com/ananana/mental-disorders

2.4   Classification Results

Results for individual disorder detection are shown in Table 2. As performance metrics we compute the F1-score of the positive class and the area under the ROC curve (AUC), which is more robust in the case of data imbalance. We show results for our model, in comparison with a baseline logistic regression model with bag-of-word features.

Table 2. F1 and AUC scores for all datasets and models trained on individual tasks.

	SELF-HARM		ANOREXIA		DEPRESSION						PTSD	
	eRisk		eRisk		eRisk		Shen et al.		CLPsych		CLPsych	
Model	F1	AUC	F1	AUC	F1	AUC	F1	AUC	F1	AUC	F1	AUC
HAN	.51	**.83**	.46	**.91**	.44	**.86**	.77	.81	.53	**.73**	.57	**.70**
LogReg	.45	.75	.49	.90	.36	.76	.71	.81	.55	.72	.49	.69

In the case of cross-disorder classification, we obtain an F1-score of 0.72 for the depression class in depression vs PTSD classification, and an AUC score of 0.75. For the eRisk datasets, we obtain an accuracy of 0.44 for discriminating between depression, anorexia and self-harm, and a macro-F1 of 0.44. The results suggest the task of cross-disorder classification is significantly more difficult than distinguishing healthy users from ones suffering from a disorder, especially in the case of depression/anorexia/self-harm classification.

3   Explaining Predictions

In this section we present different analyses meant to uncover insights into how the model arrives at its predictions, first looking at the abstract internal representations of the data in the layers of the network, and secondly providing several feature-focused analyses of misclassifications, using the lexicon-based features (emotions and LIWC categories) in order to identify particular interpretable patterns among users which the model cannot classify correctly.

3.1   User Embeddings

We start by analyzing the internal representations of the network. We can regard the final layer of the trained network as the most compressed representation of the input examples, which is, in terms of our trained model, the optimal representation for distinguishing between healthy users and those suffering from a disorder. Thus, the final layer (the output of the 32-dimensional user-level LSTM) can be interpreted as a 32-dimensional embedding for the input points, corresponding to the users to be classified.

We analyze the output of the *user embedding* layer by reducing it to 2 dimensions using principal component analysis (PCA) and visualizing it in 2D space

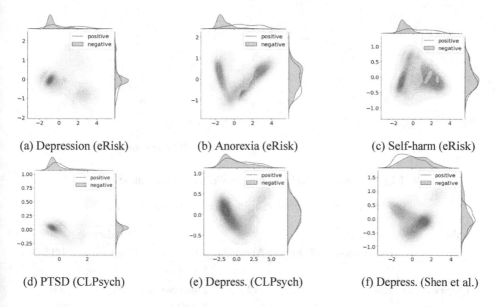

(a) Depression (eRisk)   (b) Anorexia (eRisk)   (c) Self-harm (eRisk)

(d) PTSD (CLPsych)   (e) Depress. (CLPsych)   (f) Depress. (Shen et al.)

Fig. 1. User embeddings for classification of users with a disorder vs healthy ones.

with a kernel density estimate (KDE) plot to show the distribution of scores across the 2 dimensions, separately for each dataset and disorder (Fig. 1). We make sure to train the PCA model on a balanced set of positive and negative users, then we extract 2D representations for all users in the test set. By looking at these representations, we can gain insight into the separability of the classes, from the perspective of the trained model, and better understand where it encounters difficulties in separating between the datapoints belonging to different classes. Separately, we perform the same experiments for cross-disorder classification, as shown in Fig. 2.

We notice that, in accordance with the classification performance reported previously, the highest separation in user embedding space seems to be achieved for anorexia and for depression on the Twitter (Shen et al.) dataset, while depressed users in the other datasets (eRisk and CLPsych) show higher overlap with healthy ones, as do users suffering from self-harm. Moreover, we notice an interesting pattern of multiple clusters of positive users, while healthy users' representations seem to be more compact.

In the case of cross-disorder classification, user embeddings seem highly overlapping, especially in the case of the 3-way classification of disorders in the eRisk datasets, suggesting that the model has difficulties in producing separate representations for these disorders, leading to a high misclassification rate.

In the following subsection we take a deeper dive into misclassified examples for each of the analyzed disorders and datasets. Focusing on misclassifications could also help to further explain the patterns noticed through user embedding analysis - particularly the clusters of false positives in the user embedding spaces for several disorders.

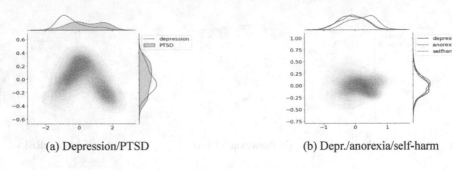

(a) Depression/PTSD   (b) Depr./anorexia/self-harm

Fig. 2. User embeddings for cross-disorder classification.

Table 3. Cross-disorder classification confusion matrices.

Label	Prediction				Prediction	
	Depression	Self-harm	Anorexia		Depression	PTSD
				Label		
Depression	139	2	113	**Depression**	126	24
Self-harm	60	67	144	**PTSD**	65	95
Anorexia	201	16	218			

3.2 Error Analysis

We provide some insight into misclassified examples for cross-disorder classification through confusion matrices, as seen in Table 3. We notice a high rate of confusion for the 3-way classification between depression, anorexia, and self-harm, and particularly that users suffering from other disorders tend to be classified as depressed. The difficulty to distinguish between these disorders might be due to their common linguistic patterns, but also to possible cases of co-morbidities.

In the case of models for detecting individual disorders, errors of classification can have serious negative impacts on the users' well-being, if such as system would be deployed into a tool for assisting social media users. False negative predictions in particular can lead to missing cases of people with high risk of suffering from mental health disorders, and, left undetected, the disorders might further develop. We attempt to understand what causes misclassifications by comparing correctly versus incorrectly classified examples in terms of different features, including words, NRC emotions and LIWC categories. We thus compare the different types of misclassified and correctly classified examples, across the four groups: true positives (TP), false positives (FP), true negatives (TN) and false negatives (FN).

In Table 4 we show the vocabulary words which are most distinctive for misclassifications for each disorder, separately for FP and FN cases. We select these words by applying the chi^2 test to extract the most discriminative features between FN and FP cases on one hand, and FP and TP cases on the other hand, and report the words with the highest scores. In some cases, these keywords can

Table 4. Top words (χ^2 test) that discriminate between incorrect and correct predictions.

Experiment	False negatives	False positives
Depression (eRisk)	Clemson, game, lemieux, team, uio play, song, pka, you, season	I, my, her, she, me, is was, the, are, trump, of
Depression (CLPsych)	Earning, mpoints, video, rewards, patientchat, thank, besties, you, gameinsight, ipadgames	Dundee, I, my, me, lol, the, vitamin, win, of, fuck, mobile, syria, love
Depression (Shen et al.)	I, rt, to, you, the, and, my, is, of, me	Rt, prayer, bestmusicvideo, iheartawards, zain, pillowtalk, location, hiphopnews, ghetsis, via
Self-harm	I, the, que, is, me, de, a, despacito, feel, myself	The, I, a, to, and, it, you, of, is, that, in, for
Anorexia	I, the, my, her, she, r, me, eating, I'm, u, senate	I, the, am, you, of, their, transfer, college, him, from, in, girls
PTSD	Mpoints, earning, reward, thank, following, you, plz, ff, ptsd, cptsd	I, besties, gameinsights, ipadgames, thatsheartgiveaway, vietnam, coins, collected

shed some light on what characterizes the sub-clusters of FN users identified with the user embedding representations. For depression, the FN group (both for eRisk and CLPsych) appear to be distinguished by discussing topics related to games. In the case of anorexia, we notice words related to college and social life in the FP group. Another interesting finding is the occurrence of "Vietnam" for FP in PTSD: the model learns to excessively associate PTSD sufferers with the topic of Vietnam, possibly showing a topic bias in the dataset.

In order to understand the effect of lexicon features on the model's prediction, we measure for each of the lexicon categories their comparative prevalence in misclassified and correctly classified examples, separately for healthy users and users suffering from a disorder. We identify four categories of features, based on their prevalence FP, FN, TP and TN examples comparatively:

Feature Bias Type 1. (FN<TP; FP>TN): features which occur to a lower degree in misclassified positive examples than in correctly classified positive examples; while for negative examples they occur more in incorrectly classified ones than in correctly classified ones. The model likely relies too much on the connection between their high prevalence and high risk scores.

Feature Bias Type 2. (FN<TP; FP<TN): generally under-represented features in misclassified examples - if they are not well represented, the model tends to make mistakes.

Feature Bias Type 3. (FN>TP; FP>TN): features which are generally over-represented in misclassified examples - when they are highly prevalent, the model is less accurate.

Feature Bias Type 4. (FN>TP; FP<TN): features which are over-represented in FN cases and under-represented in FP cases. The model likely relies too much on their low prevalence to emit high risk scores.

Table 5. LIWC features with highest differences for misclassified groups ($p < 10^{-6}$).

Experiment	Feat. bias 1	Feat. bias 2	Feat. bias 3	Feat. bias 4
Depression (eRisk)	ppron, quant, auxverb, verb, present, you, pronoun, excl, I, conj, adverb, future, cogmech, funct	-	ipron	-
Depression (CLPsych)	hear, conj, ipron, present, article,auxverb, certain, negative, verb,	-	-	-
Depression (Shen et al.,)	-	-	-	ppron, adverb, bio, funct, verb ,past,funct, article, health, present
Self-harm	conj, excl, pronoun, future, cogmech, I, funct, ppron	-	-	Incl
Anorexia	ppron, I, adverb, cogmech, auxverb, verb, pronoun, future, quant, ,excl, present, conj, funct	anxiety, health, ingest, bio	-	-
PTSD	money, number, article,work, achieve, preps	-	-	cogmech, fear, assent, pronoun, bio, I, leisure, swear, affect, feel

For each dataset, we identify misclassifications grouped into the two categories (FP and FN), and find those features for which there is a statistically significant difference of the average value between the misclassified group and the correctly classified group. We do this separately for emotions (see Fig. 3) and for LIWC features (shown in Table 5, categories with p-values below 10^{-6}). We provide more interpretations for the patterns of misclassifications in relation to emotions and psycho-linguistic categories in the following sub-section, from a deeper psychological perspective.

4   Cognitive Styles and Error Analysis: Some Interpretations

Cognitive style is a concept used in cognitive psychology to describe the way individuals think, perceive, and remember information [9]. Research in psychology suggests that some cognitive styles are more prevalent in some patients suffering from depression and anorexia. In our error analysis we find that some errors have a relation with the under-representation of these cognitive styles. Some of the features that are relevant to explain the misclassifications of the model (see Table 5) are related to cognitive styles.

For instance, for depression we find that in the case of FN, features as *cogmech*, that refers to cognitive processes (causation, discrepancy, tentative, certainty, etc.), occurs to a lower degree in FP examples. We can conclude that the model is confused when the depressed users do not express themselves in the typical pattern that refers some way of reasoning about causes, consequences, etc.

For anorexia, the under-representation of some features like *anxiety, health* or *ingest* leads to misclassifications. We can conclude that the model is relying

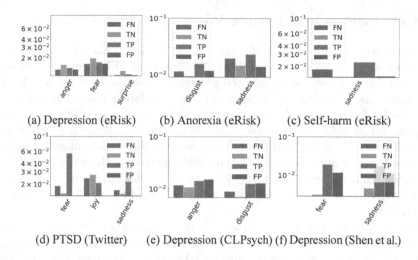

(a) Depression (eRisk)   (b) Anorexia (eRisk)   (c) Self-harm (eRisk)

(d) PTSD (Twitter)   (e) Depression (CLPsych) (f) Depression (Shen et al.)

Fig. 3. Mean values for emotions that are significantly different for misclassif. (p<0.05).

on the use of these words to detect anorexia, but there are positive cases where we do not find the typical semantics of this disorder, and these will be more difficult to detect also for clinicians.

There is an interesting result related to the use of the *future* feature of LIWC. We found that in depression (eRisk corpus), anorexia and in self-harm, if this feature that speaks about *future* occurs to a lower degree, the model tends to make more mistakes in the classification of positive examples. We can infer that the model is able to detect the mental health disorders when people speak about what life is preparing for them, but has more difficulties when users that suffer from these mental health disorders don't speak about plans and focus more on the moment.

Considering the analysis of emotions (see Fig. 3) we found also that the unclear expression of some emotions leads the model to make mistakes and that these emotions are just the ones that are relevant for each mental health disorder. For instance, we see that the model makes more mistakes when people that suffer from depression do not express anger and fear. In the case of anorexia, the FN examples are more frequent when people do not speak about disgust. This suggests that anorexia is a much more complex disorder that the one that express the development of strange eating habits. We also observe that in terms of emotions, the people with self-harm tendencies do not express their sadness emotion are more difficult to detect for the model and maybe also for clinicians. It suggests the need to explore other narratives that must being used for these people with self-harm tendencies that show a low expression of negative emotions.

5   Conclusions and Future Work

Explainability of machine learning models, especially in the domain of mental health, where automatic tools can have significant social impact, is an essential topic. In this study, we have presented several analyses for interpreting the decisions of models trained to profile users at risk of developing mental disorders from social media, going beyond more common techniques such as attention weight analysis, and including hidden layer analysis and error analysis at different levels of the language for better understanding how mental disorders manifest in social media data. In addition, we interpret our findings though the lens of psychology, identifying connections between specific topics (e.g. health, biology) or emotions (e.g. anger, fear) and certain disorders, which can lead the model to over-rely on these features.

Although we approach a novel topic in the computational research on mental disorders and present new findings, the methods used in this study could be developed into deeper and more sophisticated analyses. As future work, we intend to continue the analysis of emotion markers through applying time series analysis methods, in order to automatically detect trends and seasonal patterns in the evolution of the usage of emotion-related vocabulary for users suffering of mental health disorders. Moreover, the results of the user embedding analysis encourage us to further study the distinct patterns of symptoms for certain disorders.

Acknowledgement. The authors thank the EU-FEDER Comunitat Valenciana 2014–2020 grant IDIFEDER/2018/025. The work of Paolo Rosso was in the framework of the research project PROMETEO/2019/121 (DeepPattern) by the Generalitat Valenciana.

References

1. Abd Yusof, N.F., Lin, C., Guerin, F.: Analysing the causes of depressed mood from depression vulnerable individuals. In: DDDSM-2017, pp. 9–17 (2017)
2. Amini, H., Kosseim, L.: Towards explainability in using deep learning for the detection of anorexia in social media. NLDB 12089, 225
3. Coppersmith, G., Dredze, M., Harman, C.: Quantifying mental health signals in twitter. CLPsych **2014**, 51–60 (2014)
4. Coppersmith, G., Dredze, M., Harman, C., Hollingshead, K., Mitchell, M.: CLPsych 2015 shared task: depression and PTSD on twitter. CLPsych **2015**, 31–39 (2015)
5. De Choudhury, M., Counts, S., Horvitz, E.J., Hoff, A.: Characterizing and predicting postpartum depression from shared facebook data. In: ACM on Computer Supported Cooperative Work and Social Computing, pp. 626–638 (2014)
6. De Choudhury, M., Gamon, M., Counts, S., Horvitz, E.: Predicting depression via social media. In: AAAI (2013)
7. Eichstaedt, J.C., et al.: Facebook language predicts depression in medical records. Proc. of the Natl. Acad. Sci. **115**(44), 11203–11208 (2018)
8. Gilpin, L.H., Bau, D., Yuan, B.Z., Bajwa, A., Specter, M., Kagal, L.: Explaining explanations: An overview of interpretability of machine learning. In: IEEE DSAA, pp. 80–89. IEEE (2018)

9. Grigorenko, E.L., Sternberg, R.J.: Thinking Styles. In: Saklofske, D.H., Zeidner, M. (eds.) International Handbook of Personality and Intelligence. Perspectives on Individual Differences, Springer, Boston (1995). https://doi.org/10.1007/978-1-4757-5571-8_11

10. Holzinger, A., Biemann, C., Pattichis, C.S., Kell, D.B.: What do we need to build explainable AI systems for the medical domain? arXiv preprint arXiv:1712.09923 (2017)

11. Kaufman, J., Charney, D.: Comorbidity of mood and anxiety disorders. Depress. Anxiety 12(S1), 69–76 (2000)

12. Losada, D.E., Crestani, F., Parapar, J.: Overview of eRisk: early risk prediction on the internet. In: Bellot, P., et al. (eds.) CLEF 2018. LNCS, vol. 11018, pp. 343–361. Springer, Cham (2018). https://doi.org/10.1007/978-3-319-98932-7_30

13. Losada, D.E., Crestani, F., Parapar, J.: Overview of eRisk 2019 early risk prediction on the internet. In: Crestani, F. (ed.) CLEF 2019. LNCS, vol. 11696, pp. 340–357. Springer, Cham (2019). https://doi.org/10.1007/978-3-030-28577-7_27

14. Losada, D.E., Crestani, F., Parapar, J.: eRisk 2020: self-harm and depression challenges. In: Jose, J.M. (ed.) ECIR 2020. LNCS, vol. 12036, pp. 557–563. Springer, Cham (2020). https://doi.org/10.1007/978-3-030-45442-5_72

15. Mehltretter, J., et al.: Analysis of features selected by a deep learning model for differential treatment selection in depression. Front. Artif. Intell. 2, 31 (2020)

16. Mitchell, M., Hollingshead, K., Coppersmith, G.: Quantifying the language of schizophrenia in social media. CLPsych 2015, 11–20 (2015)

17. Mohammad, S.M., Turney, P.D.: NRC emotion lexicon. National Research Council, Canada 2 (2013)

18. Mosteller, F., Wallace, D.L.: Inference in an authorship problem: a comparative study of discrimination methods applied to the authorship of the disputed federalist papers. J. Am. Stat. Assoc. 58(302), 275–309 (1963)

19. O'dea, B., Wan, S., Batterham, P.J., Calear, A.L., Paris, C., Christensen, H.: Detecting suicidality on Twitter. Internet Interventions 2(2), 183–188 (2015)

20. Pennebaker, J.W., Francis, M.E., Booth, R.J.: Linguistic inquiry and word count: LIWC 2001. Mahway: Lawrence Erlbaum Associates 71(2001), 2001 (2001)

21. Plutchik, R.: Emotions: a general psychoevolutionary theory. Approaches Emot. 1984, 197–219 (1984)

22. Resnik, P., Garron, A., Resnik, R.: Using topic modeling to improve prediction of neuroticism and depression in college students. In: EMNLP, pp. 1348–1353 (2013)

23. Sadeque, F., Xu, D., Bethard, S.: UArizona at the CLEF eRisk 2017 pilot task: linear and recurrent models for early depression detection. In: CLEF 2017 Labs and Workshops, Notebook Papers. CEUR Workshop Proceedings. CEUR-WS.org., vol. 1866. NIH Public Access (2017)

24. Schwartz, H.A., et al.: Towards assessing changes in degree of depression through facebook. In: CLPsych, pp. 118–125 (2014)

25. Serrano, S., Smith, N.A.: Is attention interpretable? In: ACL, pp. 2931–2951 (2019)

26. Shen, G., et al.: Depression detection via harvesting social media: a multimodal dictionary learning solution. In: IJCAI, pp. 3838–3844 (2017)

27. Shen, J.H., Rudzicz, F.: Detecting anxiety through reddit. CLPsych 2017, 58–65 (2017)

28. Trotzek, M., Koitka, S., Friedrich, C.M.: Linguistic metadata augmented classifiers at the CLEF 2017 task for early detection of depression. In: CLEF 2017 Labs and Workshops, Notebook Papers. CEUR Workshop Proceedings. CEUR-WS.org., vol. 1866 (2017)

29. Trotzek, M., Koitka, S., Friedrich, C.M.: Word embeddings and linguistic meta-data at the CLEF 2018 tasks for early detection of depression and anorexia. In: CLEF 2018 Labs and Workshops, Notebook Papers. CEUR Workshop Proceedings.CEUR-WS.org., vol. 2125 (2018)
30. Uban, A.S., Rosso, P.: Deep learning architectures and strategies for early detection of self-harm and depression level prediction. In: CEUR Workshop Proceedings, vol. 2696, pp. 1–12 (2020)
31. Wang, Y.T., Huang, H.H., Chen, H.H.: A neural network approach to early risk detection of depression and anorexia on social media text. In: CLEF 2018 Labs and Workshops, Notebook Papers. CEUR Workshop Proceedings.CEUR-WS.org., vol. 2125 (2018)
32. Wiegreffe, S., Pinter, Y.: Attention is not not explanation. In: EMNLP-IJCNLP, pp. 11–20 (2019)
33. World Health Organization, W.: Depression: a global crisis. world mental health day, october 10 2012. World Federation for Mental Health, Occoquan, Va, USA (2012)
34. Yang, Z., Yang, D., Dyer, C., He, X., Smola, A., Hovy, E.: Hierarchical attention networks for document classification. NAACL-HLT **2016**, 1480–1489 (2016)
35. Yazdavar, A.H., et al.: Semi-supervised approach to monitoring clinical depressive symptoms in social media. In: IEEE/ACM in Social Networks Analysis and Mining, pp. 1191–1198 (2017)
36. Zucco, C., Liang, H., Di Fatta, G., Cannataro, M.: Explainable sentiment analysis with applications in medicine. In: IEEE BIBM, pp. 1740–1747. IEEE (2018)

Linking Documents

Using Document Embeddings for Background Linking of News Articles

Pavel Khloponin[✉] and Leila Kosseim

ClaC Lab, Department of Computer Science and Software Engineering,
Gina Cody School of Engineering and Computer Science, Concordia University,
Montreal, QC, Canada
p_khlopo@encs.concordia.ca,leila.kosseim@concordia.ca

Abstract. This paper describes our experiments in using document embeddings to provide background links to news articles. This work was done as part of the recent TREC 2020 News Track [26] whose goal is to provide a ranked list of related news articles from a large collection, given a query article. For our participation, we explored a variety of document embedding representations and proximity measures. Experiments with the 2018 and 2019 validation sets showed that GPT2 and XLNet embeddings lead to higher performances. In addition, regardless of the embedding, higher performances were reached when mean pooling, larger models and smaller token chunks are used. However, no embedding configuration alone led to a performance that matched the classic Okapi BM25 method. For our official TREC 2020 News Track submission, we therefore combined the BM25 model with an embedding method. The augmented model led to more diverse sets of related articles with minimal decrease in performance (nDCG@5 of 0.5873 versus 0.5924 with the vanilla BM25). This result is promising as diversity is a key factor used by journalists when providing background links and contextual information to news articles [27].

Keywords: Background linking · Document embedding · Proximity measures

1 Introduction

Given the sheer number of electronic sources of news available today, it is important to develop approaches for the automatic recommendation of contextual information for users to better understand a news article. In order to address this need, since 2018, the News Track at TREC has proposed two related shared tasks: background linking and entity ranking ([24–26]). The goal of the background linking task is to provide relevant background information to news articles through the identification of related articles. On the other hand, entity ranking focuses on providing a list of names, concepts, artifacts, etc. mentioned in news articles, which will help readers better understand the news. This paper focuses on the first task: background linking.

© Springer Nature Switzerland AG 2021
E. Métais et al. (Eds.): NLDB 2021, LNCS 12801, pp. 317–329, 2021.
https://doi.org/10.1007/978-3-030-80599-9_28

For the background linking task, NIST provides a large collection of news articles (see Sect. 3.1), and a set of search topics which are themselves articles from the collection. For each search topic, participants need to select up to 100 related articles from the collection and output them as a ranked list from the most related to the least related. For evaluation purposes, a 5 point rank is manually assigned to the top 5 documents by NIST assessors during the evaluation phase. The score assigned to each search topic is an integer between 0 (little or no useful information) and 4 (must appear in recommendations or critical context will be missed). The total score of the system is then computed using the nDCG@5 metric ([9]) as follows:

$$nDCG@5 = \frac{\sum_{d=1}^{5} \frac{2^{R(d)}-1}{\log(1+d)}}{IDCG@5} \tag{1}$$

where $R(d)$ is the rank that assessors gave to the document d, and IDCG@5 is the ideal nDCG@5, i.e. the best ranking possible for the query. IDCG@5 not only makes sure the backlinks with the best scores have been returned, but also that these have been ideally ranked from most relevant to least relevant. This make the nDCG@n metric harder to improve than Precision, Recall and F1-measure.

Most previous approaches to the background linking task are based on information retrieval methods, and very few have investigated to use of neural language models for the task. Given the recent successes of neural language models such as GPT2 ([22]), XLNet ([32]) and BERT ([5]), we wanted to evaluate their possible contribution to the task either as an alternative or as a complement to classic information retrieval approaches. In this paper, through our experiments for the recent 2020 TREC News Track, we show that embeddings alone do not reach the performance of the BM25 model, but combining them to BM25 can lead to more diverse sets of related articles with minimal decrease in performance (nDCG@5 of 0.5873 versus 0.5924).

2   Previous Work

News background linking can be seen as a classic information retrieval (IR) problem, where systems need to retrieve from a large document collection, a ranked list of documents related to a query; but in the case of news background linking, the query itself is a news article. For this reason, most participants in the 2018, 2019 and 2020 News TREC used approaches based on IR.

Classic IR approaches such as TF-IDF, BM25 or combinations of these are typically used for news background linking (e.g. [18,31]). Other approaches include relevance models (e.g. [12,18]) and probabilistic models (e.g. [14]). Another successful common approach is to extract named entities from articles and use them as additional features to retrieve relevant documents (e.g. [1,15]). Several models have also experimented with re-ranking the results using relevance feedback [10,20] based on the idea that the end users' behavioural information can provide additional useful information to rank relevant documents. In particular, [19] used information on the frequency of user's clicks on the initially provided links to re-rank documents; while [1] used this data for collaborative filtering. Although relevance feedback has been shown to improve background linking, in the context of the TREC News task, such information is not available. Another approach ([6]) that yielded a high performance at TREC (nDCG@5 of 0.5918 in 2019) is based on query construction using a graph-based approach. Bigrams taken from the query articles are used to construct a co-occurrence graph. After pruning the graph, keywords are extracted with weights associated to them. These weighted terms are then used to run a query on Apache Lucene.

As shown above, most previous work in background linking is based on information retrieval approaches. To our knowledge, very little work has investigated the use of novel language models for the background linking task. The expectation is that related articles would be closer to each other in vector space, and using language models for document representation, vector distance metrics would be a good approximation of document content relatedness. In that vein, [4] used BM25 for an initial retrieval of the documents, then used SBERT (Sentence BERT [23]) to perform semantic similarity reranking between the query article and the articles retrieved by BM25. For this, a list of keywords are extracted from each returned article and from the query article. These lists of keywords are then treated as sentences and SBERT embeddings are built for them. Finally, the cosine similarity is then computed between these embeddings and used to re-rank the returned articles. In [3], the authors used a similar approach by first retrieving initial results with the Elasticsearch implementation of BM25, and using SBERT to directly build embeddings for the first three paragraphs of each article before averaging them to get a final embedding for the article, and applying the cosine similarity for the final ranking. On the other hand, [17] reached promising results using BERT, ELMo and GloVe embeddings on the ad-hoc document ranking task. Based on the BM25 relevance score, they selected positive and negative pairs for training queries which were then used to fine-tune embedding models and train classifiers. Another interesting approach was used in the News2Vec system ([16]) where the authors proposed a distributed representation of news based on news-specific features such as named entities, sentiment, month, publication week and day, word count, paragraph count, etc.

Given the recent successes of neural language models such as GPT ([21]), GPT2 ([22]), XLNet ([32]), BERT ([5]) and RoBERTa ([13]), we experimented with different document embedding representations and proximity measures as an alternative or as a complement to classic information retrieval approaches.

3   Our Approach

3.1   Document Collection

The TREC News document collection consists of 671,934 articles from The Washington Post published between 2012 and 2019. This document collection is the same as the one provided in previous years (2018 and 2019) but with duplicate articles removed and with new articles from 2017 to 2019 added. NIST required participants to ignore wire articles, editorial content and opinion posts. Due to this, the initial set of 671,947 articles was reduced by 2,057 to 669,890 items. This is shown in Table 1.

Articles were pre-processed, HTML markup and other meta information was removed and only the article text was preserved. As shown in Table 1, the 671,947 documents considered have an average length of 10,391 characters prior to preprocessing, but only 4,533 after pre-processing.

Table 1. Statistics of the 2020 TREC news document collection

Original number of articles	671,947
Articles to ignore as per NIST requirements	2,057
Articles used to build the models	669,890
Average size of article before pre-processing (characters)	10,391
Average size after pre-processing (characters)	4,533
Average size after pre-processing (tokens)	945

3.2 Document Representation

After pre-processing, we experimented with 5 families of models to represent each document: GPT ([21]) & GPT2 ([22]), XLNet ([32]), BERT ([5]) and RoBERTa ([13]), PEGASUS ([33]) models trained on the Newsroom ([8]) and Multi-News ([7]) datasets. In addition, for each family of models, we experimented with a variety of specific pre-trained models without fine-tuning. In all cases, before creating the document vectors, meta-information from the text was removed and Unicode characters were normalised using the NFC form of the Unicode Standard. We used the following 19 models available from Hugging Face:

5 BERT models: `bert-base-multilingual-cased`, `bert-base-multilingual-uncased`, `bert-large-cased`, `bert-large-uncased`, and `bert-base-uncased`.
5 GPT & GPT2 models: `openai-gpt`, `gpt2`, `gpt2-large`, `gpt2-medium`, and `gpt2-xl`.
5 RoBERTa models: `roberta-large-openai-detector`, `roberta-base-openai-detector`, `distilroberta-base`, `roberta-base`, and `roberta-large`.
2 XLNet models: `xlnet-base-cased`, and `xlnet-large-cased`.
2 PEGASUS models: `google-pegasus-multi_news`, and `google-pegasus-newsroom`.

The above models have a maximum input sequence size and cannot receive entire articles as input. To overcome this limitation, the tokenized article content was split into chunks of sequential tokens. When a chunk border falls in the middle of a sentence, instead of splitting it across chunks and potentially loosing its meaning for the embedding, overlapping of 64 tokens was introduced. We built embeddings with chunks of 500 and 250 tokens. Model specific padding tokens were added to the last chunk to match the chunk size. The resulting chunks were used as input to the models. Once the embedding vectors for the chunks from an article were built, they were pooled together to create the final embedding vector for the entire article. We explored three pooling methods: min, max and mean pooling. Given the deep convolution nature of the models, we evaluated the embeddings pulled from the last hidden layer of the models as well as from the pooler output layer of the BERT and RoBERTa models.

3.3 Proximity Measures

After obtaining the embeddings for the articles, the proximity between two document vectors is computed. To do this, we explored a broad range of proximity measures. We experimented with all 62 different measures presented in [2]. These measures are grouped into 9 families:

1. L_p **Minkowsky** including Euclidean, Chebyshev ...
2. L_1 **family** including Sørensen, Gower ...
3. **Intersection** including Wave Hedges, Czekanowski, Ruzicka ...
4. **Inner product** including Jaccard, Cosine, Dice ...
5. **Fidelity or Squared-chord families** including Bhattacharyya, Matusita ...
6. **Squared L_2 or χ^2 families** including Squared Euclidean, Pearson χ^2...
7. **Shannon's entropy family** including Kullback–Leibler, Jeffreys ...
8. **Combinations** including Taneja, Kumar-Johnson ...
9. **Vicissitude** including Vicis-Wave Hedges, VicisSymmetric χ^2 ...

When taking into account the parameters in some of these measures, in total we experimented with 85 proximity measures.

3.4 Normalisation

Because the off-the-shelf embeddings are not normalised, the amplitude of the components in the computed document vectors can differ by several orders of magnitude. In that case, components with larger amplitudes dominate the distance measures and smaller components are not given an opportunity to influence the distance measures. To avoid this problem, we experimented with normalisation. Figure 1 shows scatter plots of only two components of a set of document embeddings computed from `xlnet-large-cased` embeddings. Figure 1(a) shows the original components, while Fig. 1(b) and Fig. 1(c) show the same components when the embeddings are normalised to values within the range [0.0–1.0]. As shown in Fig. 1, we experimented with two types of normalisation: amplitude normalisation and sigmoid normalisation.

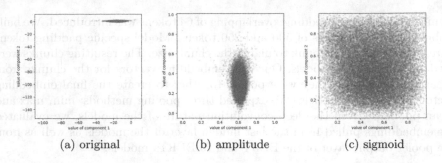

(a) original   (b) amplitude   (c) sigmoid

Fig. 1. Example of the effect of different normalisation techniques on two components of the `xlnet-large-cased` embeddings for the validation set.

Amplitude Normalisation. To generate embeddings normalised by amplitude, we calculated the minimum value (min_i) and the amplitude $(max_i - min_i)$ of each component i for each embedding and scaled each value between 0 and 1 by deducting the corresponding minimum from each component and divided it by the amplitude of the component (see Eq. 2).

$$vn_i = \frac{v_i - min_i}{max_i - min_i} \tag{2}$$

This type of normalisation ensures that all components of the vectors have a value within [0,1], but as shown in Fig. 1(b), outliers with a very large or very small component value will scale the vector components disproportionately, leaving most of the [0, 1] range unused.

Sigmoid Normalisation. In order to avoid the influence of outliers, we also experimented with sigmoid as a normalisation function. For this, we centered all components around their corresponding mean values and divided them by the component's standard deviations before applying the sigmoid function (see Eq. 3). As shown in Fig. 1(c), sigmoid normalisation allows for the components to be better spread over the [0,1] range.

$$vn_i = \frac{1}{1 + exp(\frac{v_i - \bar{v}_i}{\sigma_i})} \tag{3}$$

For each model, we used the original embedding as well as the normalised embeddings. In total, we experimented with 558 types of embeddings per document. Overall, using different embedding models, pulling methods, output layers and proximity measures gave us a total of 47,430 model configurations which we run on 2018 and 2019 validation datasets. To speed up the experiments, the proximity measures were implemented directly in Elasticsearch.

3.5   Validation

For validation purposes, NIST provided participants with the search topics and their corresponding manually evaluated results from the 2018 and 2019 TREC

News tracks. From the 2018 edition, we had 50 topics and their manually ranked (from 0 to 4) background links, and from 2019, we had 57 topics and their ranked background links. Recall from Sect. 3.1 that the past document collection was very close to this year's so they constituted a representative validation set. We used these 2018 and 2019 sets of topics (107 in total) and evaluated backlinks (22,338 in total) for validation purposes. Note that these backlinks constitute only $\approx 3\%$ of the entire document collection of 669,890 articles (see Table 1). To evaluate our 47,430 models, we generated the top 5 backlinks for each topic and computed nDCG@5 with the 2018 and 2019 datasets.

Table 2 shows the results of the top models. Although the best performances with the 2018 data are significantly lower than with the 2019 data, comparisons across years cannot be done as the queries and the backlinks are different. Comparisons should be done within the same year. Among all embedding and similarity configurations, GPT2 embeddings outperformed all embeddings and dominated the top 261 best performing configurations with the 2019 dataset, and was among the leading models with the 2018 dataset (although XLNet did perform close to the GPT2 models).

As seen from Table 2, all best configurations for 2019 and 2018 have proximity metrics from the Squared L_2 and Inner product families (see Sect. 3.3). The Pearson χ^2 distance achieved the best nDCG@5 with the 2019 data; while the cosine measure dominated the top positions with the 2018 data.

In general, normalisation seemed to improve performance. Even though for the top performing models, the improvement was small ($\approx 6\%$), for some proximity measures, normalisation did lead to a more important increase in performance. For example, the top performing model with the 2019 data, gpt2-xl with the Pearson χ^2 metric without normalization, reached only nDCG@5 of 0.0033 (not shown in the table), but with amplitude normalization it reached an nDCG@5 of 0.4790; and with sigmoid normalisation, it reached 0.5071.

Table 2. Top 5 performing models with the 2019 (top sub-table) and 2018 (bottom sub-table) validation datasets and their performance

	Embedding	Norm.	Chunk	Pooling	Distance	nDCG@5 2019	nDCG@5 2018
Best 2019	gpt2-xl	Sigmoid	250	Mean	Pearson χ^2	**0.5071**	0.3107
	gpt2-xl	Sigmoid	250	Mean	Dice	0.5067	0.2916
	gpt2-xl	Sigmoid	250	Mean	Jaccard	0.5034	0.2919
	gpt2-xl	Sigmoid	250	Mean	Vicis-Symmetric χ^2	0.5018	0.2800
	gpt2-xl	Sigmoid	250	Mean	Probabilistic Symmetric χ^2	0.5004	0.2793
Best 2018	gpt2-medium	Sigmoid	500	Mean	Cosine	0.4265	**0.3431**
	xlnet-large-cased	Sigmoid	250	Mean	Additive Symmetric χ^2	0.4345	0.3281
	gpt2-large	Sigmoid	500	Mean	Cosine	0.4868	0.3278
	gpt2-xl	Sigmoid	250	Mean	Cosine	0.4918	0.3269
	xlnet-large-cased	Sigmoid	250	Mean	Jaccard	0.4341	0.3266

A further comparison of the proximity measures is shown in Fig. 2. The figure shows the maximum nDCG@5 reached by a distance measure regardless of the embedding method used. As the figure shows, the proximity measures seem to perform relatively similarly compared to one another on both the 2018 and 2019 datasets, but the top performing proximity measures are different.

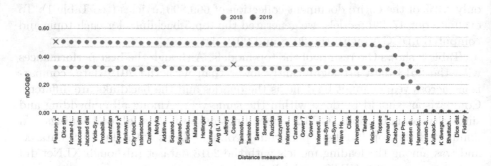

Fig. 2. Maximum nDCG@5 reached by each distance measure with the 2018 and 2019 data, independent of the embedding method. The top performing measure is marked with the a red "X". (Color figure online)

Recall from Sect. 3.2, that three pooling methods were experimented with to create the document embeddings: min, max and mean pooling. As shown in Table 3, all top models use mean pooling. In addition, except for a few PEGA-SUS models, all 47,430 configurations show significantly higher results across all proximity measures when mean pooling is applied.

Finally, we analysed the influence of the chunk size when creating the document embeddings. As indicated in Sect. 3.2, articles were split into chunks to fit the models' requirements. We expected smaller chunks to decrease performance, because each chunk contains less information, but to our surprise all but two models from Table 2 used the smallest chunk size (250 tokens).

4   Results and Analysis

Notwithstanding the results presented in Sect. 3.5, the classic BM25 model out-performed all embedding models by reaching nDCG@5 measures of 0.7418 (for 2019) and 0.5289 (for 2018). Based on this, at the recent 2020 shared task, we submitted both an embedding method along with the classic BM25.

4.1   TREC Runs

The validation of all 47,430 models of Sect. 3.5 was not ready in time for the 2020 TREC News task, therefore we used the best embedding model found from a smaller sub-set of experiments. We submitted 4 runs:

gpt2_norm is based only on GPT2 amplitude normalised embeddings with 250 token chunk size with mean pooling and Minkowski L_3 proximity measure. This configuration was chosen as it was the best one discovered among the embedding methods and proximity measures we had explored at the time of submission.

es_bm25 is based on the Elasticsearch (Lucene) implementation of the Okapi BM25 ranking algorithm.

combined is a combination of the previous two runs, where the BM25 score between the query and each target document is multiplied by the inverted distance between corresponding GPT2-embeddings.

d2v2019 is based on Doc2Vec embeddings computed from News TREC 2019 and cosine similarity as a proximity measure. This model was used because we used this approach last year in our participation to the track ([11]), and, for this year, we wished to compare last year's method to novel ones.

4.2 TREC Results

For each run and each topic, NIST provided us with our official score as well as the collective minimum, maximum and median scores. Table 3 shows the official overall scores. As shown in Table 3, two of our runs outperformed the collective median nDCG@5 of 0.5250. Among our submissions es_bm25 achieved the highest score with an nDCG@5 of 0.5924, combined performed slightly below with 0.5873, while gpt2_norm and d2v2019 performed below the collective median with nDCG@5 of 0.4541 and 0.4481 respectively.

Table 3. Overall results of our runs at TREC 2020

Run	nDCG@5
es_bm25	0.5924
combined	0.5873
gpt2_norm	0.4541
d2v2019	0.4481
TREC max	0.7914
TREC median	0.5250
TREC min	0.0660

4.3 Post-TREC Results

Once the validation of all models of Sect. 3.5 was complete, we simulated the official 2020 TREC News shared task to evaluate the top performing embedding configurations (see Table 4). We applied the methods to the entire 2020 document collection (669,890 articles, see Table 1) and used the TREC official scorer. Table 4 shows the final scores of these methods with the 2018–2020 queries.

Table 4. Performance of the top configurations with the validation set, the BM25 model and a new combined model on the entire document collection

	Model (embedding+norm+chunk+distance)	nDCG@5 2020	nDCG@5 2019	nDCG@5 2018
Best 2019	(1) gpt2-xl+sigmoid+250+mean+Pearson χ^2	0.4882	0.4157	**0.2424**
	(2) gpt2-xl+sigmoid+250+mean+Dice	0.4908	**0.4184**	0.2350
	(3) gpt2-xl+sigmoid+250+mean+Jaccard	0.4905	0.4127	0.2338
	(4) gpt2-xl+sigmoid+250+mean+Vicis-Symmetric χ^2	**0.4950**	0.4126	0.2277
	(5) gpt2-xl+sigmoid+250+mean+Probabilistic Symmetric χ^2	0.4895	0.4144	0.2269
Best 2018	(6) gpt2-medium+sigmoid+500+mean+Cosine	0.4239	0.3836	0.2394
	(7) xlnet-large-cased+sigmoid+250+mean+Additive Symmetric χ^2	0.4295	0.3539	0.2334
	(8) gpt2-large+sigmoid+500+mean+Cosine	0.4409	0.4001	0.2415
	(9) gpt2-xl+sigmoid+250+mean+Cosine	0.4409	0.4001	0.2415
	(10) xlnet-large-cased+sigmoid+250+mean+Jaccard	0.4422	0.3435	0.2206
	(11) es_bm25	**0.5924**	**0.5514**	**0.3011**
	(12) es_bm25+gpt2-xl+sigmoid+250+mean+Vicis-Symmetric χ^2	0.5737	0.5125	0.2878
	TREC median	0.5250	0.5295	N/a

As the complete validation suggested, these top-performing configurations outperformed the embedding method submitted to the shared task, gpt2_norm (gpt2 embedding with mean pooling, amplitude normalisation, chunk size of 250) which achieved an nDCG@5 of 0.4541 (see Table 3) but still performed below the overall TREC median of 0.5250 and es_bm25 (0.5924) and achieved an nDCG@5 of 0.4950. The new combined model (12), based on es_bm25 and model (4) (see Sect. 4.1), achieved an nDCG@5 of 0.5737; ranking lower than the es_bm25 model but higher than model (4) and the TREC median.

4.4 Analysis

As shown in Table 4, es_bm25 and model (12) are rather similar in terms of overall median nDCG@5, however when looking at individual topics, they return significantly different background links. Figure 3 shows this diversity graphically.

Figure 3(a) shows that es_bm25 performs better on most of the topics (see the upward bars) but significantly drops in performance on certain topics for which model (4) is very successful (see the long downward bars). For example model (4) returned the best result over all runs submitted for the topic #912 where es_bm25 missed the most relevant article. The query article and the most relevant backlink have less word overlap compared to the query article and the top backlink returned by es_bm25. Model (4), on the other hand, functioning on a different principle, was able to return the most relevant backlink in the first position.

Figure 3(b) shows the per topic difference between es_bm25 and model (12). As the figure shows, model (12) improved on most of the topics for which es_bm25 outperformed model (4) without dropping in performance on most of the topics for which es_bm25 did not yield the top results.

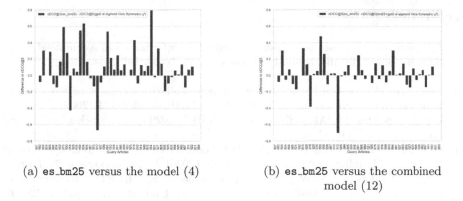

(a) es_bm25 versus the model (4)   (b) es_bm25 versus the combined model (12)

Fig. 3. Difference in nDCG@5 scores for two pairs of models showing the diversity of background links for each topic

5 Conclusions and Future Work

Through our experiments for the recent 2020 TREC News Track we have found that the best performing embedding methods are GPT2 and XLNet for the 2019 and 2018 validation sets respectively. In addition, regardless of the embedding, higher performances are reached when mean pooling, larger models and smaller token chunks are used. However, the best embedding configuration alone led to an nDCG@5 of 0.4950, which is significantly below the performance of the the classic Okapi BM25 method with an nDCG@5 of 0.5924.

This paper also showed that augmenting the BM25 model with GPT2 embeddings normalised with sigmoid funcion and using the Vicis-Symmetric χ^2 proximity measure, led to a more diverse sets of related articles with minimal decrease in performance (nDCG@5 of 0.5737 versus 0.5924).

This combination shows potential towards returning more diverse backlinks. By relying on different models with different implementations we can return topics potentially not visible to a single model system.

Many avenues of research still need to be investigated. In particular, we used the embedding models off-the-shelf with no fine-tuning. Tuning the models for our specific dataset and task itself might improve the representation of the documents in vector-space, potentially providing a better ranking for backlinks. In addition, we would like to explore different ways of combining BM25 and embedding methods, in particular to better leverage the diversity of the results, instead of favoring common results.

Acknowledgments. The authors would like to thank the anonymous reviewers for their comments on an earlier version of this paper. This work was financially supported by the Natural Sciences and Engineering Research Council of Canada (NSERC).

References

1. Adomavicius, G., et al.: Incorporating contextual information in recommender systems using a multidimensional approach. ACM Trans. Inf. Syst. **23**(1), 103–145 (2005)
2. Cha, S.H.: Comprehensive survey on distance/similarity measures between probability density functions. Int. J. Math. Model. Meth. Appl. Sci. 1 (2007)
3. Day, N., Worley, D., Allison, T.: OSC at TREC 2020 - news track's background linking task. In: TREC [30]
4. Deshmukh, A.A., Sethi, U.: IR-BERT: Leveraging BERT for Semantic Search in Background Linking for News Articles. arXiv (2020). https://arxiv.org/2007.12603
5. Devlin, J., Chang, M.W., Lee, K., Toutanova, K.: BERT: Pre-training of Deep Bidirectional Transformers for Language Understanding. arXiv (2019). https://arxiv.org/1810.04805
6. Essam, M., Elsayed, T.: bigIR at TREC 2019: Graph-based Analysis for News Background Linking. In: TREC [29]
7. Fabbri, A., Li, I., She, T., Li, S., Radev, D.: Multi-News: A Large-Scale Multi-Document Summarization Dataset and Abstractive Hierarchical Model. In: Proceedings of the ACL, pp. 1074–1084. Florence, Italy, July 2019
8. Grusky, M., Naaman, M., Artzi, Y.: Newsroom: A Dataset of 1.3 Million Summaries with Diverse Extractive Strategies. In: Proceedings of NAACL/HLT, pp. 708–719. New Orleans, June 2018
9. Järvelin, K., Kekäläinen, J.: Cumulated Gain-based evaluation of IR techniques. ACM Trans. Inf. Syst. (TOIS) **20**(4), 422–446 (2002)
10. Kashyapi, S., Chatterjee, S., Ramsdell, J., Dietz, L.: TREMA-UNH at TREC 2018: Complex Answer Retrieval and News Track. In: TREC [28]
11. Khloponin, P., Kosseim, L.: The CLaC System at the TREC 2019 News Track. In: TREC [29]
12. Lavrenko, V., Croft, W.B.: Relevance based language models. In: Proceedings of the 24th ACM SIGIR Conference, pp. 120–127. New York, NY (2001)
13. Liu, Y., et al.: RoBERTa: A Robustly Optimized BERT Pretraining Approach. arXiv (2019). https://arxiv.org/1907.11692
14. Lu, K., Fang, H.: Leveraging Entities in Background Document Retrieval for News Articles. In: TREC [29]
15. Lu, M., et al.: Scalable news recommendation using multi-dimensional similarity and Jaccard-Kmeans clustering. J. Syst. Softw. **95**, 242–251 (2014)
16. Ma, Y., et al.: News2vec: news network embedding with subnode information. In: Proceedigs of EMNLP/IJCNLP, pp. 4843–4852. Hong Kong, November 2019
17. MacAvaney, S., Yates, A., Cohan, A., Goharian, N.: CEDR. In: Proceedings of the 42nd International ACM SIGIR Conference, July 2019
18. Naseri, S., Foley, J., Allan, J.: UMass at TREC 2018: CAR, Common Core and News Tracks. In: TREC [28]
19. Okura, S., et al.: Embedding-Based News Recommendation for Millions of Users. In: Proceedings of the 23rd ACM SIGKDD Conference, pp. 1933–1942. New York (2017)
20. Qu, J., Wang, Y.: UNC SILS at TREC 2019 news track. In: TREC [29]
21. Radford, A., Narasimhan, K.: Improving language understanding by generative pre-training. Preprint (2018). https://cdn.openai.com/research-covers/language-unsupervised/language_understanding_paper.pdf

22. Radford, A., et al.: Language models are unsupervised multitask learners. Preprint (2019). https://cdn.openai.com/better-language-models/language_models_are_unsupervised_multitask_learners.pdf

23. Reimers, N., Gurevych, I.: Sentence-BERT: Sentence Embeddings using Siamese BERT-Networks. In: Proceedings of EMNLP/IJCNLP, pp. 3982–3992. Hong Kong, November 2019

24. Soboroff, I., Huang, S., Harman, D.: 2018 news track overview. In: TREC [28]

25. Soboroff, I., Huang, S., Harman, D.: 2019 news track overview. In: TREC [29]

26. Soboroff, I., Huang, S., Harman, D.: 2020 news track overview. In: TREC [30]

27. Soboroff, I., Huang, S., Harman, D.: TREC 2020 news track guidelines v2.1, May 2020. http://trec-news.org/guidelines-2020.pdf

28. TREC (ed.): NIST Special Publication: Proceedings of the 27th Text REtrieval Conference (TREC) (2018). https://trec.nist.gov/pubs/trec27/trec2018.html

29. TREC (ed.): NIST Special Publication: Proceedings of the 28th Text REtrieval Conference (TREC) (2019). https://trec.nist.gov/pubs/trec28/trec2019.html

30. TREC (ed.): NIST Special Publication: Proceedings of the 29th Text REtrieval Conference (TREC) (2020). https://trec.nist.gov/pubs/trec29/trec2020.html

31. Yang, P., Lin, J.: Anserini at TREC 2018: CENTRE, Common Core, and News Tracks. In: TREC [28]

32. Yang, Z., et al.: XLNet: Generalized Autoregressive Pretraining for Language Understanding. arXiv 1906.08237 (2020)

33. Zhang, J., Zhao, Y., Saleh, M., Liu, P.J.: PEGASUS: Pre-training with Extracted Gap-sentences for Abstractive Summarization. arXiv 1912.08777 (2019)

Let's Summarize Scientific Documents! A Clustering-Based Approach via Citation Context

Santosh Kumar Mishra[1(✉)], Naveen Saini[2], Sriparna Saha[1], and Pushpak Bhattacharyya[3]

[1] Indian Institute of Technology Patna, Patna 801103, India
{santosh_1821cs03,sriparna}@iitp.ac.in
[2] Technology Studies Department, Endicott College of International Studies, Woosong University, Daejeon, South Korea
naveensaini@wsu.ac.kr
[3] Indian Institute of Technology Bombay, Mumbai, India 400076
pb@cse.iitb.ac.in

Abstract. Scientific documents are getting published at expanding rates and create challenges for the researchers to keep themselves up to date with the new developments. Scientific document summarization solves this problem by providing summaries of essential facts and findings. We propose a novel extractive summarization technique for generating a summary of scientific documents after considering the citation context. The proposed method extracts the scientific document's relevant sentences with respect to citation text in semantic space by utilizing the word mover's distance (WMD); further, it clusters the extracted sentences. Moreover, it assigns a rank to cluster of sentences based on different aspects like similarity with the title of the paper, position of the sentence, length of the sentence, and maximum marginal relevance. Finally, sentences are selected from different clusters based on their ranks to form the summary. We conduct our experiments on CL-SciSumm 2016 and CL-SciSumm 2017 data sets. The obtained results are compared with the state-of-the-art techniques. Evaluation results show that our method outperforms others in terms of ROUGE-2, ROUGE-3, and ROUGE-SU4 scores.

Keywords: Scientific summarization · Clustering · Word mover's distance · Maximum marginal relevance

1 Introduction

The publication rate of scientific papers is increasing day by day; the availability of the massive amount of scientific literature is a big challenge for researchers in various fields to keep them up-to-date with the new developments. A recent study by bibliometric analysts shows that global scientific output doubles after every nine years [2]. Scientific document summarization aims to solve this problem by summarizing the important contributions and findings of the reference paper [5–7] and

© Springer Nature Switzerland AG 2021
E. Métais et al. (Eds.): NLDB 2021, LNCS 12801, pp. 330–339, 2021.
https://doi.org/10.1007/978-3-030-80599-9_29

thus, reducing the effort of the researchers to understand the paper. There are two approaches to scientific summarization in the literature; the first is the abstract of the document. Though the paper's abstract provides the paper's theme, but may not convey the all-important contributions and impact of the paper. The same has also been shown in recent paper [1,18]. These kinds of problems motivate the researcher to solve the scientific summarization task using the second approach, i.e., citation-based summarization [7,8,16]. Citation based summary is obtained by utilizing a set of citations referring to the original document. Citations are a short description that explains the proposed method, result, and important findings of the cited work; this description is known as citation text or citance.

This paper proposes a novel approach for scientific document summarization using an extractive summarization technique that extracts important sentences from the reference paper. Here, we extract important sentences for each citation of the reference paper-based on semantic similarity between citation text and sentences of the reference paper using word mover's distance [13]. Further, we apply clustering on all distinct important sentences. Then, we rank the clusters based on the distances between the cluster center and the document center (representative sentence of the document). Finally, we extract sentences from ranked clusters using several sentence-scoring features until the summary's desired length (i.e., 250 words) is reached. The proposed approach is evaluated on two datasets: CL-SciSumm 2016 and CL-SciSumm 2017, related to the computational linguistic domain.

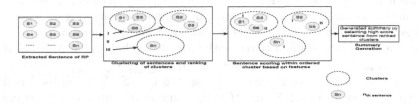

Fig. 1. Process flow chart of proposed method

2   Proposed Methodology

In this section, the steps followed in our proposed framework are discussed. The flowchart of the proposed approach is shown in Fig. 1.

2.1   Extracting the Citation Context

Initially, we have extracted the sentences from the reference paper, which is to be summarized. For this purpose, we have utilized the word mover's distance. Here, we have computed the word mover's distance (WMD) between each citation sentence and the reference paper's sentences. Then the top five sentences which are having minimum WMD are selected [11]. Let the set of distinct important sentences extracted from the reference paper (RP) after considering all citations be denoted by $\mathcal{S}$. Note that WMD calculates the similarity between the sentences in terms of distance [13], where minimum distance represents more similarity between sentences.

2.2   Grouping of Sentences Using Clustering

Sentences in $\mathcal{S}$ obtained in the previous step are grouped using the K-Medoids [19] clustering algorithm. It utilizes WMD as a distance measure between sentences instead of Euclidean sentence and, thus, is able to capture the semantic similarity present between the sentences. We have used the K-medoid clustering with the number of clusters decided by the elbow method. Let the obtained cluster centers be represented as $\{\mathcal{C}_1, \mathcal{C}_2, \ldots, \mathcal{C}_K\}$.

2.3   Ranking the Clusters Obtained

It includes two steps: representative sentence calculation and ranking of clusters, which are discussed below:

Representative Sentence Calculation: After getting clusters of sentences, it is required to build a summary. But, it is very difficult to decide which cluster should be considered first to extract the sentences. Thus, there is a need to rank the clusters. Therefore, to perform the same, firstly, we have determined the document center/representative sentence (RP) of the document. It is that sentence in the document which is the most similar to the remaining sentences. We can also call it as an document's center. In other words, among $\mathcal{S}$, the sentence having the minimum average WMD with respect to other sentences is called the RP. Mathematically, it is defined as $r = argmin \sum_{i=1}^{N} \sum_{j=1, i \neq j}^{N} \frac{wmd(s_i, s_j)}{M}$

Here, r is the index of representative sentence in $\mathcal{S}$, N is the total number of distinct sentences in set $\mathcal{S}$ and M is the number of sentence pairs, equals to $\frac{N*N-1}{2}$. s_i and s_j are the ith and j^{th} sentence in the set $\mathcal{S}$, respectively.

Ranking of Clusters: To rank the clusters, WMD distance between the cluster center, $\mathcal{C}_i$ ($1 \leq i \leq K$), and representative sentence, r, is calculated. Clusters are ranked based on their distances from the representative sentence means. The cluster closest to the representative sentence is assigned the highest priority. $d_i = wmd(\mathcal{C}_i, \mathcal{S}_r) \quad \forall i \in 1, 2, \ldots, K$

Here, d_i denotes the distance between i^{th} cluster center and representative sentence, $\mathcal{S}_r$. Then, these distances are sorted in ascending order. The cluster, which is at the lowest distance, is assigned rank-1 and so on. In other words, sentences are extracted from the higher rank to the lower rank clusters.

2.4   Calculating Sentence Scores in Each Cluster

After assigning ranks to different clusters, sentence scores are calculated in each cluster using different aspects/features. These scores help in selecting sentences from a cluster that will be part of the summary. These features are described below:

Similarity with Paper's Title (F_1): WMD between the title of the document and the sentences of the cluster has been calculated. The sentence is given the highest priority, which has minimum WMD distance with respect to the title [17].

Position of the Sentence (F_2): In most of the documents or papers, important sentences are found in the title and lead sentences of a paragraph; it is expressed as follows $m_i = \sqrt{\frac{1}{n_i}}$ where n_i is the position of a sentence in the reference paper. The sentence is given the highest priority, which lies at the starting of the paper or document [17].

Length of the Sentence (F_3): In the literature, it is shown that the longest sentences of the document are always relevant for the summaries [15,17]. The sentence is assigned the highest priority, which has the longest length.

Maximum Marginal Relevance (F_4): This feature is used to maintain anti-redundancy in the summary [3]. Sentences from each cluster are selected based on the following formula: $score(X) = \lambda Sim_1(s, D) - (1 - \lambda)Sim_2(s, Summary)$.

Here, score(X) represents linear interpolation of Sim_1 and Sim_2 where Sim_1 is the similarity of a sentence with respect to all other sentences in the cluster, and Sim_2 is the similarity of a sentence with respect to the sentences that are already included in the summary, D is the document (extracted sentences using citation context), and s is the sentence that is going to be included in the summary.

We have used WMD for the similarity between sentences. Here, $\lambda = 0.7$ which is used in [4] . The sentence is assigned the highest priority, which has the highest score of X.

2.5 Summary Generation

For the summary generation, we have considered the clusters in a rank-wise manner. Given the clusters, the summary is generated by selecting the highest ranked sentence from each cluster based on the above four features. We have generated a summary utilizing each feature and evaluated it against different types of summaries available with the datasets.

3 Experimental Setup

3.1 Datasets Used

In the current paper, we have utilized two datasets, namely, CL-SciSumm 2016 and CL-SciSumm 2017, to evaluate our method. Details of the datasets can be found at https://github.com/WING-NUS/scisumm-corpus.

3.2 Evaluation Metrics

The proposed method is evaluated with well-known evaluation metric, ROUGE score [14] for evaluating the summarization outputs.

3.3 Comparative Methods

We have compared the proposed method with the state-of-the methods of CL-SciSumm 2016 and CL-SciSumm 2017, these methods can be found in [11] and [10], respectively.

4 Results and Discussions

4.1 Results with CL-SciSumm 2016 Dataset

The results of the proposed method on the CL-SciSumm 2016 data set are shown in Table 1. This table is divided into two parts: (a) results of the proposed method using different features, (b) best results as compared with the state-of-the-art systems [11] of CL-SciSumm 2016. From these Tables it can be concluded that, the proposed method has better scores for the human summary and community summary, whereas, for the abstract summary, it lacks behind by only one system, namely, $sys8PARA7$. For the human summary, feature F2 is the most contributing feature. The proposed method has attained the highest ROUGE-SU4 score of 0.190, whereas the highest score reported in existing methods is 0.136. Our proposed approach has attained 39.70% improvement in terms of the ROUGE-SU4 score. For community summary, also, F2 is the most important feature, and our proposed approach has attained the highest ROUGE-SU4 score of 0.240, whereas the highest score reported in existing systems for CL-SciSumm 16 dataset is 0.167. Our method has obtained 43.71% improvements in terms of ROUGE-SU4. For the abstract summary, feature F3 is the best performing feature. Our proposed approach has attained a ROUGE-SU4 score of 0.308, which is the second-highest score after $sys8PARA7$.

Results of the proposed method are compared with some recent systems developed by Cohan et al. [6]; the corresponding results are shown in Table 2. It can be concluded from the table that our method performs better in terms of ROUGE-2 and ROUGE-3 scores except for one supervised model; our approach is unsupervised; this can be a reason behind the second-best performance. Note that Cohan et al. [6] have used citation contextualization and discourse facet. Our method does not use discourse facet as it needs supervised learning; our method is purely unsupervised in nature.

4.2 Results with CL-SciSumm 2017

The results of the proposed method on the CL-SciSumm 2017 dataset are shown in Table 3. Similar to Table 1 and Table 2, this table also consists of two parts: (a) results obtained using various features; (b) best results compared with the state-of-the-art system (methods) of CL-SciSumm 2017 [9]. It can be concluded from Table 3 that our proposed method performs better than all other systems for the community summary. For human summary, our proposed method has attained the highest ROUGE-SU4 score of 0.234 with 31.46% improvements over the best existing system, whereas, in terms of ROUGE-2 score, our method has attained less score in comparison to some of the systems. For community summary, our method has attained highest scores in terms of ROUGE-2 and ROUGE-SU4 metrics, which are 15.68% and 59.19% improvements in terms of ROUGE-2 and ROUGE-SU4 scores, respectively. For the abstract summary, our method has attained a better score than many methods in terms of ROUGE-2 and ROUGE-SU4 scores, but those are not the best ones. Note that the abstract is written by human authors, and our system is based on extractive summarization; therefore, this could be the reason behind poor performance by the proposed method.

Table 1. (a). Scores of generated summary in terms of ROUGE-SU4 against human summary, community Summary and abstract. (b) Comparison of performance of our proposed method with respect to state-of-the-art methods reported in CL-SciSumm 16 [12] in terms of ROUGE-SU4 metric. Here HS denotes human-summary, CS denotes community-summary and Abs denotes abstract.

Methods	HS	CS	Abs
F1	0.139	0.201	0.193
F2	**0.190**	**0.240**	0.304
F3	0.108	0.171	0.115
F4	0.176	0.228	**0.308**

(a)

Methods	HS	CS	Abs
Sys8$PARA_7	0.136	0.130	0.423
Sys3$LMKL1_CCS1	0.124	0.095	0.179
Sys3$LMEQAL_CCS2	0.121	0.102	0.214
Sys3$LMKL2_CCS3	0.114	0.095	0.158
Sys8$PARA_1	0.112	0.129	0.247
Sys8$PARA_8	0.111	0.150	0.244
Sys3$TFCCS4	0.101	0.085	0.129
Sys8$PARA_0	0.099	0.137	0.177
Sys8$PARA_4	0.094	0.162	0.170
Sys10$AUTOMATIC	0.092	0.150	0.124
Sys15$TKERN18	0.090	0.096	0.102
Sys15$TFIDF+ST+SL	0.088	0.167	0.092
Sys15$TKERN14CE	0.085	0.129	0.105
Sys10$COMMUNITY	0.085	0.149	0.111
Sys15$TKERN11CE	0.082	0.106	0.105
Sys15$TKERN11	0.081	0.103	0.107
Sys15$TKERN14	0.080	0.110	0.099
Sys15$TKERN18CE	0.071	0.103	0.093
Sys5$DEFAULT	0.065	0.082	0.087
Sys16$DEFAULT	0.048	0.107	0.053
Proposed Method	**0.190**	**0.240**	**0.308**

(b)

Table 2. (a). Scores of generated summary against human summary in terms of ROUGE-2 and ROUGE-3 for CL-SciSumm 16 dataset. (b). Comparison of performance of our method for human summary with respect to state-of-the-art methods reported in [6] in terms of ROUGE-2 and ROUGE-3 scores

Methods	ROUGE	
	ROUGE-2	ROUGE-3
F1	0.164	0.116
F2	**0.235**	**0.175**
F3	0.122	0.073
F4	0.220	0.168

(a)

Methods	ROUGE-2	
	ROUGE-2	ROUGE-3
BM25	0.152	0.130
VSM	0.148	0.127
LM	0.143	0.126
QR-NP	0.158	0.136
QR-KW	0.160	0.138
WE$_{wiki}$	0.145	0.125
WE$_{wiki}$ + retrofit	0.147	0.137
Supervised	0.175	0.150
Proposed Method	**0.235**	**0.175**

(b)

Table 3. Scores of generated summary in terms of ROUGE-2 and ROUGE-SU4 against human summary, Community Summary and Abstract. (b) Comparison of our proposed method with respect to state-of-the-art methods reported in CL-SciSumm 17 [9] in term of ROUGE-2 (R-2) and ROUGE-SU4 (R-SU4) scores; Here HS denotes human-summary, CS denotes community summary and Abs denotes Abstract.

Methods	HS		CS		Abs	
	R-2	R-SU4	R-2	R-SU4	R-2	R-SU4
F1	0.111	0.177	0.199	**0.267**	0.146	0.197
F2	**0.153**	**0.234**	**0.219**	0.264	0.108	0.151
F3	0.057	0.135	0.164	0.218	0.057	0.103
F4	0..070	0.121	0.091	0.138	0.080	0.101

(a)

Methods	HS		CS		Abs	
	R-2	R-SU4	R-2	R-SU4	R-2	R-SU4
CIST Run 4	0.156	0.101	0.184	0.136	0.351	0.185
CIST Run 1	0.171	0.111	0.187	0.137	0.341	0.167
CIST Run 6	0.184	0.110	0.185	0.141	0.331	0.172
CIST Run 3	0.275	0.178	0.204	0.168	0.327	0.171
CIST Run 2	0.225	0.147	0.195	0.155	0.322	0.163
CIST Run 5	0.153	0.118	0.192	0.146	0.318	0.178
UPF summa_abs	0.168	0.147	0.190	0.153	0.297	0.158
UPF acl_abs	0.214	0.161	0.191	0.167	0.289	0.163
UniMa Runs 1, 2, 3	0.197	0.157	0.181	0.169	0.265	0.184
NJUST Run 4	0.206	0.131	0.167	0.126	0.258	0.152
UniMa run 4, 5, 6	0.221	0.166	0.178	0.174	0.257	0.191
UniMa run 7, 8, 9	0.224	0.169	0.167	0.167	0.256	0.187
UPF summa_com	0.168	0.142	0.178	0.143	0.247	0.153
CIST Run 7	0.170	0.133	0.163	0.141	0.240	0.154
NJUST Run 2	0.229	0.154	0.152	0.114	0.214	0.138
NJUST Run 1	0.190	0.114	0.147	0.101	0.198	0.114
NJUST Run 5	0.178	0.127	0.119	0.098	0.192	0.108
Jadavpur Run1	0.181	0.129	0.132	0.119	0.191	0.133
NJUST Run 3	0.162	0.115	0.141	0.127	0.187	0.119
UPF google_abs	0.172	0.132	0.143	0.139	0.170	0.108
UPF acl_com	0.217	0.166	0.189	0.169	0.161	0.099
UPF summa_hum	0.189	0.148	0.131	0.147	0.144	0.091
UPF acl_hum	0.188	0.147	0.132	0.127	0.124	0.102
UPF google_hum	0.127	0.101	0.103	0.109	0.071	0.071
UPF google_com	0.120	0.092	0.075	0.096	0.052	0.065
Proposed Method	**0.153**	**0.234**	**0.219**	**0.267**	**0.146**	**0.197**

(b)

Table 4. Ranking based comparison with state of the art techniques for CL-SciSumm 16 (a) and CL-SciSumm 17 (b); Here AR denotes average ranking.

SOTA	HS	CS	Abs	AR
F2	1	1	1	1.66
F4	2	2	2	2
F1	3	3	7	4.33
Sys8$PARA_7	4	12	1	5.66
Sys8$PARA_8	9	8	5	7.33
Sys8$PARA_1	8	13	4	8.33
F3	10	4	14	9.33
Sys8$PARA_4	13	6	10	9.66
Sys8$PARA_0	12	10	9	10.33
Sys3$LMEQUAL_CCS2	6	19	6	10.33
Sys3$LMKL1_CCS1	5	21	8	11.33
Sys10$AUTOMATIC	14	7	13	11.33
Sys3$LMKL2_CCS3	7	22	11	13.33
Sys10$COMMUNITY	17	9	15	13.66
Sys15$TFIDF+ST+SL	16	5	22	14.33
Sys3$TF_CCS4	11	23	12	15.33
Sys15$TKERN14CE	18	22	18	16
Sys15$TKERN11CE	19	16	17	17.33
Sys15$TKERN11	20	17	16	17.66
Sys15$TKERN18	15	20	19	18
Sys15$TKERN14	21	14	20	18.33
Sys15$TKERN18CE	22	18	21	20.33
Sys16$DEFAULT	24	15	24	21
Sys5$DEFAULT	23	24	23	23.33

(a)

SOTA	HS		CS		Abs		AR	AR
	R-2	R-SU4	R-2	R-SU4	R-2	R-SU	R-2	R-SU4
F1	27	2	3	2	22	1	17.33	1.66
UNIMA Run 4, 5,6	5	5	13	4	11	2	9.66	3.66
UNIMA Run 7,8,9	4	4	16	9	12	3	10.66	5.33
CIST Run 3	1	3	2	7	4	8	2.33	6
F2	23	1	1	1	25	16	16.33	6
UNIMA Run 1,2,3	9	8	12	5	9	5	10	6
UPF acl_abs	7	7	6	8	8	11	7	8.66
CIST Run 2	3	11	4	10	5	10	4	10.33
UPF summ_abs	19	12	7	11	7	12	11	11.66
UPF acl_com	6	6	8	6	21	26	11.66	12.66
CIST Run 5	24	22	5	13	6	6	11.66	13.66
UPF summ_com	20	14	14	14	13	14	15.66	14
F3	29	15	17	3	28	25	24.66	14.33
CIST Run 7	18	16	18	16	14	13	16.66	15
CIST Run 6	12	26	10	15	3	7	8.33	16
UPF summ_Hum	11	10	25	12	23	27	19.66	16.33
CIST Run 4	22	27	11	20	1	4	11.33	17
NJUST Run 2	2	9	19	25	15	17	12	17
CIST Run 1	16	25	9	19	2	9	9	17.66
NJUST Run 4	8	18	15	23	10	15	11	18.66
UPF google_abs	17	17	21	17	20	23	19.33	19
F4	28	21	28	18	26	20	27.33	19.66
UPF acl_hum	12	13	24	22	24	24	20	19.66
Jadavpur, Run 1	14	19	23	24	18	18	18.33	20.33
NJUST, Run 3	21	23	22	21	19	19	20.66	21
NJUST Run 5	15	20	26	28	17	22	19.33	23.33
NJUST Run 1	10	24	20	27	16	21	15.33	24
UPF google_hum	25	28	27	26	27	28	26.33	27.33
UPF google_com	26	29	29	29	29	29	28	29

(b)

4.3 Ranked Analysis of the Results

It can be concluded from the previous sections, for CL-SciSumm 2016 and CL-SciSumm 2017 datasets, no system (Table 1 and Table 3) is the best suited for the human summary, community summary, and abstract summary. It can be seen from Table 1 that system $sys8PARA7$ has the best score for abstract summary (as shown in Table 1), but it is not the best system for human summary and community summary. Similarly, if we observe Table 3, system $CISTRUN4$ is the best system for an abstract summary in terms of ROUGE-2 score, but it is not the best system for human summary and community summary generations. To resolve the ties and analyze the performance of different methods, the ranking based analysis of all methods (systems) proposed in the CL-SciSumm 2016 is shown in Table 4 (a), whereas for CL-SciSumm 2017, the same is illustrated in Table 4 (b). In the ranking table, each method is assigned a rank according to its performance. Each of the systems is assigned a rank value for the human summary, community summary, and abstract summary. Finally, each system is assigned an average rank, which is the average of the ranks over the human summary, community summary, and abstract.

For CL-SciSumm 2016 and CL-SciSumm 2017 datasets, the ranking Tables are shown in Table 4 (a) and Table 4 (b), respectively. It can be concluded from Table 4 (a) that our proposed method is the best one among all the submitted systems. On the other hand, from Table 4 (b) for CL-SciSumm 2017 dataset, it can be concluded that our proposed method is the best among all the systems in terms of ROUGE-SU4 score. In terms of the ROUGE-2 score, our method is at 17th position in the overall ranking.

5 Conclusion

We present a clustering-based method for scientific document summarization. We utilize word mover's distance to extract the citation context. Incorporating different features like maximal marginal relevance, sentence position in the document, among others, helps in the summary generation process. The obtained results illustrate our proposed method's efficacy over the state-of-the-art techniques in most cases. In future, multi-objective optimization-based clustering can be used for scientific document summarization.

References

1. Atanassova, I., Bertin, M., Larivière, V.: On the composition of scientific abstracts. J. Documentation **72**(4), 636–647 (2016)
2. Bornmann, L., Mutz, R.: Growth rates of modern science: a bibliometric analysis based on the number of publications and cited references. J. Am. Soc. Inf. Sci. **66**(11), 2215–2222 (2015)
3. Carbonell, J.G., Goldstein, J.: The use of mmr, diversity-based reranking for reordering documents and producing summaries. SIGIR. **98**, 335–336 (1998)
4. Cohan, A., Goharian, N.: Scientific article summarization using citation-context and article's discourse structure. In: Proceedings of the 2015 Conference on Empirical Methods in Natural Language Processing, pp. 390–400. Association for Computational Linguistics, Lisbon, Portugal, September 2015. https://doi.org/10.18653/v1/D15-1045, https://www.aclweb.org/anthology/D15-1045
5. Cohan, A., Goharian, N.: Scientific article summarization using citation-context and article's discourse structure. arXiv preprint arXiv:1704.06619 (2017)
6. Cohan, A., Goharian, N.: Scientific document summarization via citation contextualization and scientific discourse. Int. J. Digit. Libr. **19**(2), 287–303 (2017). https://doi.org/10.1007/s00799-017-0216-8
7. Cohan, A., Soldaini, L., Goharian, N.: Matching citation text and cited spans in biomedical literature: a search-oriented approach. In: Proceedings of the 2015 Conference of the North American Chapter of the Association for Computational Linguistics: Human Language Technologies, pp. 1042–1048 (2015)
8. Hernández-Alvarez, M., Gomez, J.M.: Survey about citation context analysis: tasks, techniques, and resources. Nat. Lang. Eng. **22**(3), 327–349 (2016)
9. Jaidka, K., Chandrasekaran, M., Jain, D., Kan, M.Y.: The cl-scisumm shared task 2017: Results and key insights (2017)

10. Jaidka, K., Chandrasekaran, M.K., Jain, D., Kan, M.Y.: The cl-scisumm shared. task 2017: results and key insights. In: BIRNDL@SIGIR (2017)
11. Jaidka, K., Chandrasekaran, M.K., Rustagi, S., Kan, M.Y.: Overview of the cl-scisumm 2016 shared task. In: Proceedings of the Joint Workshop on Bibliometric-enhanced Information Retrieval and Natural Language Processing for Digital Libraries (BIRNDL), pp. 93–102 (2016)
12. Jaidka, K., Chandrasekaran, M.K., Rustagi, S., Kan, M.Y.: Insights from cl-scisumm 2016: the faceted scientific document summarization shared task. Int. J. Digit. Libr. **19**(2–3), 163–171 (2018)
13. Kusner, M., Sun, Y., Kolkin, N., Weinberger, K.: From word embeddings to document distances. In: International Conference on Machine Learning, pp. 957–966 (2015)
14. Lin, C.Y.: ROUGE: A package for automatic evaluation of summaries. In: Text Summarization Branches Out, pp. 74–81. Association for Computational Linguistics, Barcelona, Spain, July 2004. https://www.aclweb.org/anthology/W04-1013
15. Mendoza, M., Bonilla, S., Noguera, C., Cobos, C., León, E.: Extractive single-document summarization based on genetic operators and guided local search. Expert Syst. Appl. **41**(9), 4158–4169 (2014)
16. Qazvinian, V., Radev, D.R., Mohammad, S.M., Dorr, B., Zajic, D., Whidby, M., Moon, T.: Generating extractive summaries of scientific paradigms. Journal of Artificial Intelligence Research **46**, 165–201 (2013)
17. Saini, N., Saha, S., Chakraborty, D., Bhattacharyya, P.: Extractive single document summarization using binary differential evolution: optimization of different sentence quality measures. PloS one, **14**(11) (2019)
18. Yasunaga, M., Kasai, J., Zhang, R., Fabbri, A.R., Li, I., Friedman, D., Radev, D.R.: Scisummnet: a large annotated corpus and content-impact models for scientific paper summarization with citation networks. Proc. AAAI Conf. Artif. Intell. **33**, 7386–7393 (2019)
19. Zhang, Q., Couloigner, I.: A new and efficient k-medoid algorithm for spatial clustering. In: Gervasi, O. (ed.) ICCSA 2005. LNCS, vol. 3482, pp. 181–189. Springer, Heidelberg (2005). https://doi.org/10.1007/11424857_20

Multimodality

Cross-Active Connection for Image-Text Multimodal Feature Fusion

JungHyuk Im⑩, Wooyeong Cho⑩, and Dae-Shik Kim$^{(\boxtimes)}$⑩

KAIST, Daejeon, South Korea
daeshik@kaist.ac.kr
http://www.kaist.ac.kr/

Abstract. Recent research fields tackle high-level machine learning tasks which often deal with multiplex datasets. Image-text multimodal learning is one of the comparatively challenging domains in Natural Language Processing. In this paper, we suggest a novel method for fusing and training the image-text multimodal feature. The proposed architecture follows a multi-step training scheme to train a neural network for image-text multimodal classification. In the training process, different groups of weights in the network are updated hierarchically in order to reflect the importance of each single modality as well as their mutual relationship. The effectiveness of Cross-Active Connection in image-text multimodal NLP tasks was verified through extensive experiments on the task of multimodal hashtag prediction and image-text feature fusion.

Keywords: Multi-modal learning · Feature fusion · Natural language processing

1 Introduction

The development of high-performance language models has brought remarkable advance in machine learning based language tasks. As recently emerged methods [5,14] are able to represent the complex semantic properties of words regardless of tasks, comprehension abilities of word-wise encoders have been strengthened enough to tackle challenging NLP tasks. Nevertheless, high-level tasks involving both natural language understanding and text generation such as dialogue and question answering face another drawback. In real life communication between human beings, semantic representation of text is also dependent of visual information as they affect the context of words used in an utterance. Current trends of research reflect efforts to contemplate this multimodal dependency of humanlike communication tasks. A variety of models were developed to solve the challenge of Visual Question Answering [1], along with attempts to fuse image and text features for multimodal classification tasks [6,15]. However, many of the high-performance models are implemented with the ensemble of multiple networks dealing with different modalities. This shows that multimodal feature fusion is a field of research that still needs advancement.

© Springer Nature Switzerland AG 2021
E. Métais et al. (Eds.): NLDB 2021, LNCS 12801, pp. 343–354, 2021.
https://doi.org/10.1007/978-3-030-80599-9_30

Our research focuses on building a training scheme that can effectively fuse image and text features. Researchers of this field are already informed that baseline methods for image-text feature fusion involve concatenating the image and text representations separately extracted from two neural networks. But concatenation is not enough to achieve rich representation that reflects the mutual relationship between text and image information. We designed a two-phase training scheme to subdue the limitations of end-to-end models that use concatenated multimodal feature as the input. The complete architecture of the proposed model in this paper integrates two single feature extracting models with a multi-label classifier. The training scheme of the neural network multi-label classifier breaks itself down to be equivalent to training the ensemble of four individual networks; two individual single-modal networks and a set of two complementary multimodal networks.

The evaluation of our model was conducted with the task of image-text hashtag prediction. Researchers are now aware that single-modal hashtag prediction is a limited area of research as the majority of online SNS platforms deal with both image and text. It is challenging to achieve solid performance in multimodal hashtag prediction, as hashtags do not directly represent the objects in the image or written captions. To be considered as a successful approach, a multimodal predictor must not only perform better than single-modal predictors, but also be capable of handling cases when one of the two modalities does not relate to the ground truth. The results of our experiments show that the implementation of cross-active connection within the neural network is effective for building a multi-label classification model with multimodal inputs.

The main contributions of our research are summarized below:

- We present Cross-Active ConNet (CACNet), a novel network design for image-text multimodal classifier.
- Training process of CACNet involves fusing the features of two modalities within the hidden layer. As a result, the weights of the hidden layers become effective image-text feature extractor.
- The multi-level training scheme that we propose is effective for multimodal feature fusion, but too complex to implement without Batch Gradient Descent. We simplified the implementation by grouping the weight matrices into sub-sections and utilizing a virtual sigmoid output. As a result, the multi-step training scheme is reduced into the problem of training four sub-networks that add up to build CACNet, and we can apply Mini-Batch Stochastic Gradient Descent.
- Cross-Active weights are updated when the two modalities share similar latent features. This selective updating algorithm helps the network to build a complementary relationship between two modalities, making the classifer less vulnerable to cases in which one of the two inputs do not relate to the label.
- Experiments conducted in our paper show that CACNet is an effective approach for image-text multimodal classification and image-text feature fusion.

2   Related Work

Multimodal Feature Fusion. Several recently published works deal with multimodal feature fusion. Multi-modal gender prediction model [15] was implemented with Gated Multimodal Units [2]. Another recent approach [6] exploits the well-known CNN sentence classification model [9] to fuse image and text features. They have shown that the fused feature performs better than baseline models of image and text single-modal classification.

Image Based Hashtag Prediction. HARRISON [12] is a benchmark dataset for image based hashtag prediction, which is provided along with prediction experiment results using a baseline method. The authors suggest three models for evaluation, which use features extracted from VGG-Object, VGG-Scene and both of them respectively. The evaluation results of these baseline models are included in the result section of this paper for comparison of our model against single-modal classifiers.

Multimodal Hashtag Prediction. Not many published works tackle multimodal hashtag prediction. However, several online authors propose models that can handle the task. Previous work on public online repository introduces a hierarchical ensemble model of CNN [8] and word feature extractor [11,13] for image-text hashtag prediction. They have also conducted an ablation study on the importance of hashtag segmentation in terms of text pre-processing. We constructed our own dataset to conduct the experiments of our research, adapting parts of the pre-processing methods described in their works. A multimodal hashtag predictor implemented by concatenating text feature extracted from [10] and visual feature from [16] won second place on OpenResource Hackathon 2019.

3   Methods

The Overall Architecture
The complete architecture for multi-label hashtag prediction is shown in Fig. 1. The model integrates two feature extractors and a multi-label classifier. The extracted features of two modalities serve as the inputs of the multi-label classifier CACNet.

Feature Extraction
We use VGG16 model pre-trained on the 1.2 million ImageNet dataset [7] as our image feature extractor. The 1×4096 vector output is reduced into the dimension of 300, which is directly used as the image feature input of the multi-label classifier. Word2Vec model pretrained on Google News corpus of over 3 million words was used as the text feature extractor. Although there exist various methods to form representations of sentence-level texts using Recurrent Neural Network based encoding methods, previous work [3] have proven that the weighted average of word embedding can strongly represent sentences. Particularly in the task of hashtag prediction, the importance of word sequence in

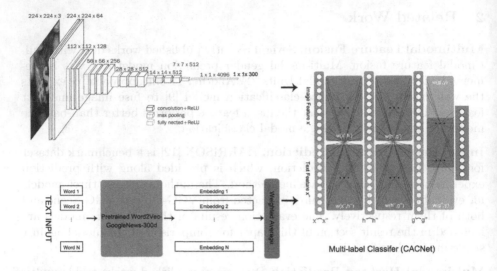

Fig. 1. The overall architecture of the multi-label classifier proposed in this paper.

the captions are reduced compared to other types of tasks such as reading comprehension. Thus our model takes the weighted average of the word vectors to represent the caption of an Instagram post as the text feature input instead of taking RNN based approaches.

Cross-Active Connection Network

The proposed network design of our research, CACNet serves as the multi-label classifier. It consists of two fully connected hidden layers of 600 dimension each and a sigmoid output for 300 categories of hashtags that our training dataset contains. The architecture of CACNet seems similar to a general Multi Layer Perceptron model with two hidden layers and the concatenated vector input of image and text features. The training algorithm of CACNet to be described later differentiates our classifier from general MLP for single-modal classification tasks. Using sigmoid function as the activation allows our classifer to perform multi-label classification.

In a Mini-Batch Stochastic Gradient Descent training scenario, CACNet updates the weights with a two-phase hierarchy. The idea of this training algorithm is to maintain the relationship between the output and each single-modality while also reflecting the complementary relationship each modality shares per single iteration. The concept is similar to adaptive dropout [4] in the sense that selective parts of the neurons are deactivated in each training phase according to a control variable. Figure 2 illustrates the structure of CACNet. Each layer is notated as a concatenation of two subsections of the neural network for convenience in mathematical formulation, and the weights that connect each subsection are grouped by the notation. The network on the right side of Fig. 2 is equivalent to the one on the left, where a slight change of arrangements

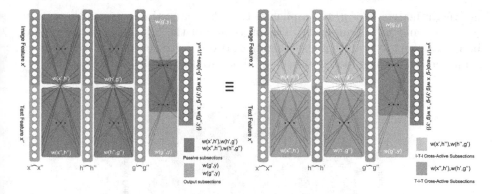

Fig. 2. Two different ways to visualize one equivalent CACNet multi-label classifier. Change of neuron alignments in the hidden layer help visualize how the weight parameters of CACNet are grouped into different sub-sections.

of weights has been made. All layers are fully connected and the weights are grouped into 10 subsections as labeled in the figures.

The first phase of training involves minimizing the cross entropy cost function for the passive subsections when the cross-active connections are deactivated. We cannot directly derive the loss function from the sigmoid output y when Cross-Active subsections are deactivated, as it is connected to both of the output subsections. We introduce the concept of creating a virtual sigmoid output which is only of temporary use for deriving the cross entropy independent of the other output subsection. Solving to minimize the error between the ground truth output and the virtual sigmoid output lets each output subsection lose dependency to the other, thus we can derive the following chain rule of partial derivatives to update passive subsections related to the image feature, where t is ground truth output.

$$y' = \sigma(g' \cdot w(g', y)) \tag{1}$$

$$E' = -y' log t - (1 - y') log (1 - t) \tag{2}$$

$$\frac{\partial E'}{\partial w(g', y)} = \frac{\partial E'}{\partial y'} \cdot \frac{\partial y'}{\partial g' \cdot w(g', y)} \cdot \frac{\partial g' \cdot w(g', y)}{\partial w(g', y)} \tag{3}$$

$$\frac{\partial E'}{\partial w(h', g')} = \frac{\partial E'}{\partial g'} \cdot \frac{\partial g'}{\partial h' \cdot w(h', g')} \cdot \frac{\partial h' \cdot w(h', g')}{\partial w(h', g')} \tag{4}$$

$$\frac{\partial E'}{\partial w(x', h')} = \frac{\partial E'}{\partial h'} \cdot \frac{\partial h'}{\partial x' \cdot w(x', h')} \cdot \frac{\partial x' \cdot w(x', h')}{\partial w(x', h')} \tag{5}$$

Notations were written as matrix multiplication for convenience. The gradients in the chain rule can all be calculated since E' is independent of the weights of the text feature related subsections and the cross-active weights. The same procedure can be processed through the text feature subsection of CACNet vice versa, by creating another virtual output y''. Notice that we maintain the notation of the weight matrix $w(g',y)$ to emphasize that the connection between the output subsection and y' is temporary. The procedure of the first phase of training is then equivalent to updating the weights of two Passive sub-networks that work as independent single-modal classifiers, illustrated in Fig. 3.

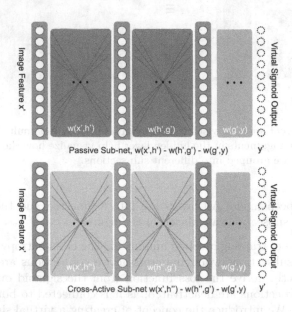

Fig. 3. The first phase of training is equivalent to training a pair of passive sub-networks to minimize the error between target output and virtual sigmoid output, when given each single-modal feature. Case 2 of the second phase is equivalent to training the Cross-Active Sub-net shown in the figure and the counterpart of it.

The second phase of training is divided into two cases controlled by the activation control variable γ,

$$\alpha = \frac{1}{N} \sum_{i=1}^{N} \frac{E_i' + E_i''}{2(-y_i log t_i - (1 - y_i) log(1 - t_i))} \tag{6}$$

$$\beta = \alpha/1 + \alpha \tag{7}$$

The activation parameter β shows how effective the whole network performs compared to the single-modal subsections. Low value of β also implies cases in which one of the two modalities do not reflect the training batch well. We define a control variable γ that ranges between 0 and 1 as a threshold that divides the high level training into two cases:

- **Case 1** If $\beta < \gamma$, update the whole network in an end-to-end manner without grouping the weights layer to minimize the cross-entropy loss of y. As described earlier, low value of β implies that the two modalities do not relate well, and it is better not to isolate the Passive subsections.

– **Case 2** If $\beta \geq \gamma$, activate the Cross-Active connection weights and deactivate passive subsections. Minimize the cross entropy cost function of a virtual sigmoid output by updating the cross-active weights.

In Case 2, where Cross-Active subsections are activated, the partial derivatives for calculating the gradients differ from the first phase training as follows.

$$y' = \sigma(g' \cdot w(g', y)) \tag{8}$$

$$\frac{\partial E'}{\partial w(g', y)} = \frac{\partial E'}{\partial y'} \cdot \frac{\partial y'}{\partial g' \cdot w(g', y)} \cdot \frac{\partial g' \cdot w(g', y)}{\partial w(g', y)} \tag{9}$$

$$\frac{\partial E'}{\partial w(h'', g')} = \frac{\partial E'}{\partial g'} \cdot \frac{\partial g'}{\partial g' \cdot w(h'', g')} \cdot \frac{\partial g' \cdot w(h'', g')}{\partial w(h'', g')} \tag{10}$$

$$\frac{\partial E'}{\partial w(x', h'')} = \frac{\partial E'}{\partial h''} \cdot \frac{\partial h''}{\partial h'' \cdot w(x', h'')} \cdot \frac{\partial h'' \cdot w(x', h'')}{\partial w(x', h'')} \tag{11}$$

Vice-versa can be done for the text-feature input involving counterpart subsection $w(x'',h')$—$w(h',g'')$—$w(g'',y)$ The loss function is independent to the deactivated weights when we minimize error between virtual output and the target output, so gradients involved in the partial derivatives are all easy to calculate. Notice that in the second phase, we are training the weights of the hidden layers connecting to the other modality, which we named Cross-Active subsections. This process is equivalent to training another complementary pair of sub-networks of structure labeled as Cross-Active Sub-net in Fig. 3.

When $\beta \geq \gamma$, the complexity of weight updates involving activating and deactivating parts of the network makes the procedure difficult to implement, especially for Mini-Batch Stochastic Gradient Descent scenarios. By utilizing virtual sigmoid outputs and grouping the weight matrix into subsections, we simplified the two-level training process into an equivalent problem of updating weights for 4 sub-networks given the same input and target output. After an iteration of Case 2 in second phase training, the 4 sub-networks jointly form CACNet.

4   Experiments

Our model was implemented with PyTorch 1.6.0, under a multi-GPU environment with 4 NVIDIA Titan Xp GPUs installed and CUDA Toolkit 10.2. The training procedure was conducted by Mini-Batch Stochastic Gradient Descent of batch size 20. Our complete dataset consists of 30k pairs of image-text multimodal inputs and text output. The CACNet classifier was trained over 500 epochs on the training dataset. The activation control variable γ described in the Methods section was set to 0.4 at the start of the training, and linearly increased up to 0.8 in the last 100 epochs.

Dataset. There are some benchmark datasets for image-based hashtag prediction [12,17], but there are no public dataset available for use in image-text multimodal hashtag prediction. For training and evaluation of our classifier, we constructed our own datasets. The process was conducted by scraping Instagram posts using Selenium over top 300 popular hashtags, as last updated on 2020-08-20. Non-english segments of the post including emoticons and special characters were removed, and the characters were converted into lower case. There has been a study about hashtag segmentation using the Viterbi algorithm to overcome the complexity caused by hashtags in Instagram posts combining multiple words into a single tag. Our multi-label classifier does not involve the ensemble of word embeddings in the prediction stage, so the pre-processing method was unnecessary. The ground truth outputs of our dataset consists of up to 10 hashtags used in a post. The details of the training and evaluation datasets are explicitly shown in Table 1 and Fig. 4.

Table 1. Details of the dataset used in our research. Train and Evaluation sets were both collected with Selenium web crawler.

	CACTrain	CACEval
# of posts	25,017	5,000
Average # of words per caption	12.82	13.1
Average # of Hashtags per post	8.71	9.11
Hashtag categories	300	300
Average # of <unk> per caption	2.51	2.27

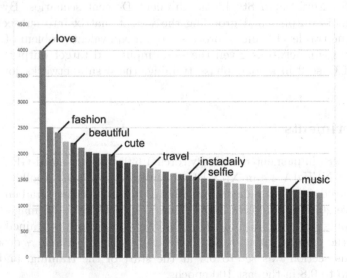

Fig. 4. The number of posts containing 40 mostly appearing hashtags in our training dataset. Single post is labeled with up to 10 multiple hashtags.

We conducted experiments to evaluate our model against baseline methods in two tasks, Image-Text Feature Fusion and Image-Text Multimodal HashTag Prediction.

Feature Fusion. The weight parameters of the trained CACNet can be extracted to serve the task of image-text feature fusion. We evaluated the feature fusion performance of CACNet against baseline methods published with UPMC Food-101, a large multimodal dataset that contains over 100k food recipes classified in 101 categories.

Hashtag Prediction. Despite the efforts of researchers on image-text multimodal tasks, there are no available published work that we can evaluate performance of CACNet on multimodal HashTag prediction against. To prove the validity of our multimodal classifier, we evaluated our model under the metrics of [12], as they provide a benchmark dataset for image-based Hashtag prediction and a baseline model. For generic evaluation, we also trained and evaluated their baseline model with our independent dataset.

5   Results

Feature Fusion. Performance in image-text feature fusion task was evaluated using the UPMC Food-101 dataset. As the authors describe, higher scores with text-only baseline method result from the bias introduced by their data crawling protocol. Evaluation was performed by comparing our results against their baseline models [18]. Our classifier CACNet achieved higher performance in classification than the baseline models. The results are shown in Table 2.

Table 2. Evaluation on the UPMC Food-101 dataset.

Methods	Avg.Precision
Very Deep (Vision only)	40.21%
TF-IDF (Text only)	82.06%
TF-IDF + Very Deep (Fusion)	85.10%
VGG16-Word2Vec300-CACNet(Fusion)	**87.63%**

Prediction examples shown in Fig. 5 show successful prediction examples in challenging cases, all of which single-modal baseline classifiers fail to predict accurately. CACNet successfully predicts hashtags in cases even when the input image does not relate to the ground truth hashtags, or when the words in the input caption are useless. **HashTag Prediction.** We used *Precision@K*, *Recall@K*, *Accuracy@K* as the evaluation measures for quantitative comparison against the baseline models introduced in the HARRISON benchmark dataset [12].

Table 3. Comparison of performance for hashtag prediction against image baseline models trained with HARRISON benchmark dataset.

Methods@Dataset	Precision@1	Recall@5	Accuracy@5
VGG-Object@HARRISON	28.30%	20.83%	50.70%
VGG-Scene@HARRISON	25.34%	18.66%	46.30%
VGG-Object + VGG-Scene@HARRISON	30.16%	21.38%	52.52%
VGG16-Word2Vec300-CACNet@CACEval	**59.7%**	**42.72%**	**71.13%**

Table 4. Generic evaluations of baseline models and our model measured with *Accuracy@K*

Methods	Accuracy@1	Accuracy@3	Accuracy@5
VGG-Object	8.41%	37.56%	48.12%
VGG-Scene	7.8%	33.74%	47.71%
VGG-Object + VGG-Scene	9.64%	38.44%	54.81%
VGG16-Word2Vec300-CACNet($\gamma=1$)	9.11%	41.47%	55.19%
VGG16-Word2Vec300-CACNet	**12.82%**	**48.91%**	**71.13%**

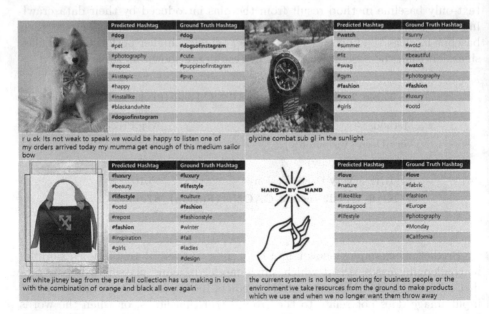

Fig. 5. Examples of successful predictions are shown above. Matching hashtags are in bold letters. Examples show cases in which our classifier was able to predict multiple matches with the ground truth when either one of caption or image are hard to relate to the hashtags.

Precision@K is the portion of top K ranked hashtags that match ground truth output. *Recall@K* is the portion of ground truth hashtags that match top K ranked hashtags. *Accuracy@K* is defined as 1 if there exists at least one match between top K ranked hashtags and the ground truth hashtags. The evaluation results for Hashtag prediction are shown in Table 3. The HARRISON benchmark contains 1,000 categories of hashtags while our dataset contains 300 categories. Thus the quantitative comparison of best results might not be reliable. For generic evaluation, we conducted further research by evaluating the baseline methods provided by HARRISON benchmark on our dataset, CACEval. Instead of using precision and recall as the metric, we evaluated the models on *Accuracy@K* only. To show the validity of our training scheme, we also trained a version of CACNet with the control variable γ set to 1. Table 4 shows the generic evaluation results.

6   Discussion

In this paper, we introduced a novel method for training a network with multimodal inputs. As far as we know, the multi-label classifier trained with our implementation, CACNet holds the state-of-the-art performance in hashtag prediction tasks. Our model has advantages over ensemble-based approaches and end-to-end approaches. The multi-phase training scheme lets the network maintain single-modal dependency as well as fusing the complimentary characteristics of two modalities. Another contribution of our research comes from introducing the concept of virtual outputs when observe the gradients from small sections of weight parameters in a whole network. This approach makes it possible to divide a network into sub-sections of weights and simplify complex training schemes. We expect our works to inspire fields of research involving image-text multimodal classification and feature fusion.

Acknowledgements. This work was supported by Institute for Information & communications Technology Promotion(IITP) grant funded by the Korea government(MSIT) (No.2016-0-00563, Research on Adaptive Machine Learning Technology Development for Intelligent Autonomous Digital Companion).

References

1. Antol, S., et al.: VQA: visual question answering. In: Proceedings of the IEEE international conference on computer vision, pp. 2425–2433 (2015)
2. Arevalo, J., Solorio, T., Montes-y Gómez, M., González, F.A.: Gated multimodal units for information fusion. arXiv preprint arXiv:1702.01992 (2017)
3. Arora, S., Liang, Y., Ma, T.: A simple but tough-to-beat baseline for sentence embeddings (2016)
4. Ba, J., Frey, B.: Adaptive dropout for training deep neural networks. In: Advances in neural information processing systems, pp. 3084–3092 (2013)
5. Devlin, J., Chang, M.W., Lee, K., Toutanova, K.: Bert: Pre-training of deep bidirectional transformers for language understanding. arXiv preprint arXiv:1810.04805 (2018)

6. Gallo, I., Calefati, A., Nawaz, S., Janjua, M.K.: Image and encoded text fusion for multi-modal classification. In: 2018 Digital Image Computing: Techniques and Applications (DICTA), pp. 1–7. IEEE (2018)
7. Goodfellow, I., Bengio, Y., Courville, A., Bengio, Y.: Deep Learning, vol. 1. MIT press, Cambridge (2016)
8. He, K., Zhang, X., Ren, S., Sun, J.: Deep residual learning for image recognition. In: Proceedings of the IEEE conference on computer vision and pattern recognition, pp. 770–778 (2016)
9. Kim, Y.: Convolutional neural networks for sentence classification. arXiv preprint arXiv:1408.5882 (2014)
10. Lan, Z., Chen, M., Goodman, S., Gimpel, K., Sharma, P., Soricut, R.: Albert: A lite bert for self-supervised learning of language representations. arXiv preprint arXiv:1909.11942 (2019)
11. Mikolov, T., Sutskever, I., Chen, K., Corrado, G.S., Dean, J.: Distributed representations of words and phrases and their compositionality. In: Advances in neural information processing systems, pp. 3111–3119 (2013)
12. Park, M., Li, H., Kim, J.: Harrison: A benchmark on hashtag recommendation for real-world images in social networks. arXiv preprint arXiv:1605.05054 (2016)
13. Pennington, J., Socher, R., Manning, C.D.: Glove: Global vectors for word representation. In: Proceedings of the 2014 conference on empirical methods in natural language processing (EMNLP), pp. 1532–1543 (2014)
14. Peters, M.E., et al.: Deep contextualized word representations. arXiv preprint arXiv:1802.05365 (2018)
15. Sierra, S., González, F.A.: Combining textual and visual representations for multimodal author profiling. Work. Notes Pap. CLEF **2125**, 219–228 (2018)
16. Simonyan, K., Zisserman, A.: Very deep convolutional networks for large-scale image recognition. arXiv preprint arXiv:1409.1556 (2014)
17. Thomee, B., Shamma, D.A., Friedland, G., Elizalde, B., Ni, K., Poland, D., Borth, D., Li, L.J.: Yfcc100m: the new data in multimedia research. Commun. ACM **59**(2), 64–73 (2016)
18. Wang, X., Kumar, D., Thome, N., Cord, M., Precioso, F.: Recipe recognition with large multimodal food dataset. In: 2015 IEEE International Conference on Multimedia & Expo Workshops (ICMEW), pp. 1–6. IEEE (2015)

Profiling Fake News Spreaders: Personality and Visual Information Matter

Riccardo Cervero[1](✉), Paolo Rosso[2], and Gabriella Pasi[1]

[1] Università degli Studi di Milano-Bicocca, Milan, Italy
r.cervero@campus.unimib.it, gabriella.pasi@unimib.it
[2] Universitat Politècnica de València, Valencia, Spain
prosso@dsic.upv.es

Abstract. Fake news are spread by exploiting specific linguistic patterns aimed at triggering negative emotions and persuading the consumers. A way to contrast this phenomenon is to analyse the psychological factors underlying consumers' vulnerabilities. This paper is situated in this research context: first, we study the correlation between psycholinguistic patterns in user's posts and the tendency to spread false information. Moreover, since online contents exploit multimedia information, a methodology aimed at profiling the authors based on the images they share is employed. The reported experiments show that the proposed method, which considers both text-related and image-related features, outperforms the results of state-of-the-art approaches.

Keywords: Author profiling · Personality traits · Visual information

1 Introduction

Social Media platforms on the World Wide Web allow to easily provide a wide range of users with potential information. However, it is evident how an irresponsible use of these open systems may cause damages to the virtual community itself. This is the case of so-called fake news, an increasingly debated phenomenon described as false articles intentionally fabricated to mislead the audience [1]; they are able, for instance, to polarize public opinion, or to deceive non-expert readers about scientific issues. The growing use of social networks as a primary source of information has created non-intermediated contexts where the evaluation of credibility is left to users' judgment, which is however compromised by the difficulty to deal with unfamiliar topics. Moreover, the Social Web also favors strong peer-to-peer connections, fostering closed and toxic virtual environments like "echo chambers", whose main characteristic is the correlation between the intensity of user engagement and the degree of negative emotional polarity [2]. Shu et al. [3] highlighted how fake news exploit consumers' vulnerabilities, triggering negative emotions and irrational reactions. Hence, effective tools turn out

© Springer Nature Switzerland AG 2021
E. Métais et al. (Eds.): NLDB 2021, LNCS 12801, pp. 355–363, 2021.
https://doi.org/10.1007/978-3-030-80599-9_31

to be algorithms able to learn the biases that penalise human judgement and to generate content that exploits them. Thus, there is an impelling need to contrast online disinformation; one possible means is to detect the users who are potential generators or sharers of fake news, by identifying the individual vulnerabilities at the basis of a lower capability to discern genuine content from fake one. Assuming that these vulnerabilities derive from psychological inclinations, this work aims to demonstrate that "fake news spreaders" are associated with specific personality traits. Therefore, after extracting personality characteristics from users' texts, we both evaluate their impact on the tendency to spread false content and test their effectiveness for the task of binary classification of users into real or fake news spreaders. However, as content flows quickly in microblogs, users' attention may be initially attracted by the visual elements of the posts. It is possible that images embedded in fake news attempt to exploit cognitive vulnerabilities, and thus they may present specific patterns. For this reason, we also report in this paper the outcomes of investigating the impact of visual features on fake news spreaders' profiling. In conclusion, the main contributions of this paper are the following. Firstly, inspired by Giachanou et al. [4], we evaluate the effectiveness of psycho-linguistic features to perform a classification of users into real and fake news spreaders. As a second task, inspired by [5], we also analyze the effectiveness of visual features - alone or mixed with personality information - to classify the authors. Lastly, we verify the feasibility of improving the effectiveness of state-of-the-art approaches for fake news detection by incorporating and/or replacing the proposed personality and visual information into the best models at the Author Profiling Task at PAN 2020[1].

The rest of the paper is organised as follows: Sect. 2 presents related works about the author profiling perspective; Section 3 introduces the PAN 2020 dataset and the two best performing solutions; Section 4 illustrates the methods of psycho-linguistic features extraction; Section 5 explains how visual information is obtained; Section 6 describes all the experiments carried out to evaluate the effectiveness of the aforementioned research contributions, whose obtained results are commented in the last Sect. 7.

2 Related Work

Fake news spreaders detection is an increasingly investigated research topic, which is more commonly tackled by means of data-driven approaches. Popular solutions are based on stylometric analysis - which aims to identify which style fits the category of "fake news spreaders", as in [6] -, or the extraction of lexicon-based emotional dimensions - the same on which Giachanou et al. [7] train an LSTM model. In particular, the employed features at the Author Profiling task at PAN 2020 can be divided into four categories [8]: *(i)* words or characters n-grams, *(ii)* stylistics, *(iii)* embeddings, *(iv)* personality and emotions, or combinations thereof. The best solutions respectively exploited a combination of n-grams and stylistic features (Buda & Bolonyai [9]) and only n-grams

---

[1] https://pan.webis.de/clef20/pan20-web/author-profiling.html.

(Pizarro [10]). Regarding the personality descriptors, the reference point of this work is what has been done by Giachanou et al. in [4], covered in the Sect. 4. Previous alternatives were the Myers-Briggs Type Indicator [11], or the manual compilation of questionnaires. Images are less frequently considered for the author profiling task, and the combination of visual and personality information is still under-explored. The reference point, in this case, is the approach aimed at extracting the visual features in [5] (Sect. 5).

3   PAN 2020: Profiling Fake News Spreaders on Twitter

The 2020 edition of the PAN event came with a shared task [8] aimed to inquire the feasibility of detecting authors who shared fake news in their past timeline in a bilingual perspective, i.e. considering both English and Spanish tweets. Two datasets, provided separately for each language, were generated as explained below. After selecting news labelled as fake on debunking websites, the organizers downloaded and manually labelled the tweets related to them as content supporting the false information, or vice versa. Thus, users in the sample who had shared at least one tweet supporting a fake news were labelled as "fake news spreader", and only the ones with the highest count were included in the final dataset, together with the same number of randomly selected "real news spreaders". In the end, the datasets are generated by collecting the last 100 tweets from each user's timeline, discarding those directly related to the fake news considered above, so as to avoid biases. On these datasets, the best average performance has been achieved, with equal merit, by Buda & Bolonyai [9] and by Pizarro [10]. In details, Buda-Bolonyai's model provided the highest accuracy on the English dataset (0.75), while Pizarro obtained the best result on Spanish tweets (0.82). Buda & Bolonyai's solution [9] is structured as follows. Firstly, four baseline classifiers (Logistic Regression, Support Vector Machine, Random Forest, and the gradient boosting algorithm XGBoost) undergo a training process consisting in an extensive grid search of the optimal combination among text pre-processing methods, vectorization techniques and baseline parameters. In details, the authors experimented different ranges for words n-grams. Then, another XGBoost algorithm is trained on user-wise statistical indicators: (i) minimum, maximum, mean, standard deviation and range of the length - both in words and in characters - of the tweets; (ii) number of retweets and mentions by the author; (iii) count of additional elements: URLs, hashtags, emojis and ellipses; (iv) lexical diversity calculated as the type-token ratio of lemmas. Buda & Bolonyai have preferred cross-validation techniques to prevent overfitting while optimizing the parameters of the baselines, instead of a single hold-out. Lastly, the five sub-models are trained to determine the probability of being a fake news spreader, and then they are stacked together through the best ensemble method chosen among (i) Majority Voting, (ii) Linear Regression, and (iii) Logistic Regression, which turned out to be the most reliable. The training of the ensemble model has been performed on the approximation of the predictions distribution, obtained by refitting the sub-models on different chunks of the training set. Pizarro [10] performed an optimization of the parameters of a Linear Support Vector Classifier trained on combinations of word and character n-grams, and experimenting with twelve pre-processing pipelines, based on mixtures of four basic operations: *(i)* downcase all the letters; *(ii)* replace numbers, URLs, users' name and hashtags with tokens; *(iii)* replace emojis with word representation; *(iv)* reduce number of repeated characters. The final linguistic features derive from the calculation of the Term Frequency - Inverse Document Frequency (TF-IDF) weight for each n-gram.

4   Personality Information

Giachanou et al. [4] originally proposed a method to classify users into "fake news spreaders" and "fact checkers", i.e. those interested to share posts that refute false information with evidences. In this work, instead, we aim at a classification into fake news "speaders" and "non spreaders". Their CheckerOrSpreader architecture is composed of a Convolutional Neural Network built upon two components: *(1)* one aimed at defining word-embeddings vectors from a pre-trained GloVe model, and *(2)* one aimed at eliciting the psycho-linguistic information that can describe users' personality. To obtain the latter, two approaches have been used simultaneously: *(i)* use of the LIWC software [12], mapping the text into 73 "psychologically-meaningful categories"; *(ii)* the Five-Factor Model (FFM) [13], which quantifies the evidence of a particular trait or disorder in user's text, considering five basic factor: openness to experience, conscientiousness, agreeableness, extraversion, and neuroticism. A personality score is then derived with Neuman and Cohen's method [14], computing the semantic similarity between the context-free embedding representations of both input text and a set of benchmark adjectives empirically observed as to be able to encode the essence of personality. The aforementioned components are then combined with further sets of features: *(1)* eight emotional dimensions and two related to the sentiment polarization (both through the NRC lexicon)[2]; *(2)* Bag-Of-Words vectors. The CheckerOrSpreader model was trained and tested on the two PAN datasets, obtaining an accuracy repectively equal to 0.52 and 0.51 for English and Spanish datasets. This result will be useful for subsequent comparisons.

5   Visual Information

In [5], the authors mined features from the images embedded in the tweets, by using pre-trained neural networks. The work presented in this paper follows a similar extraction methodology, although from an author profiling perspective. This new approach offers an average description of all the images posted by each user in the sample, and subsequently it evaluates to which of the two target classes this description can correspond. In details, the set of non-duplicated images scraped from each user's texts are passed to five models, pre-trained

---

[2] https://saifmohammad.com/WebPages/NRC-Emotion-Lexicon.htm.

on the popular ImageNet dataset - VGG16, VGG19, ResNet50, InceptionV3, Xception. After applying an average pooling operation, five vectors per image are obtained. Finally, the five compressed representations - one per neural network -, from which it is possible to draw the typical characteristics of the visual contents posted by a user, are obtained by averaging all the vectors per image. In case no images were available for a particular author, vectors of zero have been assigned.

6   Experiments and Results

All the experiments have been performed on the same training and test sets provided at the Author Profiling Task at PAN 2020. All the results, organized with reference to the considered research issues, are available at the following link: github.com/results. The evaluation metric considered is the Accuracy.

First of all, to verify the effectiveness of psycho-linguistic information, we tested all the possible combinations between the LIWC features and the Five Factor model features - separately or jointly -, mixing with the emotional dimensions and the BOW vectors mentioned in Sect. 4, as well as the statistical features implemented by Buda-Bolonyai (Sect. 3). The predictive architectures tested are: Logistic Regression (LogReg), Convolutional Neural Network (CNN) - like the CheckerOrSpreader model requires - and a Long Short-Term Memory (LSTM). The last two also consider as input the vector representation of the tweets provided respectively by the pre-trained GloVe model and a pre-trained FastText model for the English and Spanish datasets, both fine-tuned to better capture the semantic contexts. Looking at the results displayed in the Tables A (available here) and B (available here), the best solution for both languages appears to be a Logistic Regression trained on the mix of personality scores and TF-IDF values, offering an accuracy of 0.69 for English and 0.75 for Spanish. Despite the fact that these values are respectively lower w.r.t. Buda-Bolonyai's performance (0.75 on English users) and Pizarro's result (0.82 on the Spanish dataset), the difference w.r.t. Buda-Bolonyai's accuracy is not statistically significant with a confidence level set at 95%. This confirms that personality scores derived by FFM - in combination with a BOW approach - are powerful enough to significantly conform the state-of-the-art performances on the author profiling task in case of English text. Personality scores without BOW vectors always offer poorer results, but, in the English case, still better than the accuracies produced by the LIWC features alone. In contrast, this latter software-generated representation outperforms the FFM on Spanish text. It is then also possible to conclude that emotional and Buda-Bolonyai's features, in combination with personality information, make a little contribution to the accuracy result. Finally, it is important to note that deeper models like CNN or LSTM always give worse results than Logistic Regression, probably because this latter is able to intrinsically manage the strong collinearity among variables in a better way than the two others architectures, and in general its performance is not penalised by a small amount of data, as is the case with neural networks.

The evaluation of the usefulness of a user's "visual profile" for the given task - second goal of the project - consisted in the experimentation of all possible combinations among the five vector representations from each truncated neural network, and, at a later time, mixing also with the psycho-linguistic components, emotional dimensions and Bag-Of-Words. The tests have been carried out by training a Logistic Regression, since, as aforementioned, this ensures efficient management of the strong multicollinearity among the visual features. Observing the results in Tables C (available here) and D (available here), in both cases the best solution remains the union of the linguistic patterns extracted from the LIWC software with the visual information, even if the results on the two datasets (0.675 for English and 0.706 for Spanish) are lower than the best performances reported for the PAN task in 2020. It is important to note that in the Spanish case the VGG16 vector representation appears to be the only useful one. Finally, although it may seem that visual information alone offers poor accuracy (0.59 with a VGG16-Xception combination on English users and 0.553 with VGG16 vector for Spanish ones), it is necessary to consider that these solutions, actually, still manage to exceed the result achieved by the original CheckerOr-Spreader model on the same PAN test sets (respectively 0.52 and 0.51 for English and Spanish datasets), even without considering any textual information at all. In the first case, this difference is even statistically significant with a 95% confidence level.

Regarding improvements to state-of-the-art models, it is worth mentioning that, to reduce computational weight and training time, only the best performing combinations between visual and psycho-emotional information have been tested. As far as variations on Buda-Bolonyai's model, we maintained the simultaneous training of the four baselines on word n-grams. The variations, instead, concerned the features set the XGBoost algorithm is trained on, and the trial of both Logistic Regression and Linear Regression as an ensemble method. Focusing on English dataset, we can see that any replacement and integration of visual/personality information in the ensemble model improves the original result (as visible in Table E, available here). The maximum accuracy (0.775) is reached with the combination of the baselines trained on N-grams plus an XGBoost model fed with personality scores and VGG16-Xception vectors. The only exception - an accuracy worse than the original one - is found when only integrating the Five Factor Model representation. In the Spanish sample, the opposite is observed: any modification worsens the original result (as seen in Table F, available here). With regard to Pizzaro's system, since it was iteratively trained only on mixtures of n-grams extracted from pre-processed text, the modifications consisted in simple concatenations of the new features sets - including visual, statistical and psycho-emotional features - to the TF-IDF weights originally considered. However, it has been necessary to estimate the personality scores only once with the original text preparation performed by Giachanou et al. [4]. Since the FFM paradigm compresses input text to compute the similarity with the vectors of the benchmark adjectives, variations in text preparation – searching for an optimal pipeline, as Pizarro's original system does - could penalize the result. From Tables G (available here) and H (available here), it appears that, for both datasets, integrations almost always lead to a performance worsening. The exception on the English dataset occurs with the only addition of personality scores (with an increases from 0.735 to 0.76). On the Spanish dataset, the only improvement in the accuracy is due to the concatenation of the VGG16 vector: the result rises from 0.82 to 0.832.

7 Conclusions

Fake news is an increasingly debated phenomenon due to the dramatic influences it has on both virtual and real communities. It is, thus, of primary importance that scientific research is concerned with countering the spread of false information. In this paper, we test the effectiveness of personality information and visual features for profiling fake news spreaders on Twitter. To summarise the results obtained from the performed experiments, Tables 1 and 2 show the accuracy measures achieved by the respective best combinations of features sets and predictive models, on both English and Spanish corpora. From these Tables, it appears that the fake news spreader detection task can be addressed more effectively with a combination of N-grams, personality information and visual features, in both datasets. Therefore, the obtained results demonstrate the relevance of the visual and personality information proposed. In both languages, the second best solution combines visual information with textual features. A second consideration can thus be made on the effectiveness of visual features: although they offer worse results if used alone, when combined with N-grams they always obtain a better performance w.r.t. the mix of text and personality information. In general, even in combination with psycho-linguistic features, visual information offers good results.

Table 1. Best overall solution on the English dataset.

Combination	Model	Features	Accuracy
TXT+PERS+IMG	LinReg Ensemble	N-grams + FFM + VGG16, XNC	**0.775**
TXT+STAT	*Buda-Bolonyai's*	N-grams + Stat.	0.75
TXT	*Pizarro's*	N-grams	0.735

Table 2. Best overall solution on the Spanish dataset.

Combination	Model	Features	Accuracy
TXT+PERS+IMG	Linear SVC	N-grams + FFM + VGG16	**0.832**
TXT	*Pizarro's*	N-grams	0.82
TXT + STAT	*Buda-Bolonyai's*	N-grams + Stat.	0.805

Then, personality scores, modeled together with BOW by a Logistic Regression, offer a result statistically not inferior to state-of-the-art solutions for English only. In particular, we observe that the most powerful psycho-linguistic features in both languages are offered by the Five Factor Model. However, when combined with the LIWC patterns, it often penalizes the result. Moreover, in this context it was noted that it is advisable to use less complex models like Logistic Regression. Finally, it is possible to conclude that the integration of visual/personality information allows to improve the performance of state-of-the-art models in many cases.

Acknowledgements. This work is funded by Project 2020-ATE-0632, "Definition of models and systems for the representation, management and analysis of information and knowledge", University of Milano Bicocca, and supported by the MISMIS-FAKEnHATE research project on Misinformation and Miscommunication in social media: FAKE news and HATE speech (PGC2018-096212-B-C31).

References

1. Allcott, H., Gentzkow, M.: Social media and fake news in the 2016 election. In: National Bureau of Economic Research (2017)
2. Del Vicario, M., et al.: Echo chambers: emotional contagion and group polarization on facebook. In: Scientific Reports 6 (2016)
3. Shu, K., Sliva, A., Wang, S., Tang, J., Liu, H.: Fake news detection on social media: a data mining perspective. In: ACM SIGKDD Explorations Newsletter 19 (2017)
4. Giachanou, A., Ríssola, E.A., Ghanem, B., Crestani, F., Rosso, P.: The role of personality and linguistic patterns in discriminating between fake news spreaders and fact checkers. In: Métais, E., Meziane, F., Horacek, H., Cimiano, P. (eds.) NLDB 2020. LNCS, vol. 12089, pp. 181–192. Springer, Cham (2020). https://doi.org/10.1007/978-3-030-51310-8_17
5. Giachanou, A., Zhang, G., Rosso, P.: Multimodal fake news detection with textual, visual and semantic information. In: Sojka, P., Kopeček, I., Pala, K., Horák, A. (eds.) TSD 2020. LNCS (LNAI), vol. 12284, pp. 30–38. Springer, Cham (2020). https://doi.org/10.1007/978-3-030-58323-1_3
6. Afroz, S., Brennan, M., Greenstadt, R.: Detecting hoaxes, frauds, and deception in writng style online. In: ISSP 2012 (2012)
7. Giachanou, A., Rosso, P., Crestani, F.: Leveraging emotional signals for credibility detection. In: Proceedings of the 42nd International ACM SIGIR Conference on Research and Development in IR, pp. 877–880 (2019)
8. Rangel, F., Giachanou, A., Ghanem, B., Rosso, P.: Overview of the 8th author profiling task at PAN 2020: profiling fake news spreaders on Twitter. In: Cappellato, L., Eickhoff, C., Ferro, N., Névéol, A. (eds.) CLEF 2020 Labs and Workshops, Notebook Papers. CEUR-WS.org, vol. 2696 (2020)
9. Buda, J., Bolonyai, F.: An ensemble model using N-grams and statistical features to identify fake news spreaders on Twitter. In: Cappellato, L., Eickhoff, C., Ferro, N., Névéol, A. (eds.) CLEF 2020 Labs and Workshops, Notebook Papers. CEUR-WS (2020)
10. Pizarro, J.: Using N-grams to detect fake news spreaders on Twitter. In: Cappellato, L., Eickhoff, C., Ferro, N., Névéol, A. (eds.) CLEF 2020 Labs and Workshops, Notebook Papers. CEUR-WS.org (2020)

11. Briggs-Myers, I., Myers, P.B.: Gifts Differing: Understanding Personality Type. Davies-Black Publishing, Mountain View (1995)
12. Pennebaker, J.W., Boyd, R.L., Jordan, K., Blackburn, K.: The Development and Psychometric Properties of LIWC 2015. Technical report (2015)
13. John, O.P., Srivastava, S.: The big-five trait taxonomy: history, measurement, and theoretical perspectives. In: Handbook of Personality: Theory and Research, pp. 102–138 (1999)
14. Neuman, Y., Cohen, Y.: A vectorial semantics approach to personality assessment. Sci. Rep. 4(1), 1–6 (2014)

Applications

Comparing MultiLingual and Multiple MonoLingual Models for Intent Classification and Slot Filling

Cedric Lothritz$^{(\boxtimes)}$ ⓘ, Kevin Allix ⓘ, Bertrand Lebichot ⓘ, Lisa Veiber ⓘ, Tegawendé F. Bissyandé ⓘ, and Jacques Klein ⓘ

University of Luxembourg, 6 rue Richard Coudenhove-Kalergi, 1359 Luxembourg, Luxembourg
cedric.lothritz@uni.lu

Abstract. With the momentum of conversational AI for enhancing client-to-business interactions, chatbots are sought in various domains, including FinTech where they can automatically handle requests for opening/closing bank accounts or issuing/terminating credit cards. Since they are expected to replace emails and phone calls, chatbots must be capable to deal with diversities of client populations. In this work, we focus on the variety of languages, in particular in multilingual countries. Specifically, we investigate the strategies for training deep learning models of chatbots with multilingual data. We perform experiments for the specific tasks of Intent Classification and Slot Filling in financial domain chatbots and assess the performance of mBERT multilingual model vs multiple monolingual models.

Keywords: Chatbots · Multilingualism · Intent classification · Slot filling

1 Introduction

Chatbots usually operate in a single language depending on where they are deployed (e.g., a chatbot for a British bank will only handle requests written in English). While deploying a single monolingual chatbot is usually sufficient in countries where the entire population speaks one language, this strategy presents challenges in multilingual areas where people do not necessarily speak the same language at a high level. In multilingual countries, such as Switzerland, Luxembourg, India, South Africa, etc. with two or more national languages, companies and banks need to be able to communicate with their clients in the language of the latter's choosing in order to stay competitive. The same holds true for client support chatbots, which have to support multiple languages to stay viable in a multilingual environment. This requirement presents a challenge as companies have to decide on a strategy for implementing a multilingual chatbot system. Two such strategies are as follows: (S1) For n languages, employ n chatbots, each

© Springer Nature Switzerland AG 2021
E. Métais et al. (Eds.): NLDB 2021, LNCS 12801, pp. 367–375, 2021.
https://doi.org/10.1007/978-3-030-80599-9_32

of which is trained to handle requests in a single language. (S2) For n languages, employ one chatbot which is trained using data written in n languages. There are some immediate advantages for training a chatbot using mixed-language data as one would have to train only a single chatbot and maintain only one database as opposed to multiple. However, it is unclear how the performance of a singular multilingual chatbot (S2) compares to a combination of multiple monolingual chatbots (S1). In this paper, we explore these two strategies for chatbots in a multilingual environment. Specifically, we investigate the performance of S1 and S2 on two tasks that represent fundamental blocks for chatbot systems: Intent Classification (IC), which is the task of identifying a user's intent based on a piece of text, and Slot Filling (SF), the task of identifying attributes that are relevant to a given intent. For this study, we use the Rasa chatbot framework, which uses the Dual Intent and Entity Transformer Classifier [2] for both the IC and SF tasks. Furthermore, we compare two techniques for text representation, namely bag-of-words (BOW) and multilingual BERT (mBERT) [5].

We aim to answer the following research questions:

- RQ1: How does the distribution of data samples per language influence the performance of multilingual chatbots?
- RQ2: How do S1 and S2 compare in terms of Intent Classification and Slot Filling?

For this study, we use a novel dataset for IC and SF in the financial domain, which we name the *Banking Client Support* (BCS) dataset. We also use the MultiATIS++ dataset published by Xu et al. [11].

This paper is structured as follows: In Sect. 2, we explain the datasets we use, the chatbot framework, and give a detailed description for S1 and S2. In Sect. 3, we present the results of our experiments, answer the research questions, and show the merits of multilingual chatbots. Section 4 shows various papers related to this study, and we finally conclude our findings in Sect. 5.

2   Experimental Setup

2.1   Datasets

For this study, we use two multilingual datasets to evaluate the performance of multilingual chatbots. We created one dataset for client support bots in the banking domain as there are no public datasets available to the best of our knowledge. We also use a multilingual version of the well-known ATIS dataset to verify the results using a larger dataset.

Banking Client Support Dataset: The first dataset (which we refer to as banking client support dataset (BCS) throughout this paper) is based on a toy dataset provided by Rasa[1]. The original dataset contains 337 samples divided

---

[1] https://github.com/RasaHQ/financial-demo

into 15 intents. We removed three of the intents together with 93 samples as they seemed too vague (*inform*) or were not directly related to the banking domain (*help&human_handoff*), and added 763 samples and introduced 16 new intents, resulting in 1003 samples across 28 intents with each intent being distributed quite equally. The intents cover basic conversational phrases such as *greet* or *affirm* and requests specific to the banking domain such as *make_bank_transfer*, *block_card* or *search_atm*. Additionally, the set contains 253 entities, divided into 6 unique entity types such as *account_type* or *credit_card_type*. We then translated the dataset into three languages (German, French and Luxembourgish) with Google Translate and manually corrected translation errors, resulting in a total of four distinct, but parallel datasets[2]. For this study, we use these four base datasets to construct mixed-language datasets containing equal numbers of samples from the base datasets, e.g., the English-French dataset consists of 50% English samples and 50% French data samples. There are 11 possible language combinations: six combinations with two languages, four with three languages, and one combination with all four languages, which gives us a total of 15 different datasets containing varying numbers of languages.

MultiATIS++ Dataset: The second dataset is based on the popular Airline Travel Information System (ATIS) dataset [4]. The original dataset contains a total of 5871 sentences divided into 26 intents. Furthermore, it contains 19 356 samples for slot filling, divided into 128 slot types. MultiATIS++ is a multilingual version of ATIS created by Upadhyay et al. [8] and Xu et al. [11]. For this study, we use the English, German and French versions of the MultiATIS++ dataset. Furthermore, we reduced the number of intents by removing intents with fewer than five samples, resulting in a total of 5860 sentences divided into 17 intents. It is to note that the distribution of the intents is highly imbalanced with 73.6% of the samples having the intent *atis_flight*. There are four possible language combinations, resulting in a total of seven datasets.

2.2 Chatbot Framework Used in This Study

Rasa: Bocklisch et al. introduced the Rasa NLU and Rasa Core tools [2], with the objective of making a framework that is more accessible for creating conversational software. The modular design of a chatbot made with Rasa allows to swap out configuration files and training data. For this study, we created two different configurations: (C1) a bag-of-words (BOW) pipeline consisting of a Whitespace-Tokenizer, RegexFeaturizer, LexicalSyntacticFeaturizer, and a CountVectorsFeaturizer. (C2) an mBERT pipeline which consists of the HFTransformersNLP model initializer using the cased multilingual BERT-base as its pretrained model as well as its accompanying tokenizer and featurizer[3].

mBERT: For this study, we will use the multilingual BERT [5] (mBERT) model as our datasets contain texts written in English, French, German, and Luxem-

---

[2] Available at https://github.com/Trustworthy-Software/BCS-dataset.
[3] Further information on Rasa models: https://rasa.com/docs/rasa/components/.

bourgish. However, as the number of Wikipedia articles varies greatly for every language of mBERT, there are significant disparities between the datasets used to train the different language components. Specifically, the English dataset is the largest with around 6 million articles, the German and French datasets have comparable sizes with 2.5 and 2.2 million articles respectively, and the Luxembourgish dataset is the smallest with only 59 000 articles.

For this study, we use the cased mBERT model with 12 transformer blocks, 768 hidden layers, attentions heads and 110 trainable parameters provided by Devlin et al.[4] [5].

2.3   Implementation Strategies

S1: Pseudo-multilingual Chatbots. For each monolingual dataset, we train two chatbots: one using an mBERT model, and one without. By combining a language-selector (LS) and monolingual chatbots, we can create pseudo - multilingual chatbots. This allows us to directly compare the performance between monolingual chatbots and multilingual chatbots. For the LS, we use langid[5].

S2: Multilingual Chatbots. Based on the monolingual datasets, we construct mixed-language datasets. For every language combination, we extract a stratified subset from each monolingual dataset and combine them to create multilingual datasets. For each of these new datasets, we train two multilingual chatbots, one using a BOW model, and one using an mBERT model.

3   Experimental Results

In this section, we will answer the two research questions that we formulated for this study (cf. Sect. 1) as well as discuss the results in Sect. 3.5.

3.1   RQ1: How Does the Distribution of Data Samples per Language Influence the Performance of Multilingual Chatbots?

In order to answer this question, we create chatbots trained on bilingual datasets, vary the distribution of both languages in the sets, and evaluate their performance on various test sets. Specifically, we train 11 chatbot models on 11 mixed-language datasets where dataset 0 contains 0% samples from language A and 100% samples of language B, dataset 1 contains 10% samples of language A, 90% samples of language B, etc. These models are tested on three test sets: (1) a monolingual test set containing samples from language A, (2) a test set containing samples from language B, (3) a stratified test set containing an equal number of samples from both languages A and B.

---

[4] https://github.com/google-research/bert/blob/master/multilingual.md.
[5] https://github.com/saffsd/langid.py.

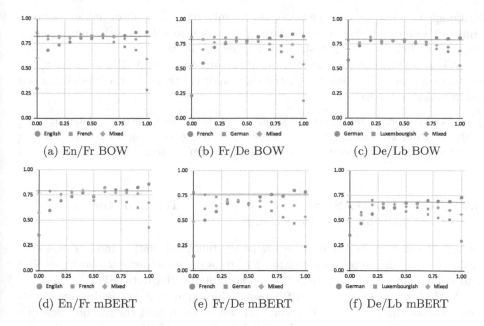

(a) En/Fr BOW   (b) Fr/De BOW   (c) De/Lb BOW

(d) En/Fr mBERT   (e) Fr/De mBERT   (f) De/Lb mBERT

Fig. 1. Evolution of the F1 score for bilingual chatbots for IC task when varying the distribution of data samples per language. The horizontal line represents the performance of the LS+monolingual chatbots models.

Intent Classification. Figure 1 shows the performances of three language combinations in terms of F1 score. These combinations are: English/French (En/Fr), French/German (Fr/De) being two languages that are very dissimilar in terms of syntax and vocabulary, and German/Luxembourgish (De/Lb) being syntactically very similar. When varying the distribution of per-language data samples, we can make several observations: (1) when tested on a monolingual test set, we tend to observe very low performances if the training set does not contain the tested language at all, while we can see very high performances for the opposite case. This performance drop is less apparent for the De/Lb combinations (cf Fig. 1c and Fig. 1f). Furthermore, the Fr/De combinations (cf Fig. 1b and Fig. 1e) show the highest performance drop for these extreme cases. (2) When testing on the mixed-language test set, we can observe comparable performances for every training set, except for the models that were trained on monolingual training sets. (3) Models that are trained on sets containing 50% samples from each language tend to perform similarly for each test set. When performing the same experiment on the MultiATIS++ dataset, we observed that the performance remained stable except for the models trained on monolingual data.

Slot Filling. For the task of slot filling, we can make similar observations as we did for IC: very high and low performances for chatbots that were trained on monolingual datasets, with less noticeable drops for the German/Luxembourgish language combinations. When tested on the mixed test sets, most models perform similarly well except for the monolingual ones. It is to note that this performance drop is smaller for the SF task than it is for the IC task.

When performing the same experiment on the MultiATIS++ dataset, the performance of the models fluctuated only slightly except for the models trained on monolingual data.

> **RQ1 Answer:** There is a noticeable drop in performance if a language is absent from the training set. A 50/50 split in the training set tends to lead to the highest performances on the mixed-language test sets.

3.2   RQ2: How Do S1 and S2 Compare in Terms of Intent Classification and Slot Filling?

In order to answer this question, we reuse the bilingual chatbot models that were trained on the datasets which contain 50% data samples from each language (S2) and compare their performance to pseudo-bilingual chatbots (S1).

Table 1 compares F1 scores for pseudo-bilingual chatbot models and bilingual chatbot models for the IC task. Our results show that the combination of a language selector and two monolingual chatbots yields higher performances with regard to every performance measure used. It is to note that the English/French variant is an exception to the rule as the model with the S2 strategy significantly outperforms the S1 model. This trend can be observed for both the chatbot models with an mBERT and the ones with a BOW model. The performance differences between S1 and S2 models with mBERT are usually larger when compared to the performance differences between the models that do not use pretrained models. Furthermore, the models based on BOW consistently outperform the models with mBERT by several percentage points.

Table 2 shows the results of the same task on the MultiATIS++ datasets. In contrast to the BCS sets, the results are in favour of the S2 strategy. When comparing the MCC scores, we observe that the performance of the bilingual models either exceeds or matches that of the combinations of LS+monolingual chatbots.

Table 1. Test results for bilingual chatbots (S2) vs monolingual chatbots with language selector (S1) on Intent Classification task on BCS set.

	BOW						mBERT					
	Bilingual			LS + Monolingual			Bilingual			LS + Monolingual		
	Prec	Rec	F1	Prec	Rec	F1	Prec	Rec	F1	Prec	Rec	F1
En/Fr	0.851	0.835	**0.833**	0.864	0.805	0.823	0.779	0.753	0.745	0.830	0.771	**0.791**
En/De	0.810	0.801	0.797	0.867	0.835	**0.843**	0.744	0.708	0.706	0.796	0.766	**0.769**
En/Lb	0.807	0.797	0.794	0.845	0.810	**0.819**	0.712	0.679	0.676	0.747	0.697	**0.703**
Fr/De	0.787	0.764	0.761	0.835	0.788	**0.796**	0.691	0.664	0.654	0.800	0.753	**0.763**
Fr/Lb	0.805	0.780	0.778	0.824	0.777	**0.787**	0.703	0.677	0.662	0.728	0.674	**0.679**
De/Lb	0.794	0.788	0.783	0.826	0.784	**0.797**	0.668	0.640	0.638	0.725	0.678	**0.683**

Table 2. Test results for bilingual chatbots(S2) vs monolingual chatbots with language selector(S1) on Intent Classification task on MultiATIS++ set

| | BOW | | | | | | | | mBERT | | | | | | | |
| | Bilingual | | | | LS + Monolingual | | | | Bilingual | | | | LS + Monolingual | | | |
	Prec	Rec	F1	MCC	Prec	Rec	F1	MCC	Prec	Rec	F1	MCC	Prec	Rec	F1	MCC
En/Fr	0.973	0.967	0.969	**0.929**	0.976	0.961	0.967	0.914	0.971	0.970	0.97	**0.941**	0.979	0.968	0.973	0.929
En/De	0.977	0.974	0.975	**0.942**	0.930	0.966	0.968	0.924	0.973	0.972	0.972	**0.937**	0.978	0.972	0.974	**0.937**
Fr/De	0.964	0.959	0.961	0.911	0.971	0.962	0.966	**0.916**	0.974	0.97	0.971	**0.933**	0.974	0.965	0.968	0.922

In order to determine if pseudo-bilingual (S1) significantly outperform bilingual (S2) models, we perform a Wilcoxon test for both strategies over every dataset used. We find that the differences in performance for mBERT models are indeed significant, but in the case for BOW models, only the difference in precision is clearly significant.

For the SF task. We generally see better results for the mBERT model. Similarly to the IC task, the combination of monolingual chatbots and a language selector almost consistently outperforms the chatbots trained on bilingual datasets by a large margin. This is true for both the BCS and the MultiATIS++ datasets. We once again determine statistical significance of the obtained results through a Wilcoxon test. The resulting p-values show that the performance differences are significant except for recall and F1 score for the BOW models.

> **RQ2 Answer:** In most cases, S1 performs better than S2, with IC on MultiATIS++ being a notable exception.

3.3 Discussion

When using a small dataset, the results of the conducted experiments are generally in favour of strategy S1 and by a significant margin. This is true for both the IC and the SF tasks. The results are less conclusive when training the chatbots on the larger MultiATIS++ dataset. For the IC task, neither strategy is consistently outperforming the other. On the other hand, strategy S1 is superior regardless of the dataset as it outperforms S2 for the BCS dataset as well as the MultiATIS++ dataset. The performances of the investigated models were significantly dependent on the task. While BOW-models generally performs better for the IC task, mBERT-models seems to be the favourable choice for the SF task, as strategy S1 with mBERT generally largely outperformed the BOW-models when compared directly.

4 Related Work

Multilingual IC and SF: Previous multilingual text classification systems are usually based on two different approaches: (1) machine translation systems that translate training data into the target language [10] or (2) parallel corpora that are used to learn embeddings jointly from multiple languages [6]. Such crosslingual embeddings prove useful for binary classification tasks such as sentiment classification [12,13] and churn intent detection [1]. Abbet et al. [1] use multilingual embeddings for the task of churn intent detection in social media. They show that bilingual embeddings trained on an English and German dataset outperform monolingual embeddings for this binary IC task. Furthermore, they show that models trained on social media data can be applied to chatbot conversations as well. Schuster et al. [7] evaluate three methods for multilingual IC and SF, namely translating the training data into the target language, using pretrained crosslingual embeddings, and using a novel pretrained translation encoder to generate embeddings.

Multilingual Datasets: One major challenge for multilingual IC and SF is the lack of textual data in languages other than English. Schuster et al. created a dataset containing 57 000 utterances divided into three languages [7]: 43 000 utterances in English, 8600 in Spanish and 5000 in Thai. Their data is annotated for 12 intent types, and 11 slot types in total. They use their dataset to evaluate various crosslingual transfer methods for IC and SF. The ATIS dataset [4] is one of the most popular datasets for IC and SF. Originally available only in English, it was partially translated into Hindi and Turkish [9], creating MultiATIS. Xu et al. further extended MultiATIS to six more languages [11], resulting in MultiATIS++, consisting of nine versions of the original ATIS dataset. Datasets related to banking are difficult to find as most of them are proprietary [3], making our BCS dataset one of the few public datasets related to that domain.

5   Conclusion

In this paper, we presented a study on multilingual chatbots, specifically on the Intent Classification and Slot Filling tasks.

We compared two implementation strategies and two embedding techniques. We noticed that training a chatbot on mixed-language data decreases the overall performance. We concluded that, in the case of two languages, the combination of a language selector and two monolingual chatbots (S1) usually outperforms chatbots that are directly trained on bilingual datasets (S2). While the BOW models almost consistently outperform the mBERT models in the Intent Classification tasks, the mBERT models usually perform better in the Slot Filling tasks when using the S1 strategy.

References

1. Abbet, C., et al.: Churn intent detection in multilingual chatbot conversations and social media. arXiv preprint arXiv:1808.08432 (2018)
2. Bocklisch, T., Faulkner, J., Pawlowski, N., Nichol, A.: Rasa: Open source language understanding and dialogue management. arXiv preprint arXiv:1712.05181 (2017)

3. Costello, C., Lin, R., Mruthyunjaya, V., Bolla, B., Jankowski, C.: Multi-layer ensembling techniques for multilingual intent classification. arXiv preprint arXiv:1806.07914 (2018)
4. Dahl, D.A., et al.: Expanding the scope of the atis task: the atis-3 corpus. In: HUMAN LANGUAGE TECHNOLOGY: Proceedings of a Workshop held at Plainsboro, New Jersey, 8–11 March 1994 (1994)
5. Devlin, J., Chang, M.W., Lee, K., Toutanova, K.: Bert: pre-training of deep bidirectional transformers for language understanding. In: 2019 Conference of the North American Chapter of the ACL: Human Language Technologies (2019)
6. Lauly, S., Larochelle, H., Khapra, M.M., Ravindran, B., Raykar, V., Saha, A., et al.: An autoencoder approach to learning bilingual word representations. arXiv preprint arXiv:1402.1454 (2014)
7. Schuster, S., Gupta, S., Shah, R., Lewis, M.: Cross-lingual transfer learning for multilingual task oriented dialog. In: Proceedings of the 2019 Conference of the North American Chapter of the Association for Computational Linguistics: Human Language Technologies, vol. 1 (Long and Short Papers), pp. 3795–3805 (2019)
8. Upadhyay, S., Faruqui, M., Tür, G., Dilek, H., Heck, L.: (Almost) zero-shot cross-lingual spoken language understanding. In: 2018 IEEE ICASSP, pp. 6034–6038 (2018). https://doi.org/10.1109/ICASSP.2018.8461905
9. Upadhyay, S., Faruqui, M., Tür, G., Dilek, H.T., Heck, L.: (Almost) zero-shot cross-lingual spoken language understanding. In: 2018 IEEE International Conference on Acoustics, Speech and Signal Processing (ICASSP), pp. 6034–6038. IEEE (2018)
10. Wan, X.: Co-training for cross-lingual sentiment classification. In: Joint Conference of the 47th Annual Meeting of the ACL, pp. 235–243 (2009)
11. Xu, W., Haider, B., Mansour, S.: End-to-end slot alignment and recognition for cross-lingual NLU. In: Proceedings of EMNLP 2020, November 2020, pp. 5052–5063. ACL, Online (2020). https://doi.org/10.18653/v1/2020.emnlp-main.410
12. Zhou, G., He, T., Zhao, J.: Bridging the language gap: learning distributed semantics for cross-lingual sentiment classification. In: International Conference on Natural Language Processing and Chinese Computing, pp. 138–149. Springer, Heidelberg (2014)
13. Zhou, H., Chen, L., Shi, F., Huang, D.: Learning bilingual sentiment word embeddings for cross-language sentiment classification. In: 53rd Annual Meeting of the ACL and the 7th International Joint Conference on NLP, pp. 430–440 (2015)

Automated Retrieval of Graphical User Interface Prototypes from Natural Language Requirements

Kristian Kolthoff[1]([⊠]), Christian Bartelt[1], and Simone Paolo Ponzetto[2]

[1] Institute for Enterprise Systems, University of Mannheim, Mannheim, Germany
{kolthoff,bartelt}@es.uni-mannheim.de
[2] Data and Web Science Group, University of Mannheim, Mannheim, Germany
simone@informatik.uni-mannheim.de

Abstract. High-fidelity Graphical User Interface (GUI) prototyping represents a suitable approach for allowing to clarify and refine requirements elicitated from customers. In particular, GUI prototypes can facilitate to mitigate and reduce misunderstandings between customers and developers, which may occur due to the ambiguity and vagueness of informal Natural Language (NL). However, employing high-fidelity GUI prototypes is more time-consuming and expensive compared to other simpler GUI prototyping methods. In this work, we propose a system that automatically processes Natural Language Requirements (NLR) and retrieves fitting GUI prototypes from a semi-automatically created large-scale GUI repository for mobile applications. We extract several text segments from the GUI hierarchy data to obtain textual representations for the GUIs. To achieve ad-hoc GUI retrieval from NLR, we adopt multiple Information Retrieval (IR) approaches and Automatic Query Expansion (AQE) techniques. We provide an extensive and systematic evaluation of the applied IR and AQE approaches for their effectiveness in terms of GUI retrieval relevance on a manually annotated dataset of NLR in the form of search queries and User Stories (US). We found that our GUI retrieval performs well in the conducted experiments and discuss the results.

Keywords: Automatic Prototyping of Graphical User Interfaces (GUIs) · GUI retrieval · GUI prototypes from natural language requirements

1 Introduction

Effective requirements elicitation techniques play a vital role in early development stages [18], in order to mitigate or eliminate misunderstandings of requirements between customers and developers, which might occur due to the ambiguity and vagueness inevitably encompassed in Natural Language (NL) communication [4]. GUI prototyping poses a meaningful technique to visualize the developers' understanding of the requirements and enable their verification by

© Springer Nature Switzerland AG 2021
E. Métais et al. (Eds.): NLDB 2021, LNCS 12801, pp. 376–384, 2021.
https://doi.org/10.1007/978-3-030-80599-9_33

the customer as a tangible artifact. Moreover, GUI prototypes can provide the foundation for incorporating the customer early into the application development and lead to productive discussions and clarification of requirements [16].

In this work, we propose an ad-hoc GUI retrieval approach that is based on a semi-automatically created large-scale GUI repository for mobile apps. Kolthoff et al. [12] showed how such a GUI retrieval system can be useful to support rapid prototyping and it could be used as part of a virtual prototyping assistant [11].

2   Approach: GUI2R

The main goal of our *GUI2R* approach is to retrieve matching GUI prototypes for a NL query provided by a user. In order to achieve that, we employ Natural Language Processing (NLP) and Information Retrieval (IR) techniques to compute a ranking over a large-scale GUI prototype repository for the given NL query. In addition, we experiment with various Automatic Query Expansion (AQE) techniques to tackle the vocabulary mismatch problem [13]. In our approach, we employ a GUI repository of mobile applications (Android). Figure 1 shows an overview of the system architecture of *GUI2R*, which in general follows the extended Boolean model [13]. First, a user specifies a NLR in the form of a search query or in the structured form of a US. The input is processed *(A)* by a NLR parser that detects US and extracts only specific parts for further processing. Subsequently, a pipeline of text preprocessing techniques is applied on the NL input. As a foundation for the GUI repository, we employ *(B)* the large-scale mobile app GUI dataset *Rico* [7]. This dataset consists of Android apps crawled from Google Play. To make the GUI prototypes searchable for NL input, we extract particular text segments from the corresponding GUI hierarchy data and represent the GUI prototypes as text documents. Afterwards, *(C)* an inverted index is computed from the GUI text documents and applied to match GUI documents that contain at least a single query term. The matched GUI documents are then scored by a retrieval model and the top-ranked documents are used for AQE to compute (D) the final ranking of the GUI prototypes. In the following, we describe the individual components of *GUI2R* in more detail.

2.1   NLR Parsing and Preprocessing

We employ several text preprocessing methods on the NL input. First, we lowercase the NL input and apply tokenization. Tokens are then excluded by several filters: We remove basic English stopwords, words comprising numeric or non-ASCII characters and out-of-vocabulary words based on a dictionary derived from the textual representation of the GUIs. We initially apply our US parser that is based on pattern matching to detect US and extract the *user-role, user-task* and *user-goal* from the US template (based on the Connextra format). From the parsed US, we only use the *user-task* description as NL input to our GUI retrieval system, but apply previously discussed preprocessing steps beforehand.

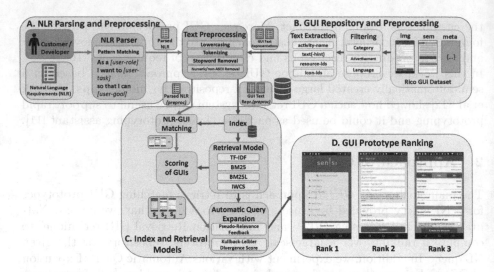

Fig. 1. Overview of *GUI2R* with (A) NLR parsing of US and text preprocessing, (B) GUI repository and its preprocessing, (C) Matching of NLR and GUIs with index and GUI scoring with multiple retrieval models and (D) final GUI ranking

2.2   GUI Repository and Preprocessing

Recent research on data-driven design published several GUI datasets suitable for our retrieval system such as *ReDraw* [14], *ERICA* [8] and *Rico* [7]. All of these GUI datasets are gathered from mobile applications crawled from the app store Google Play. We decided to use *Rico* and the reasons for employing *Rico* as our GUI repository are manifold: *(i)* the large scale, making a retrieval system particularly valuable, *(ii)* the wide spectrum and diversity of mined applications available for retrieval, covering potentially many reusable GUIs and *(iii)* the provided rich textual information, including component identifiers and semantically tagged GUI components. *Rico* mines GUI screenshots, GUI hierarchy data, application meta data and interaction traces with both human-based and automatic exploration techniques and constitutes the largest design dataset of the discussed ones with 72,219 GUIs collected from 9,772 unique Android apps.

Since *Rico* crawls applications and extracts GUIs partly in an automatic fashion, incorporating noisy GUIs in the dataset is inevitable. First, (1) we filter all GUIs that belong to applications of the entertainment category, since we are not interested in game GUIs. Second, (2) we remove GUIs covered with advertisement overlay screens by checking for particular patterns of the component labels in the semantically tagged GUI document. Third, (3) we apply language detection on the extracted text segments in order to remove non-English GUIs from the repository. To achieve that, we employ a language detection framework that computes language probabilities by accumulating character-level n-gram spelling feature probabilities [17].

To enable NL-based search queries on the mobile app GUI repository, we first require to represent the GUIs as text documents. From the GUI hierarchy data provided in *Rico*, we extract several text segments through XPath expressions. We extract text from all components that are explicitly marked as `text` or `text-hint` and displayed to the user. In addition, we extract the full activity name of the GUI and the resource identifier of each individual GUI component. Developers often provide semantically rich descriptive names for their activities and GUI component identifiers, thus we consider them as a valuable resource for retrieval. To make these special strings such as *"com.sample.sens.register.-CreateNewAccountActivity"* searchable, we apply a pipeline of various tokenizers. First, we apply punctuation, basic camel case and snake case tokenization. On top, we use a custom probabilistic tokenizer based on English Wikipedia unigram frequencies to split remaining concatenated words. On the resulting tokens, we apply a specially created stopword list to remove non-descriptive general terms (e.g. *"com"*, *"main"* and *"activity"*). From the semantically tagged GUI representation, we extract the textual descriptions of the detected icons (for example *"add"* and *"search"*) since they often provide descriptive terms that may similarly be used in NL queries. The text segments are preprocessed identically to the query and represent the GUI as a text document as the basis for GUI retrieval.

2.3   Information Retrieval Models

To retrieve matching GUIs from NL queries from our GUI repository, we adopt retrieval methods that have a long history in IR research [13]. In particular, we employ TF-IDF, BM25 and BM25L [15]. These models provide a strong baseline to many other specific IR tasks and have shown their effectiveness in other domains before. In addition, we evaluate a more recent method that exploits TF-IDF weighted pre-trained dense word embeddings (based on 300-dimensional *word2vec* embeddings) for similarity scoring (IWCS) [9]. Another way to enhance the retrieval performance of IR systems is the introduction of AQE techniques to tackle the vocabulary mismatch problem through Pseudo-Relevance Feedback (PRF) [13]. Many expansion candidate scoring methods based on PRF have been proposed [1]. These methods follow a similar underlying notion. First, a ranking over the terms contained in the relevant documents D_R (top-k documents initially retrieved by a base model) is computed. Here, terms that are special for the relevant documents D_R and distinguish them from the rest of the document collection D_C should receive a higher ranking score. This can be achieved through comparing the term distributions between the D_R and D_C documents. The initial user query is then expanded with the top-n words from the ranked candidate terms. For our retrieval system, we compute the Kullback-Leibler Divergence (KLD) score [5] for each term $t \in D_R$ as

$$Score_{KLD}(t) = p(t|D_R) \cdot \log \frac{p(t|D_R)}{p(t|D_C)}$$

with $p(t|D_R)$ and $p(t|D_C)$ being the probability of term t occurring in D_R and D_C, respectively. The probabilities are computed as the Maximum Likelihood

Table 1. Evaluation dataset overview and examples for the three different NLR

#	NLR Type	Size	Examples
(1)	KB Queries	30	(1) "daily log" (2) "watch video" (3) "image blog with search" (4) "export data" (5) "select clothing size between s and xl" (6) "select my age" (7) "image grid" (8) "new price old price product" (9) "show training statistics"
(2)	US (int.)	20	(1) "As a user I want to see the product price, product image and product description" (2) "As a user I want to choose my favorite language" (3) "As a user I want to see the number of votes of a post (4) "As a user I want to create a new account"
(3)	US (ext.)	10	(1) "As an OlderPerson, I want to maintain my contact list in my phone." (2) "As a user, I want to be able to search any dataset published and publicly accessible by their title and metadata, So that I can find the datasets I'm interested in." (3) "As a User I want to set my own username, So that my data is more easily discoverable."

Estimates i.e. $p(t|D_X) = \frac{f_{t,x}}{|D_X|}$. We decided to apply and evaluate the KLD score in our experiments since it showed its effectiveness compared to other scoring methods before [5]. We also evaluated two other variants that include the KLD score as a weight for the expanded terms in the retrieval model to control their influence and by computing expansion terms for each text segment separately.

3   Experimental Evaluation

In order to evaluate the proposed approach, we investigate two research questions that relate to the retrieval performance of the discussed IR and AQE models. In the following, we describe our evaluation dataset, the annotation schema, the employed evaluation metrics and discuss the obtained evaluation results. In particular, we investigate the following research questions:

– **RQ$_1$**: Are traditional Information Retrieval (TF-IDF, Okapi BM25, BM25L) and more modern scoring functions (IWCS) suitable for GUI prototype retrieval from Natural Language Requirements (NLR)? Which method performs best for GUI retrieval from NLR?
– **RQ$_2$**: Can pseudo-relevance feedback methods based on the Kullback-Leibler divergence score improve the retrieval performance using Okapi BM25 as a base model? Which AQE method performs best for GUI retrieval from NLR?

3.1   Experimental Setup

In order to evaluate the proposed research questions, we created a requirement collection consisting of 60 Natural Language Requirements (NLR), since there is no evaluation dataset available for evaluating the GUI retrieval systems performance. This dataset provides the foundation for our evaluation and is separated into three sub-datasets: (1) Keyword-based search queries that represent the typical format for conducting searches with 30 examples, (2) User Stories (US) (internal) that we created to investigate different US types and their application in our GUI retrieval approach with 20 examples and (3) User Stories (US) (external) that we gathered from an external resource with 10 examples [6]. Table 1 shows an overview of the evaluation dataset and provides concrete examples for all three sub-datasets. For the keyword-based search queries sub-dataset (1), we attempted to include many diverse topics from rather broad queries such as (1.1) requiring daily log functionality and (1.2) requiring functionality to watch a video to more specific queries such as (1.5) requiring a particular clothing size selection range. For the internal User Story sub-dataset (2), we included US that represent typically reusable requirements and occur among many applications such as (2.2) requesting functionality to choose the favorite language or (2.4) requiring functionality to create a new account, but also US that are more specific to a particular domain and difficult such as (2.1) containing many details. For the external User Story sub-dataset (3), we employed a publicly available US requirement dataset [6] and gathered 10 US that are related to GUIs from two applications (*openspending* and *alfred*). These requirements are generally more specific and custom to their particular application such as example (3.2) requiring a dataset search functionality with very specific search parameters. To evaluate the retrieval performance in terms of relevancy of the returned GUI prototypes and since there is no goldstandard available for this particular problem, we annotated the retrieval results manually. We annotated the top-k retrieved GUI prototypes for all requirements from our evaluation dataset. In our experiments in particular, we retrieved and annotated the top-15 GUI prototypes for each requirement and method ($k = 15$). For a particular evaluation requirement, we annotated each retrieved GUI prototype on a relevancy scale of 0 (*not relevant*), 1 (*related*), 2 (*relevant*) through a web-based evaluation application. Finally, we computed the following standard IR metrics: Precision ($P@k$), Average Precision (AP) and Normalized Discounted Cumulative Gain ($NDCG@k$).

3.2   Results and Discussion

The evaluation results for our different experiments are shown in Table 2. For our first experiment (**RQ$_1$**), we observe that BM25 outperforms all other evaluated IR models by a large margin for datasets (1) and (2), and only for dataset (3) TF-IDF outperforms all other models. IWCS can outperform TF-IDF on the search query dataset (1) but performs worse on both US datasets (2) and (3). During the annotation of the results, we observed some typical retrieval errors. GUI prototypes with an opened menu overlapping most of the screen were often ranked

Table 2. Evaluation results overview of the different experiments

	(1) Search queries					(2) User Stories (int.)					(3) User Stories (ext.)				
	P@1	P@5	P@15	AP	N@15	P@1	P@5	P@15	AP	N@15	P@1	P@5	P@15	AP	N@15
TF-IDF	.467	.427	.344	.496	.763	.500	.430	.407	.530	.777	**.200**	**.240**	**.193**	**.376**	**.532**
BM25	**.767**	**.633**	**.513**	**.710**	**.902**	**.600**	**.580**	**.503**	**.676**	**.839**	.000	.180	.167	.207	.493
BM25L	.333	.320	.256	.417	.714	.400	.310	.300	.398	.752	.100	.140	.153	.196	.412
IWCS	.600	.507	.427	.608	.825	.500	.400	.383	.518	.747	.100	.140	.087	.179	.336
BM25	**.767**	.633	.513	.710	.902	.600	.580	.503	**.676**	**.839**	.000	.180	.167	.207	.493
+PRF	.667	.647	.509	.687	.901	.650	.510	.460	.571	.837	**.400**	.180	.147	**.331**	.494
+PRF(c)	.633	.647	.484	.670	.887	.500	.550	.487	.601	.829	.300	.200	.160	.315	**.541**
+PRF(w)	.733	**.673**	**.527**	**.718**	**.905**	**.650**	.550	.543	.637	.823	.200	**.220**	.160	.257	.495
+PRF(cw)	.600	.673	.493	.672	.892	.650	**.590**	**.550**	.656	.831	.300	.200	**.180**	.264	.520

among the top-15 results since the underlying GUI contained relevant text that was not marked as non-visible. We also observed GUIs that were represented properly as textual documents, however, have erroneous GUI screenshots due to GUI capturing errors. Often, semantic retrieval errors occurred, for example, *login* screens which are retrieved for requirement (1.1) *"daily log"*.

For our second experiment (**RQ₂**), we observe that BM25-PRF (w) and BM25-PRF (cw) outperform the BM25 model for most of the cases, however, often only on small margins. During the annotation, we observed that the base model performance could be improved especially for requirements that are less ambiguous and where it is in general easier to find matching GUIs for. For example, for requirements such as *"login"* or requirement (2.4) requesting functionality for creating a new account, the AQE method could filter out some incorrect GUIs by extracting relevant expansion terms from the top-ranked results.

4   Related Work

Guigle [3] automatically crawls and extracts GUI screenshots and GUI hierarchy data from Android apps harvested from Google Play [14]. Their approach indexes multiple parts of the hierarchy such as the app name, screen color, GUI component text and type and employs a basic Boolean query language to quickly retrieve relevant GUIs. However, our GUI retrieval system *GUI2R* particularly focuses on customer-friendly NLR input, proposes are more sophisticated retrieval architecture including IR methods based on word embeddings and AQE techniques and provides an in-depth evaluation of these methods. In contrast, *Swire* [10] and *GUIFetch* [2] enable mobile app GUI retrieval not through simple NL input, however, using basic Android apps or hand-drawn sketches. In particular, *Swire* employs a neural network-based joint embedding space between the GUI screenshots and the hand-drawn sketches for retrieval, whereas *GUIFetch* computes similarities between GUIs to rank applications based on an app sketch.

5 Conclusion

In this work, we presented a GUI retrieval system for Android applications that ranks GUIs from a semi-automatically created large-scale GUI repository based on NLR to facilitate GUI prototyping with customers. Our experimental results showed that standard IR models can be employed to effectively retrieve GUIs from NLR formulated as search queries or US. We also showed that AQE techniques could slightly improve the retrieval effectiveness of the BM25 base model.

Acknowledgements. This work is supported by the German Federal Ministry of Education and Research (BMBF).

References

1. Azad, H.K., Deepak, A.: Query expansion techniques for information retrieval: a survey. Inf. Process. Manag. **56**(5), 1698–1735 (2019)
2. Behrang, F., Reiss, S.P., Orso, A.: Guifetch: supporting app design and development through gui search. In: Proceedings of the 5th International Conference on Mobile Software Engineering and Systems, pp. 236–246 (2018)
3. Bernal-Cárdenas, C., et al.: Guigle: a gui search engine for android apps. In: 2019 IEEE/ACM 41st International Conference on Software Engineering: Companion Proceedings (ICSE-Companion), pp. 71–74. IEEE (2019)
4. Berry, D.M., Kamsties, E.:Ambiguity in requirements specification. In: do Prado Leite, J.C.S., Doorn, J.H. (eds) Perspectives on Software Requirements. The Springer International Series in Engineering and Computer Science, vol. 753. Springer, Boston, MA (2004). https://doi.org/10.1007/978-1-4615-0465-8_2
5. Carpineto, C., De Mori, R., Romano, G., Bigi, B.: Brigitte: an information-theoretic approach to automatic query expansion. ACM Trans. Inf. Syst. (TOIS) **19**(1), 1–27 (2001)
6. Dalpiaz, F.: Requirements data sets (user stories). Mendeley Data, vol. 1 (2018)
7. Deka, B., et al.: Rico: a mobile app dataset for building data-driven design applications. In: Proceedings of the 30th Annual ACM Symposium on User Interface Software and Technology, pp. 845–854 (2017)
8. Deka, B., Huang, Z., Kumar, R.: Erica: interaction mining mobile apps. In: Proceedings of the 29th Annual Symposium on User Interface Software and Technology, pp. 767–776 (2016)
9. Galke, L., Saleh, A., Scherp, A.: Word embeddings for practical information retrieval. INFORMATIK 2017 (2017)
10. Huang, F., Canny, J.F., Nichols, J.: Swire: sketch-based user interface retrieval. In: Proceedings of the 2019 CHI Conference on Human Factors in Computing Systems, pp. 1–10 (2019)
11. Kolthoff, K.: Automatic generation of graphical user interface prototypes from unrestricted natural language requirements. In: 2019 34th IEEE/ACM International Conference on Automated Software Engineering (ASE), pp. 1234–1237. IEEE (2019)
12. Kolthoff, K., Bartelt, C., Ponzetto, S.P.: GUI2WiRe: rapid wireframing with a mined and large-scale GUI repository using natural language requirements. In: 35th IEEE/ACM International Conference on Automated Software Engineering (ASE 2020). ACM (2020)

13. Manning, C.D., Raghavan, P., Schütze, H.: Introduction to information retrieval. Cambridge University Press, Cambridge (2008)
14. Moran, K., Bernal-Cárdenas, C., Curcio, M., Bonett, R., Poshyvanyk, D.: Machine learning-based prototyping of graphical user interfaces for mobile apps. arXiv preprint arXiv:1802.02312 (2018)
15. Robertson, S.E., et al.: Okapi at TREC-3. Nist Special Publication Sp, 109:109 (1995)
16. Rudd, J., Stern, K., Isensee, S.: Low vs. high-fidelity prototyping debate. Interactions **3**(1), 76–85 (1996)
17. Nakatani, S.: Language detection library for java. Retrieved 7 July 2016 (2010)
18. Zowghi D., Coulin C.: Requirements elicitation: a survey of techniques, approaches, and tools. In: Aurum, A., Wohlin, C. (eds) Engineering and Managing Software Requirements. Springer, Heidelberg. (2005). https://doi.org/10.1007/3-540-28244-0_2

Author Index

Printed in the United States
by Baker & Taylor Publisher Services